Cereals and Pseudocereals

UNDERUTILIZED CROP SERIES

Series co-ordinator: N. Haq

Interests in the use of underutilized crop species are governed by many different concerns in developed and underdeveloped economies, these interests forming a continuum related to crop diversification, production of value-added products, rural development and increased income for growers and processors.

For too long, the world has relied heavily on a relatively small number of crops grown across wide areas and there have been many justifiable environmental concerns expressed. There is an increasing international awareness in the importance of investigating options for the development of new and alternative crops, set against a backdrop of worrying statistics on the loss of species diversity, in particular the loss of species which have not been looked at with a view of commercial exploitation as new crops.

The Underutilized Crop Series is a short series of books, each covering a range of crops within a particular group which have received both local and international attention over the past decade and where enhanced research attention has been justified. Each book in the series has been drawn together by the International Centre for Underutilized Crops (ICUC), the series co-ordinator is Dr N. Haq of ICUC. The first two books in the series are edited by Professor J.T. Williams an adviser to a number of international research networks and projects.

This series is an invaluable source of reference for agricultural and plant scientists, nutritionists, development agencies and all those involved in research and policy decisions concerning agricultural development, biodiversity exploitation and so on.

Titles in this series

Pulses and Vegetables
Edited by J.T. Williams

Cereals and Pseudocereals
Edited by J.T. Williams

Forthcoming titles

Fruit and Nuts

Industrial Crops

Cereals and Pseudocereals

Edited by

J.T. Williams

International Consultant in
Agricultural Research,
Washington D.C., USA

CHAPMAN & HALL

London · Glasgow · Weinheim · New York · Tokyo · Melbourne · Madras

Published by Chapman & Hall, 2–6 Boundary Row, London SE1 8HN, UK

Chapman & Hall, 2–6 Boundary Row, London SE1 8HN, UK

Blackie Academic & Professional, Wester Cleddens Road, Bishopbriggs, Glasgow G64 2NZ, UK

Chapman & Hall GmbH, Pappelallee 3, 69469 Weinheim, Germany

Chapman & Hall USA, One Penn Plaza, 41st Floor, New York NY 10119, USA

Chapman & Hall Japan, ITP-Japan, Kyowa Building, 3F, 2-2-1 Hirakawacho, Chiyoda-ku, Tokyo 102, Japan

Chapman & Hall Australia, Thomas Nelson Australia, 102 Dodds Street, South Melbourne, Victoria 3205, Australia

Chapman & Hall India, R. Seshadri, 32 Second Main Road, CIT East, Madras 600 035, India

---

First edition 1995

Typeset in 10/12pt Palatino by Acorn Bookwork, Salisbury, Wiltshire
Printed in Great Britain by St Edmundsbury Press, Bury St Edmunds, Suffolk

ISBN 0 412 46570 1

A catalogue record for this book is available from the British Library

∞ Printed on permanent acid-free text paper, manufactured in accordance with ANSI/NISO Z39.48-1992 and ANSI/NISO Z39.48-1984 (Permanence of Paper).

Contents

List of contributors ix

Introduction: Development of underutilized cereals and pseudocereals xi
J.T. Williams
Background xi
Priorities xiv
Context in which underutilized crops fit xvii
Institutional frameworks xix
Useful lessons for further research and development xx
References xxii

Part One
Crops Benefiting from Major Research

1 *Quinoa* (Chenopodium quinoa) 3
J.E. Fleming and N.W. Galwey
1.1 History, distribution and uses 3
1.2 Morphology 7
1.3 Taxonomy 10
1.4 Physiology of growth and development 15
1.5 Biochemistry 21
1.6 Agronomy, yield and production 37
1.7 Genetics and breeding 51
1.8 Post-harvest technology 64
1.9 Market development 68
1.10 Future potential 69
References 71

2 *Buckwheat* (Fagopyrum esculentum) 85
B.D. Joshi and R.S. Rana
2.1 History and uses 85
2.2 Taxonomy and ecology 89

2.3 Morphology and physiology 93
2.4 Biochemical composition 98
2.5 Agronomy 103
2.6 Post-harvest technology 110
2.7 Genetics and breeding 112
2.8 Trade 119
2.9 Himalayan farmers' perception of the crop 120
2.10 Future tasks 121
References 121

3 *Grain amaranth* (Amaranthus *species*) 129
J.T. Williams and D. Brenner
3.1 History and uses 129
3.2 Taxonomy and ecology 137
3.3 Morphology and anatomy 141
3.4 Physiology of growth and development 145
3.5 Biochemical composition 147
3.6 Agronomy, yield and production 154
3.7 Post-harvest technology 162
3.8 Genetics and plant breeding 169
3.9 Future prospects 177
References 178

4 *Triticale* (x Triticosecale) 187
Anne Hörlein and J. Valentine
4.1 Introduction 187
4.2 Classification 191
4.3 Morphology 192
4.4 Physiological adaptation 194
4.5 Chemical composition and nutritive value 198
4.6 Agronomy, yield and production 199
4.7 Post-harvest technology 205
4.8 Genetics and breeding 207
4.9 Octoploid triticales 212
4.10 Future prospects 213
Acknowledgement 214
References 214

Part Two
A Neglected Cereal and Two Promising Ones

5 *Fonio* (Digitaria exilis *and* D. iburua) 225
N. Haq and F.M. Dania Ogbe
5.1 History and origins 225

5.2 Taxonomy 230
5.3 Biochemistry 235
5.4 Agronomy 237
5.5 Genetics 242
5.6 Gaps in knowledge 243
References 244

6 *Intermediate wheatgrass* (Thinopyrum intermedium) 247
Peggy Wagoner
6.1 Introduction 247
6.2 Crop history 248
6.3 Taxonomy 248
6.4 Origin and ecology 251
6.5 Crop morphology 251
6.6 Nutritional composition and food use potential 252
6.7 Agronomy 252
6.8 Economics of intermediate wheatgrass grain production 254
6.9 Breeding and varietal development 255
6.10 Market development 256
6.11 Future potential 257
References 258

7 *Wild rice* (Zizania palustris) 261
M.R. Duvall
7.1 History 261
7.2 Taxonomy and ecology 262
7.3 Morphology and anatomy 263
7.4 Physiology of growth and development 265
7.5 Biochemical composition 266
7.6 Agronomy, yield and production 267
7.7 Diseases and pests 268
7.8 Post-harvest technology 268
7.9 Genetics and breeding 269
7.10 Future potential 269
References 270

***Postscript* 272**
N. Haq

Index 275

Contributors

D. Brenner
USDA/ARS,
Plant Introduction Station,
Iowa State University,
Ames,
Iowa 50011,
USA.

M.R. Duvall
Department of Botany and Plant Sciences,
University of California,
Riverside,
CA 92521-0124,
USA.

J.E. Fleming
Department of Genetics,
Cambridge University,
Cambridge,
CB2 3EH,
UK.

N.W. Galwey
Department of Genetics,
Cambridge University,
Cambridge,
CB2 3EH,
UK.

N. Haq
International Centre for Underutilized Crops,
Department of Biology,
Southampton University,
Southampton,
SO9 3TU,
UK.

Anne J. Hörlein
UK Research and Higher Education European Office,
83 Rue de la Loi,
1040 Brussels,
BELGIUM.

B.D. Joshi
National Bureau of Plant Genetic Resources,
Regional Station,
Phagli,
Shimla-171 021,
INDIA.

Folu M. Dania Ogbe
Department of Botany,
University of Benin,
PMB 1154,
Benin City,
NIGERIA.

R.S. Rana
National Bureau of Plant Genetic Resources,
Pusa Campus,
New Delhi 110 012,
INDIA.

J. Valentine
AFRC Institute of Grassland and Environmental Research,
Aberystwyth,
SY23 3EB,
Wales,
UK.

Peggy Wagoner
Rodale Institute Research Center,
611 Siegriedale Road,
Kutztown,
PA 19530,
USA.

J.T. Williams
1, Scott's Circle NW
Apartment 102
Washington DC 20036
USA.

Introduction: development of underutilized cereals and pseudocereals

J.T. Williams

BACKGROUND

During the past 20 years there has been renewed interest in economic botany. This stems from a number of reasons and developments, both scientific and those related to international development.

First of all, in relation to major crops, in many developing areas of the world the carrying capacity of the land was rapidly being exceeded because of increasing population growth. Not only was the application of technology needed to modernize agriculture in the 1960s and 1970s (the process now referred to as the Green Revolution) but somewhat later international attention also became focused on the needs to change cropping cycles and to incorporate other crops, including more minor ones, into the cropping cycles along with the major staples.

For those cereals which feed the majority of people – wheat, rice and maize – our depth of evolutionary understanding was greatly broadened in the 1960s and 1970s by the results of research on the origins and diversity of the crop genepools and the relationships between the species of the genepools. Concerns, during the same period, about genetic erosion and also on genetic vulnerability as a potential hazard in mono-cropping and continuous cultivation, led to concerted action on crop genetic resources. Although it was clear that priority attention had to be given to the major cereals, other cereals and pseudocereals began to receive some attention. Much of the progress reported on for some crops in this book would not have been possible without the collection and description of germplasm and I feel proud to have been

Cereals and Pseudocereals. Edited by J.T. Williams. Published in 1995 by Chapman & Hall, 2–6 Boundary Row, London SE1 8HN. ISBN 0 412 46570 1.

instrumental in this as the Director of the International Board for Plant Genetic Resources.

Apart from wheat, rice and maize, in the 1960s international action to expand production of other cereals was also mobilized: barley, sorghum and pearl millet, all three being ancient domesticates, and also triticale, a new man-made crop.

Second, food quality, in terms of protein content and amino acid composition, had become an important international issue in view of widespread malnutrition. Unusually high protein levels were noted in a number of underutilized crop species, e.g. in the pseudocereals quinoa, and grain amaranths, or in legumes such as the winged bean. The major cereals, of course, provide protein in the diet (as much as 50%), and they also become feed for livestock, and the production of meat, eggs, milk and cheese for human food adds to the protein intake. Enhancing the quality protein in major cereals has not been easy, as witness the 'Opaque-2' maize development identified in 1963. The International Maize and Wheat Improvement Center (CIMMYT) found a number of difficulties in yields, an unacceptable kernel, greater susceptibility to pathogens and a shift by nutritionists in their perceptions of the magnitude of protein malnutrition in poorly fed populations. These problems caused a set-back in the research in the 1970s which is now re-organized and continuing (National Research Council, 1989; Hardin, 1990). Whilst such efforts have been ongoing to improve nutrition of major cereal grains, consensus opinions have emerged on the value of some of the more minor crops in the human diet. There was even a tendency to 'over-sell' some of the minor crops. Such promotion could hardly be justified due to the need for lengthy research and development in an integrated multidisciplinary crop improvement framework when other research priorities were more important. The latter were necessitated by the expected increase in human population and hence needs to implement systems for greatly increased agricultural productivity and food delivery systems. These systems have to be in place by around the year 2000.

Third, the international emphasis on conservation of biodiversity and wiser use of resources towards sustainable development has spurred increased attention to actual uses of plants by traditional communities and to the potential uses of many wild plants as well as the use of crops which are very minor in terms of their acreage. Associated with these interests are those focused on alternative agriculture. Reading the literature one almost feels that we stand on the eve of a new crop revolution, but this is a complex issue fraught with misunderstandings. In fact, any new crop development will probably be more easily implemented in the developed world, rather than in the developing world, because the former may be able to absorb the risk.

Certainly, the grower, who is probably a resource-poor farmer, is in no position to absorb any risk.

Rather than being about to see many new crops coming on the scene, there will be a limited number, and such development has to be based on good science, backed by significant research and development over a time horizon of more than a few years. We have, in recent years, witnessed some successful examples: wheat in Bangladesh, cassava in north-east Thailand and several food legumes introduced to new areas of the tropics and subtropics. In the case of cereals and pseudocereals, triticale and amaranths, dealt with later in this book, also provide success stories.

Fourth, international trade and pricing have a major effect on agricultural production, as do the fluctuations in value of currencies. If a drop in value of a currency makes the cost of an imported commodity too high then politicians begin to urge the production of domestic alternatives. Similarly, if the international price of a commodity falls then a nation may start looking for possible alternative crops. Usually these will be those associated with high value-added prices rather than staple crops. Low agricultural prices of major commodities in recent years have in fact opened the door in some instances to new or underutilized crops through increased interest in diversification.

Against the background resulting from these four trends it has to be firmly recognized that world agriculture has reached a stage of global surplus for several key commodities. The question needs to be asked: why invest more effort on less important crops when so much is known about the major staples and many researchers and institutions are involved with their production and improvement?

Taking the numerous cereals embracing millets in the broadest sense, we can address that question by looking at the range of species and the importance of each one. Clearly, the most important, pearl millet, is a staple cereal in the drier parts of tropical Africa and in India, where it is the fourth most important cereal after rice, wheat and sorghum. Finger millet assumes importance in tropical areas of Africa and India and it survives when there is well-distributed rainfall during the growing season and the absence of droughts. Hence finger millet and pearl millet are not mutually exclusive, and each has an ecological niche. As well as finger millet there are a number of other small millets which likewise are grown as rain-fed crops under more marginal conditions and with low inputs, in areas where they produce a more dependable harvest compared with the other more major crops. As a result, several of the small millets are now receiving more research attention because this latter attribute has been the reason for their continued cultivation since they were domesticated.

After all, out of the thousands of grass species, only a limited number have been domesticated and those that were domesticated either spread widely or remained of significant regional or local importance. Table 1 lists the small millets which have significance except for those that are only important locally. Of those listed, research attention is already being paid to many of them. *Digitaria*, or fonio, stands out as one which has been neglected, and in answering the question posed, research is indeed justified because of the crop's agro-ecology; small research inputs are likely to provide substantial returns in terms of productivity. Suitable research inputs would consider the crop's role in intercropping, selection and multiplication of superior genotypes, multilocational trials and modest use of fertilizers, all relevant to the situation in the African savanna, debts of nations, their population increases and their struggles for food self-sufficiency. Since yields of fonio can be 600–800 kg/ha and these yields compare extremely well with the normal yields of pearl millet under similar conditions (no fertilizers and cultivation by hand), fonio has been selected for inclusion in this book, even though, by contrast to some of the other crops described, there has been minimal research.

Other justifications for enhanced research are based upon a wider view of rural development than is often taken. The industrialized countries with a minute proportion of the population involved in agriculture, do not often appreciate that many countries which have over two-thirds of the people involved in farming have not only a low level of production from the agricultural sector but also policies rarely appropriate to those farmers who are expected to lead the country to self-sufficiency in food. Experience worldwide with numerous development assistance projects in integrated rural development has pointed the way to new partnership needs between farmers, extension workers and the research scientist. All too often, research, especially on underutilized crops with potential, has not been well slotted into this framework. Hence there is a feeling of frustration by promoters of such research. However, some crops discussed in this book have fitted into this framework, notably triticale and grain amaranths. There are many lessons from international agricultural research which can be applied to the minor crops, and some are discussed later in this introduction.

PRIORITIES

Faced with feeding a burgeoning world population, many scientists and politicians, not least a number of donor agencies, feel that attention to anything other than the major cereals has no future. Whilst this may be true for the productive agricultural lands it is certainly not so for the more harsh marginal lands nor for degraded lands where a number of the minor cereals and pseudocereals can produce more efficiently.

Table 1 Small millets

Species		*Origin*	*Cultivation*	*Remarks*
More important				
Eleusine coracana[a]	Finger millet	Africa	Widely grown in Africa and Asia, some in Europe and Latin America	Substantial literature base. Researched by ICRISAT and numerous countries
Panicum miliaceum[a]	Proso millet	Eurasia	Widespread across Eurasia, some in East Africa	Moderate knowledge base. Active research in India, Russia, Kenya and Bangladesh
Setaria italica[a]	Foxtail millet	Eurasia	Widespread across Eurasia also in North and East Africa, fodder in the New World	Moderate knowledge base. Active research in India, China, Bangladesh, Sri Lanka and Kenya
Less important				
Echinochloa colona[b]	Barnyard millet	Eurasia	Extensive in Central India	Adequate knowledge base. Research in India
E. frumentacea[b]	Barnyard millet	Eurasia	Extensive in South and East Asia	Moderate knowledge base. Some research in Asia
Paspalum scrobiculatum[b]	Kodo millet	India	Widespread in Old World tropics	Moderate knowledge base
Coix lacryma-jobi[b]		Asia	South, South-east and East Asia. Introduced to central Africa, Caribbean and Brazil.	Moderate knowledge base. Major research in India
Eragrostis tef[b]	Teff	Ethiopia	Cultivated in Ethiopia, used as hay in Kenya, South Africa, Australia. New in USA.	Reasonable knowledge base. Suitable research ongoing
Digitaria spp.[c]	Fonio	West Africa	Cultivated across semi-arid West Africa.	Poor knowledge base. Underutilized in view of potential

[a] Germplasm held for the global community by ICRISAT (International Crops Research Institute for the Semi-Arid Tropics).
[b] Reasonably representative germplasm collections in national programmes.
[c] Germplasm collections not representative of geographical range.

Even with this myopic view there are tendencies to reduce investments in productivity research even of the major staple crops. This is largely as a result of the need to increase research on natural resources. This is unwise, since productivity research has to sustain past yields through maintenance research, close yield gaps and raise the yield ceilings (Plucknett, 1993).

We should also not lose sight of the fact that the reason the major cereals are now so successful is because, although they originated under primitive farming systems, in very recent times scientific innovation has transformed them. The yield take-offs, i.e. when rapid advances were made, in cereals in Europe, North America and other advanced countries, were mostly after World War II. Those of the Green Revolution were 20 years later. Such scientific innovation produced a new crop such as semi-dwarf wheat in the 1960s, a crop that has made a massive contribution to economic growth and development. With such experience to draw upon, even if not fully appreciated by politicians or others, the scientific community can probably condense the time span in which the major cereals underwent slow improvement – millennia in all cases – and combine this phase with that of more recent scientific innovation for the improvement of minor crops especially if they possess the required ecological adaptations. In some cases this process may be more successful and cost-effective than trying to adapt the major staple cereals to difficult environments.

Although the comments above relate to food production, particularly in the less affluent parts of the world, there are rapidly changing consumer diets in the industrialized countries which fully justify the production of a number of minor crops to satisfy speciality needs.

In some cases, the two driving forces – enhanced food production and satisfying market demands – overlap. In the case of grain amaranths, aims are production for food in Latin America, Africa and Asia, and potential production for industrial starch purposes and production for speciality foods, as well as part of an option to reduce surplus agricultural production, in Europe (Denmark and the UK) and North America. However, in Mexico production is also being commercialized to produce for export, conversely the value of amaranths on hill-lands is under investigation in the mid-south USA.

The need for priorities is essential to mobilize limited research funding. The background to international priority setting for underutilized crops has been largely missing from the relevant publications of the past 20 years. Nonetheless these have emerged at the national level, for example, in the USA (Conrad, 1993), Netherlands (van Soest, 1993), India (Mal and Joshi, 1991) and Nigeria and neighbouring countries (Diversity, 1990). The extreme situations between industrialized countries and the traditional subsector of agriculture in economic dualism in poorer countries may seem so diverse that there can be no common background to priority setting. However, Williams and Haq (1993) have itemized a number of criteria relevant to such a consideration. These include the following.

1. Development of sustainable production systems can be justified on a number of fragile resources areas (Ruttan, 1991). These include the semi-arid tropics of Africa, arid areas such as the south-western USA and others. Such developments are likely to be adjuncts to food supply systems rather than important components.
2. New crop production can be justified on degraded lands which are marginal to food production. In many cases crops are needed to generate off-farm income and full consideration needs to be given if new institutional arrangements are needed (Schuh, 1990).
3. Since many of the sustainability and environmental issues will be addressed through sustainable crop production on the robust agricultural lands, due attention to multiple cropping and farming systems research will identify some role for minor species.
4. Much as it should be possible to propose public/private sector partnerships this is likely to remain difficult for some time outside the developed world. In less developed areas, food and food processing ranks low in the interests of the private sector. Private interest is rarely mobilized until a crop is commercial, hence the degree to which the programme will move towards that is an important criterion.
5. Underutilized crops may be the most obvious and cost-effective way of satisfying market demands for particular products.

Initially the selection of crops for this book was to have been based on a group which had received major research attention and yet still needed continued research and development and on a smaller group where less research had shown excellent potential but which was further away from wider adoption. Both groups satisfied the criteria outlined above. The former group includes quinoa, buckwheat, grain amaranths and triticale but to these we have added fonio because of its neglect and its known potential. The second group is limited to two specific examples where research has been carried out and where further work is fully justifiable.

CONTEXT IN WHICH UNDERUTILIZED CROPS FIT

Whereas there is always the possibility that a crop can be developed to supply a specific product in demand internationally, a chemical for industry or a flavouring for example, development of a minor cereal or pseudocereal will rarely fit into this category. At most they will supplement normal cereal production, maybe providing steady yields in more difficult environments, excepting when they provide a niche for speciality foods in more affluent societies.

In the context of the imperatives of global agricultural research, the world has learned how to produce enough food and how to increase agricultural production. The debate that remains is how to reduce poverty (Mellor, 1990) and increase returns from technical change. As land pressure increases there is an urgent need to intensify smallholder agriculture (Lele and Mellor, 1990) and it is usually this group of farmers who have continued for generations the cultivation of many of the underutilized crops.

Rather than sweeping away traditional cultural practices the recent emphasis on sustainability could, through integrated action, lead to the rational development of some of the crops in suitable areas. But we must not automatically retain traditional crops and traditional cultivation practices for the sake of retaining them; crops have been lost from areas many times in the past and agricultural systems do not remain static in perpetuity.

Sustainable agriculture has become the watchword for both countries and donor agencies, and appropriate land utilization and protection of the environment and the natural resource base is now widely recognized. Traditionally the productivity of the land has been sustained by different technologies – bush fallow, mixed cropping, shifting cultivation and many others in wide areas of the tropics and subtropics. These were compatible with the ecological environment when there was low population pressure. More recent research (as opposed to unwise changes in traditional systems which lead to soil erosion, nutrient imbalance, desertification, and other negative results) is now being applied to modernization of traditional cultivation systems to provide continual cropping on tropical soils. For instance, slash and burn which allows cropping only one year or less in 10 can be replaced with adjusted tillage systems. Better agro-ecological approaches to the uplands and marginal soils might well include, with benefit, a number of underutilized crops.

Maybe the most striking example of the more recent use of more minor crops has been within the framework of diversified cropping in rice-based farming systems where concentrated research is finding better ways to grow 'upland' rice crops during the dry season. The aims are not to replace rice but to find ways to grow other crops and manage irrigation. Attention to multiple cropping in the rice-based system in Asia led to the rapid development of wheat in Bangladesh as a new crop.

Crop diversification, in whatever farming system or agro-ecological zone, has the effect of reducing the risk in crop production. For instance, the Philippines programme in this area is clearly a market-led crop diversification and that in India is aimed at technologically feasible and economically viable changes in existing cropping systems (dominated by the rice-wheat system) towards more balanced farming systems to meet the needs for food, fibre, fodder and fuel (Rao, 1989).

Use of underutilized cereals and pseudocereals clearly fits this perspective; the key to their adoption will be the ready availability of knowledge and suitable germplasm. Such knowledge is gathered together for a handful of species in this book.

Some scientists, other than those actively promoting the use of underutilized species, go even further, and include in the key to sustained or increased agricultural production 'explorations for new crops among the world's existing vegetation' (Wittwer, 1988).

INSTITUTIONAL FRAMEWORKS

The underutilized cereals and pseudocereals described in the book have been researched through a variety of mechanisms. In some cases the knowledge has been amassed through the research of individual scientists, fonio being an example. Another crop has been promoted because it is a 'mandate' crop of an International Agricultural Research Centre (triticale, researched by CIMMYT, the International Maize and Wheat Improvement Center, since 1963). For others, research has advanced through donor support, and in some cases research is fostered through associations of interested scientists or growers.

Unless stable funding is provided by a national government the international centre approach is very attractive in ensuring that the research progresses and has short- to medium-term pay-offs. This is largely because international centres tend to use a co-ordinated networking approach, involving all interested scientists and institutes as partners. However, a number of agricultural networks has been established which act as proxies for international centres; some have become absorbed, others have remained separate. Much has been written about the networking approach recently (Faris, 1991; Plucknett *et al.*, 1990; Winkelmann, 1987; SPAAR, 1987), but essentially networks aim to bring together effectively an interdisciplinary critical mass of research effort. In the case of underutilized crops, it is critical that those scientists involved give due thought to strategic planning for this type of activity. This is especially true when germplasm is being screened for length of growing season, or time to maturity, as well as for adaptations for yields. Williams and Haq (1993) stated that for new crops 'quantum jumps in establishment of policy will emerge when attempts are made, on the one hand, to develop adequate information and data, and on the other, to effectively work together'.

USEFUL LESSONS FOR FURTHER RESEARCH AND DEVELOPMENT

In order to plan research, development and utilization of a new crop or an underutilized crop a number of lessons can be drawn from experience with two crops described in this book, namely triticale and grain amaranths.

In the case of triticale, the first appeared as a botanical curiosity last century. In this century it became the subject of attention by breeders in Europe, primarily in Sweden and Russia, but by the 1950s it was the subject of research in other countries, primarily Canada and Japan. Hardin (1990) details the start of the internationalization of research through a joint University of Manitoba, Canada – CIMMYT programme and by 1970 major funding was mobilized by two Canadian agencies: the International Development Research Centre (IDRC) and the Canadian International Development Agency. Hardin notes that, despite the scientific advances, enthusiasts and private firms pushed the crop prematurely and a backlash developed from disenchanted farmers, seed companies and university scientists. This had to be overcome by the efforts of CIMMYT. Triticale, at present, represents the product of new crop research started in the 1950s and is a crop that has been developed without a comprehensive initial plan to include production, processing and marketing (Johnson, 1990). Triticale has not yet reached the stage of rivalling other common bulk grains but it has become a successful specialized-use crop and still has potential to be realized.

International action on grain amaranths was spurred by the recognition of the high lysine content of the seed, by M.T.S. Downton of the Australian National University, Canberra, and by J. Robson of the University of Michigan, USA. Rather as for triticale, a champion was found, in this case the Rodale Research Center in Pennsylvania. Nonetheless the research on amaranth differed from that on triticale, by virtue of the material. Rather than having to make thousands of crosses initially, germplasm had to be assembled, tested and selected and new varieties developed. This all started in 1974 and research on breeding went on in parallel with research on agronomy and genetics in several institutions and also in India. International collaboration developed by 1982 with programmes in Guatemala, Kenya, Mexico, Peru and Thailand through donor support, and an international testing network developed.

Nor was the amaranth work developed through initial planning or international collaboration, although strategy was well defined at the Rodale Research Center since pressure was on for the development of a commercial crop. However, the research was successful, even though it might have been better planned (see flow chart). Despite enthusiasts promoting the crop the real push came from a series of seminars and

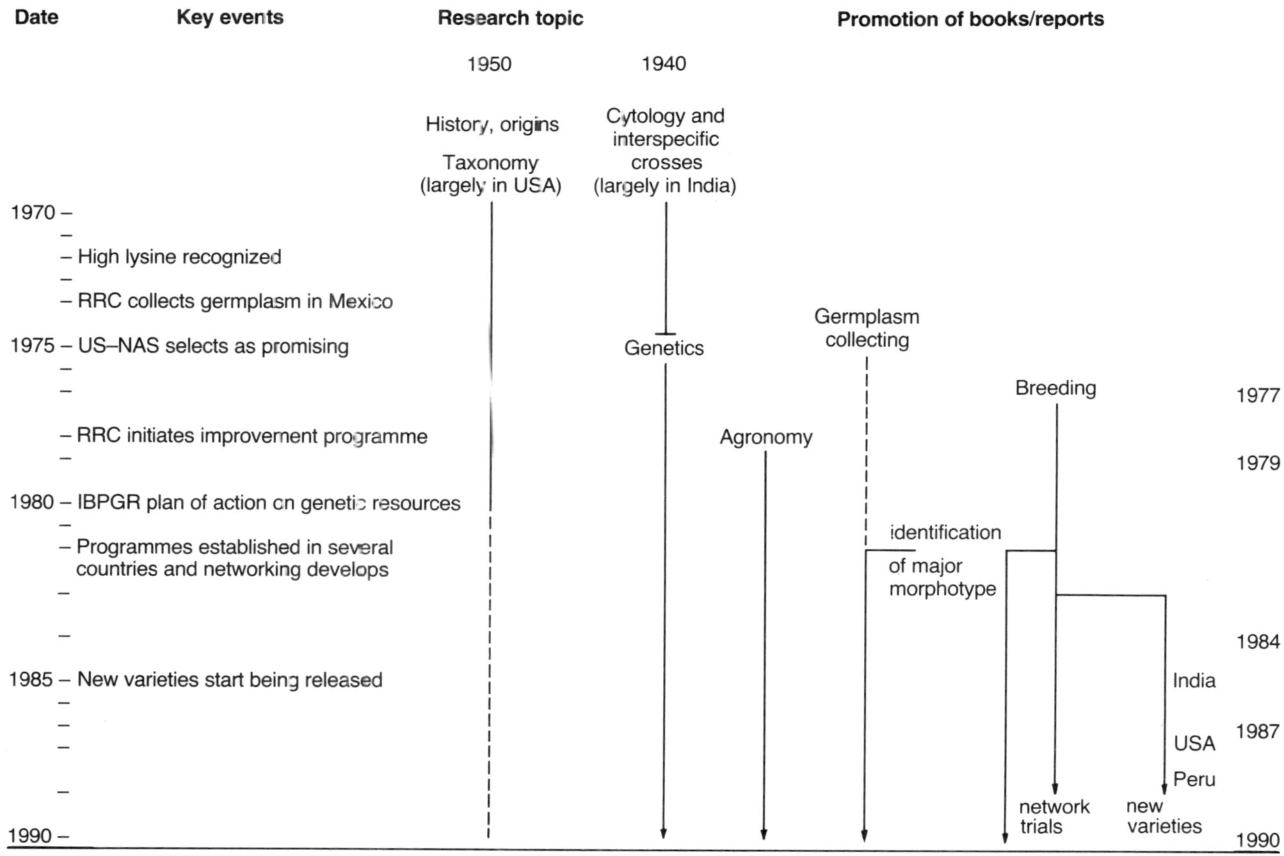
Date
Key events
Research topic
Promotion of books/reports
1950
1940
History, origins
Taxonomy (largely in USA)
Cytology and interspecific crosses (largely in India)
1970
High lysine recognized
RRC collects germplasm in Mexico
1975
US–NAS selects as promising
Genetics
Germplasm collecting
Breeding
1977
RRC initiates improvement programme
Agronomy
1979
1980
IBPGR plan of action on genetic resources
identification of major morphotype
Programmes established in several countries and networking develops
1984
1985
New varieties start being released
India
1987
USA
Peru
network trials
new varieties
1990
1990

publicity articles combined with reviews of progress and identification of constraints.

Both these crops perhaps suffered from the lack of an agro-ecological approach early in the research and the need of socio-economic evaluation at the producer level. The latter is being identified as a most important factor following analysis of research programmes. There is a need to include socio-economic research as an integrated part of the crop improvement process. Apart from experience at international centres, this has also been vividly shown by a number of co-operative crop improvement programmes between US universities and institutions in the developing world supported by US-AID and it has also emerged as one of the most important areas for the International Network on Bamboo and Rattan, currently a network supported by IDRC and the International Fund for Agricultural Development.

Analysing the progress and the results obtained for both triticale and grain amaranths provides a basis for any of the other crops described in this book to be developed according to a strategic plan. Even for those crops where the research dates back hardly a decade, such as quinoa, there is benefit in the approach and it should not be difficult for the one neglected crop, fonio, to be developed in a sound manner through strategic planning and networking.

REFERENCES

Conrad, J.W. (1993) National new crops policy, in *New Crops*, (eds J. Janick and J.E. Simon), Wiley and Sons, New York, pp. 2–4.

Diversity (1990) New plant diversity research effort launched at Nigerian symposium. *Diversity*, **6** (3/4): 7–8.

Faris, D.G. (1991) *Agricultural Research Networks as Development Tools*, IDRC, Ottawa and ICRISAT, Patancheru, India.

Hardin, L.S. (1990) International repercussions of new crops, in *Advances in New Crops*, (eds J. Janick and J.E. Simon), Timber Press, Portland, OR, pp. 47–51.

Johnson, D.L. (1990) New grains and pseudograins, in *Advances in New Crops*, (eds J. Janick and J.E. Simon), Timber Press, Portland, OR, pp. 122–7.

Lele, U. and Mellor, J.W. (1990) Agricultural growth, its determinants, and their relationship to world development: an overview, in *Agriculture and Governments in an Interdependent World*, (eds A. Maunder and A. Valdés), Int. Assoc. Agric. Economists, pp. 47–62.

Mal, B. and Joshi, V. (1991) Under-utilised plant resources, in *Plant Genetic Resources. Conservation and Management*, (eds R.S. Paroda and R.K. Arora), IBPGR, New Delhi, pp. 211–19.

Mellor, J.W. (1990) Dramatic poverty reduction in the third world, in *Agricultural Prospects and Challenges in Developing Countries*, (ed. J. Ryan), ACIAR/ANU, Canberra, pp. 17–20.

National Research Council (1989) *Triticale. A Promising Addition to the World's Cereal Grains*, National Research Council, National Academic Press, Washington, D.C.

Plucknett, D.L. (1993) Science and Agricultural Transformation. Lecture at IFPRI, Washington, D.C.

Plucknett, D.L., Smith, N.J.H. and Ozgediz, S. (1990) *Networking in International Agricultural Research*, Cornell University Press, Ithaca, NY.

Rao, T.H. (1989) Irrigation management for diversified cropping in rice-based systems in Andra Pradesh, India, in *Research Network on Irrigation Management for Diversified Cropping in Rice-based Systems*, IIMI, Colombo, Sri Lanka, pp. 41–9.

Schuh, G.E. (1990) Sustainability, marginal areas, and agricultural research. *Development*, **1990** (3/4), 138–43.

SPAAR (1987) Collaborative research networks: desirable characteristics, Special Program for African Agricultural Research, World Bank, Washington, D.C. (mimeographed).

van Soest, L.J.M. (1993) New crop development in Europe, in *New Crops*, (eds J. Janick and J.E. Simon), Wiley and Sons, NY, pp. 30–8.

Williams, J.T. and Haq, N. (1993) International new crops policy, in *New Crops*, (eds J. Janick and J.E. Simon), Wiley and Sons, NY, pp. 5–11.

Winkelmann, D.L. (1987) Networking: some impressions from CIMMYT, in *The Impact of Research on National Agricultural Development*, (eds B. Webster, C. Valverde and A. Fletcher), ISNAR, The Hague, pp. 125–34.

Wittwer, S. (1988) Issues of concern in land use in developing countries, in *Sustaining Land Productivity in Developing Countries*, OICD/USDA and US-AID, Washington, D.C., pp. 15–34.

Part One

Crops Benefiting from Major Research

1

Quinoa (*Chenopodium quinoa*)

J.E. Fleming and N.W. Galwey

1.1 HISTORY, DISTRIBUTION AND USES

Quinoa (*Chenopodium quinoa* Willd.) is a grain crop of Andean origin. Until recently its cultivation was restricted to subsistence farming in some parts of this region of South America. Since about 1975, however, the crop has received considerable attention both within and outside South America, due to a number of attractive features. It is a versatile plant and a prolific biomass producer. It is able to grow well under poor environmental conditions and can be combine-harvested. Being a dicotyledon, it would be suitable for use as a break crop in cereal rotations, allowing better control of Graminaceous weeds and preventing build-ups of cereal pests and diseases. It is good at extracting nutrients left in the soil by other crops and so could contribute to a reduction in agrochemical pollution of the environment. Its excellent amino acid balance and high content of protein (typically 14–18%) make it attractive for human consumption and animal feed. Other attributes of the grain, such as its small starch granules and the presence of saponins, may have industrial applications. These features suggest that quinoa is a crop which merits wider exploitation, and research on quinoa has been carried out in an estimated 22 countries outside South America (National Research Council, 1989). Obstacles to the wider cultivation of quinoa include the lack of an established market and potential antinutritional effects of the saponins. Current yields are too low to enable it to compete economically with cereals except, perhaps, in regions with agriculturally marginal soils. Plant breeding research is underway to improve yields and reduce saponin content in varieties for nutritional use. At the same time market development is underway in several

Cereals and Pseudocereals. Edited by J.T. Williams. Published in 1995 by Chapman & Hall, 2–6 Boundary Row, London SE1 8HN. ISBN 0 412 46570 1.

countries and is aided by new developments in processing the grain into food products.

1.1.1 **History and distribution**

Quinoa is thought to have been domesticated in the Altiplano region surrounding Lake Titicaca, an ancient centre of civilization and crop domestication. This region is now a centre of diversity for the crop. The earliest archaeological records of quinoa date to 5000 BC in Ayacucho, Peru (Uhle, 1919, reported by Tapia, 1979) and to 3000 BC in Chinchorro, Chile (Nuñez, 1970, reported by Tapia, 1979). Quinoa was grown by the Incas throughout their Empire, by the Araucanian Indians of Argentina and Chile, and by the Chibcha Indians of northern Colombia (Tapia, 1979). The exact distributional boundaries of quinoa are unknown (Wilson, 1990), but during Inca and pre-Inca times it is thought to have been grown widely from Bogotá in Colombia (5 °N), southwards through Peru and Bolivia to the island of Chiloe in Chile (42 °S) and south-eastwards to Córdoba in Argentina (Tapia, 1979), being grown from sea-level to altitudes of about 3800 m.

Quinoa was held in an honoured position in Inca society and religion, and was referred to as the 'Mother grain', being second in importance only to the potato in the highest areas of cultivation (Risi and Galwey, 1984). At the time of the Spanish conquest of South America, the Inca Empire was marked by the prolific nature of its agriculture: it is thought that the Incas cultivated as many species as the farmers of all of Asia and Europe combined. It has been estimated that the Andean Indians domesticated as many as 70 species, including root crops, grains, legumes and fruits (Cook, 1925). However, Pizarro and the later Spaniards who conquered Peru suppressed the Indian traditions and destroyed much of their intricate agricultural system (National Research Council, 1989). Crops such as quinoa underwent a decline and were replaced by European species such as wheat and barley (Cardozo and Tapia, 1979). Quinoa, unlike potatoes and tomatoes, was not exploited outside South America. An understanding of the reasons for this decline would help in assessing whether quinoa has potential benefit for modern agriculture. It is thought that the replacement of quinoa with wheat and barley may have been a deliberate strategy on the part of the Spanish, who believed that quinoa was a source of strength and resilience for the Incas. They also wished to discredit the Inca cultural practices and traditions, and undermine their religion. However, the decline of quinoa cannot be entirely attributed to the prejudices of the conquistadors: the European species which the Spaniards introduced were well suited to the environments where quinoa was cultivated, and largely displaced it. The cultivation of about a dozen root crops, two other grains, three legumes and more than a dozen fruits also declined. It is possible that the Spaniards did attempt to introduce quinoa in the Old World: the Inca Garcilaso de la Vega referred to the first exportation of the grain to the Old World, but reported that the crop could not be cultivated as the seeds 'arrived dead' (Tapia, 1979).

The decline in cultivation of quinoa continued almost to the present time. In Peru, for example, the area under quinoa declined rapidly between 1947 and 1975, from 47 000 to 15 000 ha (Tapia, 1979). Since 1975 there has been a resurgence of interest in quinoa, both within and outside South America. This interest has developed due to a recognition of its high nutritional and agricultural value, and due to growing economic pressures within many South American countries to reduce food imports, thus improving national economies. The opening of markets for quinoa outside South America has further helped to halt the decline.

Subsistence farming has remained the predominant agricultural system in the Andean region, and consequently small amounts of quinoa have been continually produced, and the genetic diversity of landraces has been maintained. Today, quinoa continues to be grown in southern Peru and Bolivia as a sole crop. It is also grown on a smaller scale in central Peru, Ecuador, southern Colombia, Argentina and Chile, usually at the household level where it is intercropped. Recently, a sole-crop production area has been developed in Ecuador, to supply the export market to North America. Quinoa is also reported to be grown in south-central Mexico (Ma, Heinstein and McLaughlin, 1989). Figure 1.1 shows the presumed pre-conquest distribution and the present distribution of quinoa in South America.

The Altiplano region surrounding Lake Titicaca, both to the Peruvian north and the Bolivian south, remains the area where quinoa is most genetically diverse and most widely cultivated, 75% of the quinoa produced in Peru being grown on the Altiplano around Puno. It is grown as a sole crop, but its cultivation is largely confined to marginal land (Risi, 1989). In the Mantaro Valley it is also grown as a sole crop, in mechanized agricultural systems, using fertilizers and pesticides in fields of up to 5 ha (Narrea, 1977). In the more northerly Andean valleys it is intercropped with maize or faba beans, or grown as a border crop in potato fields (Narrea, 1976). The area under quinoa cultivation in Peru increased from its minimum of 15 000 ha in 1975 to 25 000 ha in 1980. Thereafter about 20 000 ha has been cultivated annually (Risi, 1989). Quinoa retains greatest importance in Bolivia, with an estimated 40 000 ha being cultivated annually between 1980–1985 (Risi, 1989). It is cultivated on the southern side of Lake Titicaca and in the Salares region to the south of the country. A small amount of quinoa is grown in the Chilean region of the Altiplano, but the main Chilean cultivation area is at sea-level near Concepción in central Chile (Risi and Galwey,

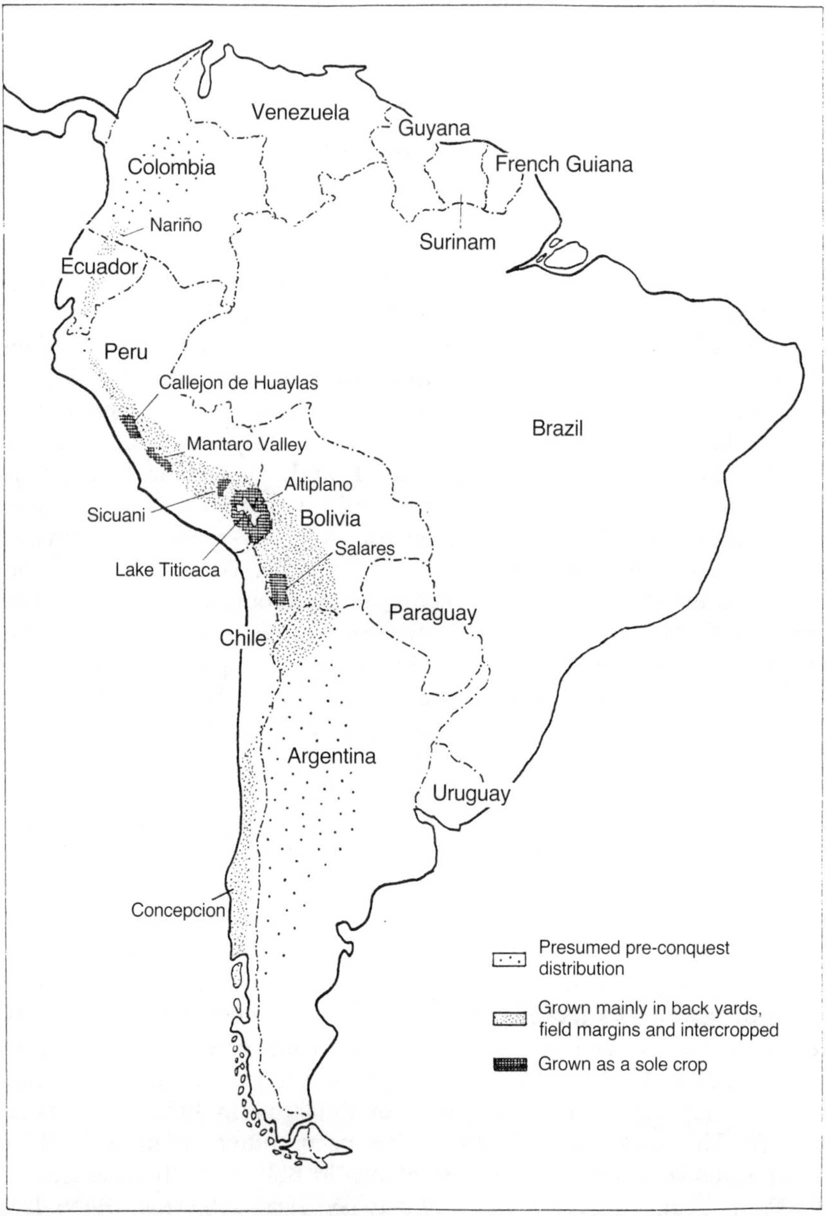

Figure 1.1 Presumed pre-conquest and present distribution of quinoa in South America.

1984). Recent reports suggest that quinoa is being cultivated more widely in Chile, possibly as a result of the promotion of indigenous species. Quinoa is now grown as far south as the Mt Cochrane area (47 °S) (Forster, cited by Wilson, 1990).

There are scattered reports of the cultivation and use of quinoa outside South America. Church (1886, cited by Singh, 1961) reported that quinoa was introduced into India from Peru. Cultivation of quinoa is still reported to be practised in some of the north-eastern regions of India (Partap and Kapoor, 1987) and in Pakistan. Quinoa was also introduced to Kenya where it showed some promise as a grain plant for land between 7000–8000 ft (Elmer, 1942). Quinoa was cultivated in Germany during World War I (Anon,. 1930) and it is reported to have been an essential staple in Japan during World War II (F. Fabricant, undated newspaper cutting). However, it is said to be unpopular there today as it is still associated with hard times and poverty. Seeds of *Chenopodium album* have also been used as food in Russia, Denmark, Greece and northern Italy. In most areas where Chenopods have been used they are seen as a low status food due to their association with difficult times.

In the last decade a more concerted effort has been made to introduce the cultivation of quinoa as a grain and forage crop outside South America. Early in the 1980s, the value of using quinoa as a break crop in cereal rotations in eastern Britain was recognized and research was initiated to adapt it to the agricultural and climatic conditions of this region. At about the same time, Sierra Blanca Associates in Colorado, USA, began to import grain from South America for sale in the North American health food market (Cusack, 1984). A breeding programme for adapting the crop for cultivation in the Rocky Mountains was also initiated (Cusack, 1983). In addition to its cultivation in Britain and the USA, quinoa is also being grown in Sweden (where it is grown on an experimental basis for forage (Carlsson, 1984)); in Finland (Carmen, 1984); in Holland (Sahin, reported by Galwey, 1989); in West Germany (Ritter, reported by Galwey, 1989), and in Denmark (for both grain (Jacobsen and Jorgensen, in press) and forage (Haaber, 1991)). Recently, the potential of quinoa starch for specialized industrial applications has attracted attention. It can be expected that additional uses will be found for this crop and that interest in it, both within and outside South America, will grow.

1.2 MORPHOLOGY

Quinoa is an annual gynomonoecious plant, varying in height from 0.4–3.0 m, with an erect stem which may be either branched or unbranched and which bears alternate leaves (Figure 1.2). Its growth cycle varies from 150–220 days. The root system is highly ramified and its

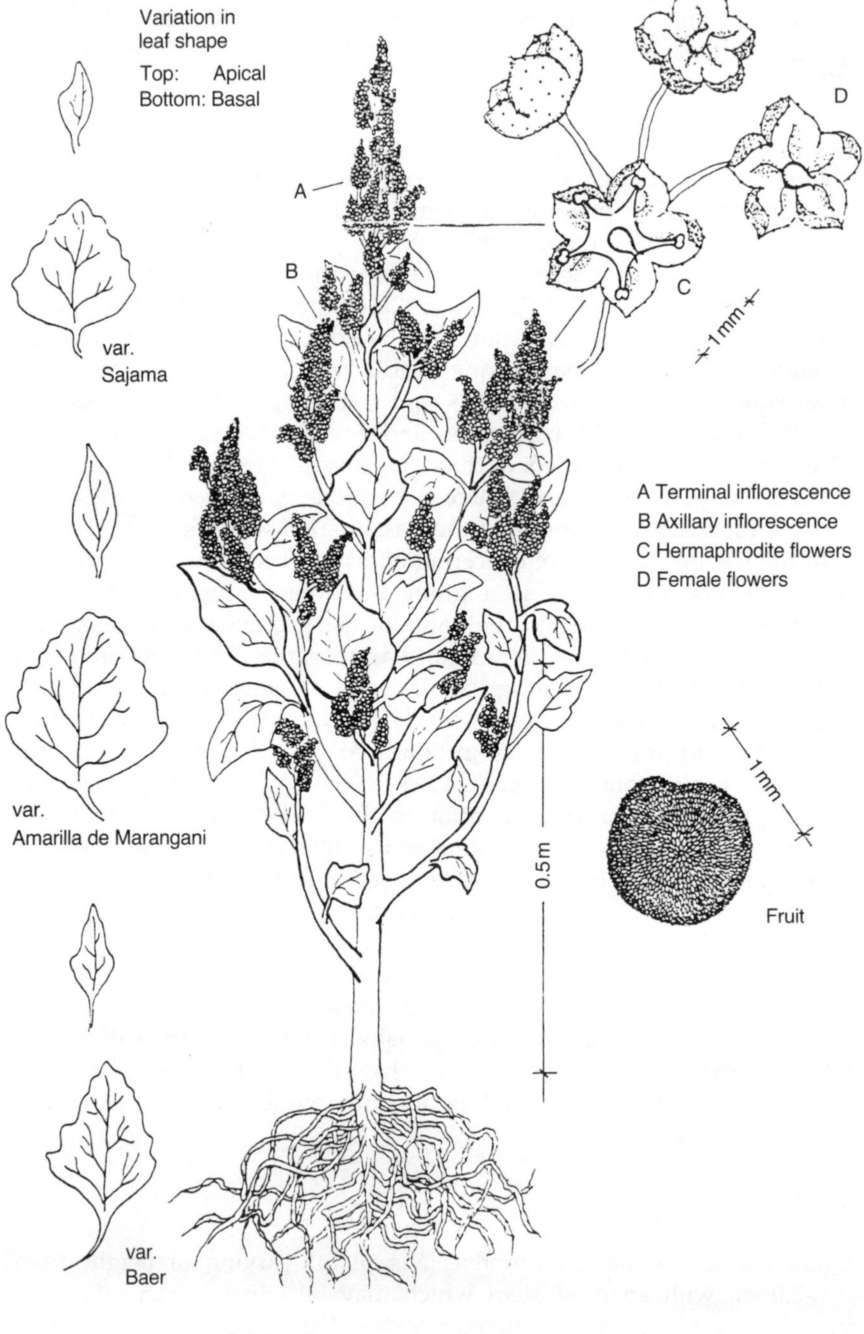

Figure 1.2 Morphological details of quinoa.

depth is roughly equivalent to plant height. Plant height and the amount of stem branching depend on variety and sowing density (Risi and Galwey, 1991b); branched plants tend to have thicker stems (Risi and Galwey, 1989b). Stem colour is variable and may be uniformly green, yellow, red, purple or orange, or may be green with stripes of another colour (Lescano, 1980). At maturity the colour changes to pale yellow or red.

The leaf petiole is long and narrow, and is grooved on its upper surface. Leaf lamina shape is polymorphic within a single plant: lower leaves tend to be rhomboidal, whereas upper leaves tend to be lanceolate (Hunziker, 1943; Nelson, 1968). Leaf axils may have a purple pigmentation. Young leaves are covered with a granular vesiculate pubescence which varies in colour from white to purple (Gandarillas, 1979a). The number of indentations on the lamina border varies from 0–27 (Risi and Galwey, 1989a), being characteristic of the variety (Gandarillas, 1968). As the leaves mature, they change colour from green to yellow, red or purple.

The inflorescence of quinoa is a panicle, having a principal axis from which secondary axes originate. The form of the inflorescence varies considerably among varieties. In the amaranthiform type, the groups of flowers are directly inserted into the secondary axes, whereas in the glomerulate type, they are inserted into the tertiary axes. The length of the inflorescence also varies from about 6–70 cm, depending on variety and environment. Inflorescence diameter varies from about 1–13 cm (Risi and Galwey, 1989a).

Quinoa fruits are disc-shaped and range in diameter from 1–3 mm. The major anatomical parts of the fruit are the pericarp, the perisperm and the embryo. After harvesting, the fruit is often still enclosed in the perianth, which is easily removed by rubbing. The pericarp may be translucent, white, yellow, orange, pink, red, brown, grey or black (Lescano, 1981). It consists of a dense compact layer of cells about 10 μm thick. Beneath this there are two more seed-coat layers, one about 20 μm thick, containing polygonal starch granules and electron-dense regions. The other layer, 3 μm thick, is cemented to the perisperm and can be considered as the cuticle. The next layer of the grain is the episperm, which may also be pigmented. The embryo takes up about 60% of the volume within the episperm, the remainder being endosperm. The storage reserves for the developing embryo are located in the perisperm, in the centre of the fruit. The cells of the perisperm are filled with starch granules which occur either singly or as compound structures consisting of spherical or oblong aggregates. The granules are surrounded by matrix protein (Varriano-Marston and de Francisco, 1984).

Risi and Galwey (1989a, 1989b) investigated the relationships between various morphological characters (and saponin content) in a

Table 1.1 Correlations between plant and inflorescence dimensions[a]

	Plant height	Stem diameter	No. of leaf protrusions[b]	Inflorescence length	Inflorescence diameter
Stem diameter	0.501 ***				
No. of leaf protrusions[a]	0.430 ***	0.222 ***			
Inflorescence length	0.862 ***	0.461 ***	0.421 ***		
Inflorescence diameter	0.792 ***	0.409 ***	0.374 ***	0.832 ***	
Saponin content[b]	0.283 ***	0.223 ***	0.257 ***	0.322 ***	0.257 ***

*** $P < 0.001$
[a]From Risi and Galwey (1989a).
[b]Transformed to square root.

wide range of genotypes, and found strong correlations between plant and inflorescence dimensions (Table 1.1). The glomerulate inflorescence form is very common among the types of quinoa grown in the Andean valleys north of the Altiplano (Aguirre and Tapia, 1982), and a high proportion of plants with glomerulate inflorescences have a branched plant habit. This combination is commonly observed in accessions from Colombia, Ecuador and Peru, with the exception of the department of Puno. Laxness of inflorescence is also common among such accessions. Compact inflorescences and unpigmented axils are strongly associated, as are green plants and unpigmented axils. White seeds are associated with green plants, unpigmented axils, glomerulate inflorescences and/or compact inflorescences. The association between this group of characteristics probably reflects breeders' selection criteria. The glomerulate and lax inflorescence characters are highly significantly associated; this presumably represents the ancestral form of the inflorescence (Risi and Galwey, 1989a).

1.3 TAXONOMY

Quinoa is a member of the family Chenopodiaceae. This family also contains crops of Old World origin, namely sugar beet, beetroot and mangold (all *Beta vulgaris* L.) and spinach (*Spinacea oleracea* L.). The genus *Chenopodium* contains about 250 species (Giusti, 1970) and is divided into 16 sections. The domesticated species of the genus are placed in subsections Lejosperma and Cellulata of section *Chenopodium* (Aellen, 1960). Several *Chenopodium* species are cultivated as food plants. *Chenopodium quinoa* Willd. (quinoa) and *Chenopodium pallidicaule* Aellen. (cañihua) are cultivated for grain in South America, the latter being cultivated particularly in the highest regions of Peru and Bolivia (Risi,

1989) where little else will grow. *Chenopodium berlandieri* Moq. subsp. *nuttalliae* Safford (hauzontle) is cultivated in Mexico, mainly for use as a vegetable, though it is sometimes also grown for grain (Wilson, 1990). A chenopod, designated *Chenopodium album* L., is cultivated in the western Himalayas (especially the state of Himachal Pradesh), mainly as a vegetable and fodder crop. In times of scarcity dried stems are also used as fuel (Partap and Kapoor, 1987). The exact taxonomic position of this chenopod is uncertain, as *Chenopodium album* is often used as a 'convenient taxonomic receptable' for species that are otherwise difficult to assign (Wilson, 1980). There are also several important weed species in the genus *Chenopodium*. Another form of *Chenopodium album* is a weed with a worldwide distribution, and is commonly known in Britain as fat hen. South American weed species include *Chenopodium hircinum* Schrader (known as *quinua silvestre*, 'wild quinoa'), *Chenopodium murale* L. (*hierba de gallinazo*, 'vulture herb') and *Chenopodium ambrosioides* L. (*paico*).

The somatic chromosome number of quinoa is 2n = 36. Quinoa is an allotetraploid, though it is functionally diploid due to the presence of genes which suppress pairing between homologous chromosomes (Risi and Galwey, 1984). Diploid species in the subsection Cellulata are found only in North America, so it is likely that tetraploidy occurred there, giving rise to species such as the free-living forms of *Chenopodium berlandieri*. Free-living tetraploid species occur in both North and South America, and it is therefore likely that they spread to South America prior to domestication, and that quinoa was domesticated there. If this were the case, its immediate ancestor would have been a free-living South American tetraploid similar to *C. hircinum*. There is evidence to support this hypothesis. Electrophoretic analysis of *C. hircinum* populations from eastern South America has shown that allozyme variation among Andean cultivated types of *C. quinoa* and the weed types associated with them is simply a subset of that observed in *C. hircinum* (Wilson, 1988a). It is thus possible that current Andean populations were derived from lower elevations in the eastern foothills, subsequent structural differentiation occurring at higher elevations. This would suggest that quinoa was domesticated in the southern Altiplano region surrounding Lake Titicaca. From here, small relatively uniform groups of founders would have spread both northward to the Andean valleys and southwards to the Chilean coast, giving rise to the distinct populations of these regions. Alternative theories concerning the origin of quinoa have suggested that it spread from north to south. Aellen (1929) tentatively placed Ecuadorian weed forms as the linking element between a putative North American tetraploid progenitor (possibly *C. berlandieri* subsp. *zschackei* (Murr) Zobel) and South American elements of the section. The distribution of peptidase alleles also supports this theory

(Wilson and Heiser, 1979). There is currently insufficient evidence for a conclusive choice between these two theories, though the theory that quinoa is derived from *C. hircinum* and was domesticated around Lake Titicaca seems the more likely.

Domesticated types of quinoa have been classified according to conspicuous variation in morphological and agronomic characters. Among the characteristics which vary are the length of the life cycle, plant height, degree of branching, the compactness of the inflorescence and its form (glomerulate, with seeds in clusters, or amaranthiform, with seeds in long tapering spikes), the presence of purple or red betacyanin pigment throughout the plant, or of purple pigment in the axils, and the colour of the seed, which varies from black to white, through red, orange and other colours. Gandarillas (1968) classified samples of quinoa collected from Peru, Bolivia and Ecuador into 17 races on the basis of seed, inflorescence, leaf and whole-plant characteristics. The most widely used classification is that of Tapia *et al.* (1980), who defined five ecotypes on the basis of agronomic and morphological characters and geographical origins. Details of this classification are given below, supplemented in some cases by information from a study of the pattern of genetic diversity in a large germplasm collection (Risi, 1986; Risi and Galwey, 1989a, 1989b; Galwey, 1989):

1. *Valley type*
   From the Andean valleys at altitudes between 2000 and 4000 m in central Peru and places further north. The plants are tall (2–3 m), branched, have lax inflorescences and are late-maturing (growth cycles longer than 7 months). They are very diverse, but many valley varieties have large white seeds with a low saponin content.
2. *Altiplano type*
   From the area around Lake Titicaca, at an altitude of about 4000 m. Shorter (1.0–1.8 m) and earlier maturing (4–7 months) than the Valley type. Tapia (1979) reported that this type is mostly unbranched, whereas Risi and Galwey (1989a) report that branched plants are common in this ecotype. The plants are mostly pigmented and have small inflorescences. The seeds of the varieties studied by Risi and Galwey had a low saponin content, surprising in heavily pigmented landraces from the centre of the crop's distribution.
3. *Salar type*
   From the Salares, the salt flats of the southern Altiplano in Bolivia, at an altitude of about 4000 m. These varieties are grown in soils with a pH above 8.0, and mostly produce red-pigmented plants with amaranthiform inflorescences. Most varieties have black seeds with a very high saponin content and sharp edges. In other respects they resemble the Altiplano types.

4. *Sea-level type*
   From central and southern Chile, at latitudes of about 40 °S. The earliest-maturing type when grown in England (6 months). They produce short (about 1.5 m), unbranched, green plants, long compact inflorescences, and small, yellow, translucent seeds of medium saponin content. They form a particularly homogeneous group, mainly because they come from an isolated, peripheral part of the geographical range and from a distinctive environment. They have also been influenced by plant breeders to a minor extent.
5. *Subtropical type*
   Tapia (1982) reported that a plant was found in the subtropical Yungas region of Bolivia. It had an intense green coloration that turned orange at maturity and produced very small yellow–orange seeds. This plant may be considered to constitute a fifth group.

Risi and Galwey (1989a, 1989b) classified the accessions in a large germplasm collection into these ecotypes according to information on their origins. A canonical variate analysis on a range of morphological and developmental characteristics provided criteria for distinguishing between ecotypes, but there was considerable overlap between different ecotypes, suggesting that this classification is rather artificial. There was more overlap between the Altiplano, Salar and Valley ecotypes than between these and the Sea-level ecotype. The Subtropical ecotype was found to be very different from the other accessions; however, there was only one accession of this ecotype in the collection.

Electrophoretic analysis of the variation in the crop and its related weed species has also suggested that the division into ecotypes is artificial. However, classification of crop and weed forms of quinoa has not been entirely consistent, and weedy types and domesticates have responded to human selection in different ways. Weeds have responded through physiological adaptation to human disturbance, but do not show marked morphological differentiation from their wild relatives. Domesticates, on the other hand, show many adaptive morphological modifications, such as inflorescence condensation, increased plant and fruit size, and loss of fruit dispersal mechanisms, as well as an important physiological adaptation, loss of seed dormancy. Unlike the crop, weed forms of quinoa have been classified according to the established fruit characters of the genus. Weed forms are generally designated as *Chenopodium quinoa* subsp. *milleanum*, or as Andean taxa of *Chenopodium hircinum*. Although the weeds and domesticates have thus been classified as different species or subspecies, biosystematic studies of the Andean crop/weed complex (Wilson, 1976, 1980, 1981, 1985; Wilson and Heiser, 1979; Wilson *et al.* 1983) produced no evidence of specific differentiation between domesticated and free-living populations in the Andes. Comparative analysis of allozyme and leaf blade data from 187 population samples (Wilson, 1988a, 1988b, 1988c) clearly separated free-living *Chenopodium hircinum* populations from eastern South America from a complex of domesticated and free-living populations in the Andean highlands and along the Chilean coast (*Chenopodium quinoa*). Within this latter group, however, the differentiation found was not between free-living and domesticated groups. Free-living populations of the Andes, known locally as 'ajara' or 'aspha quinua', showed a closer affinity to Andean quinoa than either showed to the coastal quinoa populations of Chile.

The Chilean coastal populations, known locally as 'quingua' and originally cultivated by the Araucanian Indians, are quite homogeneous and are probably early derivatives from the highlands. All have translucent perisperm, glomerulate inflorescences and trullate leaf blades with dentate–lacinate margins, similar to free-living *Chenopodium* species. All cultivated Andean varieties, on the other hand, express the floury-perisperm character to some extent, and many have amaranthiform inflorescences. Acute terminal angles of leaves and toothed margins are very rare in Andean populations. Diffusion to the Chilean coast must therefore have occurred quite early, before these characteristics developed. The uniformity among the Chilean types is probably due to pre-Inca founder events. Cultural isolation could explain why the more advanced types from the highlands did not continue to diffuse down to the coastal plains. Looking at both morphological and electrophoretic information, Wilson has suggested that the variation in the crop is essentially bimodal. However, in addition to the basic distinction between highland and coastal types, the following groupings can be recognized:

1. *Northern populations*
   These occur from Nariño in southern Colombia through to central Peru. They produce tall, bushy plants, characterized by obtuse leaf-blade angles, with shallow marginal sinuations or toothing, compressed fruit with sharp marginal truncation, 'marbled' perisperm and a distinct lack of allozyme variation. This electrophoretic uniformity suggests that these populations, like the coastal populations, were derived from the southern highland populations at a fairly early stage.
2. *Southern populations*
   These extend from southern Peru through the southern Andes and exhibit a wider range of both morphological and electrophoretic variation than the northern populations. They are characterized by acute leaf angles with deeper marginal sinuations and lobing, spherical fruits with floury perisperm and a relatively compact, condensed plant and inflorescence morphology.

3. *Cochabamba Valley populations*
   These are relatively uniform mid-elevation types, morphologically similar to northern populations, but allied to the southern populations with regard to allozyme variation. It is possible that they are present elsewhere in the eastern Andes.
4. *Yungas populations*
   This type is grown at lower elevations (about 1800 m) in a subtropical region northwest of Cochabamba. It is likely to be the most distinctive domesticated element of the Andean complex.

Attempts to classify quinoa should take into consideration the genetic affinity between the domesticated crop and its companion weeds. Weedy types show a wide range of phenotypic variation that corresponds to local habitat and is geographically congruent with variation in the domesticated populations. Thus, northern ajaras have the same morphology as the 'Valley' type quinoa, while ajara plants found in the Altiplano resemble the 'Altiplano' type quinoa. The common agricultural weed form of ajara is similar to *Chenopodium album*, and is thus easily spotted and removed from fields of quinoa prior to anthesis. Some plants, however, have the black seeds typical of free-living types, but the high levels of pigmentation, inflorescence compactness and fruit retention characteristic of domesticated quinoa. They are thus harvested and carried into the next generation: free-living elements, or at least genes derived from them, enter the selective domain of the domesticate and produce progeny. The Andean crop/weed complex should thus be viewed as a single clade. It is inappropriate to view either the crop or the weed as being derived from the other: both elements have moved together to the same phyletic distance from the progenitor taxon (Wilson, 1990).

1.4 PHYSIOLOGY OF GROWTH AND DEVELOPMENT

1.4.1 **Phenology**

Developmental processes in quinoa have not been investigated extensively. Flores (1977) divided the life cycle of quinoa into five growth periods:

Growth period 1:	from sowing to emergence
Growth period 2:	from emergence to the formation of the first two true leaves
Growth period 3:	from the two-true-leaf stage to the appearance of the inflorescence

Growth period 4: from the appearance of the inflorescence to anthesis

Growth period 5: from anthesis to maturity.

Flores found that growth periods 3 and 5 were the longest and that there was little variation between varieties in the length of growth periods 1 and 2. Risi and Galwey (1989a), studying the relationships between growth periods in a large germplasm collection, found that the associations between them, though often significant, were usually not strong (Table 1.2). Differences in adaptation to the long days of British summers were expressed from growth period 3 onwards. The ranges, means and coefficients of variation for the durations of growth periods are given in Table 1.3. Risi and Galwey (1989a) also investigated the relationship of growth periods with various morphological characters. The correlations obtained are given in Table 1.4.

Table 1.2 Correlations between durations of growth periods in quinoa

	Growth period 1	Growth period 2	Growth period 3	Growth period 4
Growth period 2	0.589 ***			
Growth period 3	0.346 ***	–0.085 n.s.		
Growth period 4	0.389 ***	–0.078 n.s.	0.786 ***	
Growth period 5	0.121 n.s.	–0.0368 ***	0.328 ***	0.085 n.s.

n.s. not significant
* P < 0.05
** P < 0.01
*** P < 0.001

Source: Risi and Galwey (1989a).

Table 1.3 Ranges, means and coefficients of variation of the durations (in days) of the growth periods

Growth period	*Range*	*Mean*	*C.V. (%)*
1	9–15	11.23	9.0
2	11–19	15.72	7.6
3	41–89	51.22	2.6
4	7–53	14.18	10.6
5	65–137	116.39	2.0

Source: Risi and Galwey (1989a).

Table 1.4 Correlations between durations of growth periods and morphological characteristics

Growth period 1	0.351 ***	0.344 ***	0.387 ***	0.362 ***	0.220 ***	0.316 ***
Growth period 2	−0.088 n.s.	−0.040 n.s.	0.198 **	−0.149 *	−0.058 n.s.	−0.060 n.s.
Growth period 3	0.248 ***	0.179 **	0.271 ***	0.040 n.s.	0.042 n.s.	0.114 n.s.
Growth period 4	0.302 ***	0.190 **	0.297 ***	0.143 *	0.128 *	0.129 *
Growth period 5	0.005 n.s.	0.160 *	−0.251 ***	−0.047 n.s.	0.007 n.s.	−0.068 n.s.
	Plant height	Stem diameter	No. of leaf protrusions[a]	Inflorescence length	Inflorescence diameter	Saponin content[a]

n.s. not significant
* $P < 0.05$
** $P < 0.01$
*** $P < 0.001$
Source: Risi and Galwey (1989a).
[a]Transformed to square root.

1.4.2 **Photoperiod sensitivity**

Reports in the literature have suggested that quinoa is a photoperiod-sensitive crop which is only adapted to high altitude Andean conditions (Cardenas, 1949; León, 1964; Simmonds, 1976) and would not, therefore, be suited to low altitudes or to temperate latitudes with long summer days. However, these conclusions were drawn from observations of a restricted range of genotypes and do not apply to genotypes of the Sea-level ecotype from southern–central Chile (Risi and Galwey, 1984). Throughout most of its geographical range, quinoa is grown at low latitudes and is not subject to much variation in photoperiod, but in southern–central Chile, summer days are up to 14 h in length. Quinoa ecotypes may be expected to be adapted to the length of the growing season in their place of origin, and, therefore, to vary in photoperiod sensitivity with latitude of origin as other *Chenopodium* species are reported to do (Cumming, 1959). Genotypes from high latitudes would be expected to be less sensitive to photoperiod, or to have different photoperiodic requirements from those from low latitudes.

There has been some disagreement in the literature concerning the type of flowering response exhibited by quinoa. Fuller (1949) found that Bolivian quinoa plants would flower under a broad range of photoperiods, and concluded that quinoa is photoperiodically indeterminate, though shorter photoperiods generally favoured the earlier appearance and opening of flowers. Nelson (1968) found that quinoa varieties from near the equator tended to flower later than those from high latitudes and the same altitude, even under fairly short photoperiods, and Simmonds (1965) found that altitude of origin is also important in determining flowering response. He suggested that flowering in quinoa is controlled by nutritional status rather than by photoperiod and that the importance of nutritional status varies among genotypes.

Risi (1986) found large differences in flowering date between varieties when grown at high latitudes in England, possibly due to differences in their sensitivity to long days. However, the duration of the vegetative period was shorter following later sowings and when within-row sowing density was increased. This indicates that, although genetic effects largely determine when flowering occurs, the environment also exerts some influence. Analysis of segregating generations from a diallel cross (Kenwright, 1989) suggested that some aspects of development determining flowering date are under close genetic control, additive effects being more important than dominance effects. The number of days to the initiation of flowering was shown to be under close genetic control, but the actual flowering date was influenced by prevailing weather conditions, anthesis being promoted by warm weather.

Experiments are currently under way to study the pattern of photoperiod sensitivity in more detail in varieties from a range of geographical origins. There is variation between varieties in the length of the juvenile stage and in the magnitude of the effects of photoperiod on development. Long days generally retard reproductive development, but to a varying extent, the most sensitive varieties being those from low latitudes. Some varieties are virtually photoperiod insensitive with respect to flowering. The number of short days required to induce flowering to the maximum extent also varies between varieties. Plants continue to be sensitive to photoperiod after flowering: long days have a strong delaying effect on maturity, again particularly in varieties from low latitudes. However, all varieties examined flowered eventually, even in continuous light. The pattern of flowering in quinoa thus involves at least two processes: there is the basic age-related flowering process on which are superimposed processes to hasten flowering in response to suitable environmental stimuli, of which photoperiod is probably only one. Temperature, nutrient availability and competition are probably also involved. It should therefore be possible to select varieties that are photoperiod insensitive, or varieties which, although photoperiod sensitive, are still early maturing.

1.4.3 **Response to environmental extremes**

Quinoa is grown during the southern hemisphere summer – from September/October to May/June. The rains are concentrated during these months, the mean temperature rises and frosts become less frequent. At altitudes of 3000–4000 m, where quinoa is typically grown, the mean temperature ranges from 7–10 °C. However, the extremes of temperature vary depending on the distance from Lake Titicaca, which has a buffering effect. Both frosts and droughts are common. In most areas where quinoa is grown, the climate is very harsh. It is regarded as a crop for dry years, being able to withstand high insolation, high evapotranspiration, low soil moisture and low dew temperatures at night. It is usually cultivated on marginal soils which are stony and are either poorly or excessively freely drained. These soils are generally low in natural fertility and often have extreme pH values – 4.8 in parts of Peru and Ecuador (Narrea, 1976), and 8.5 in the Bolivian Salares (Tapia, 1979).

There are contradictory reports of frost resistance in quinoa. Romero *et al.* (1977) reported that trials of 5981 and 3597 accessions in successive seasons at an altitude of 3850 m in Catac, Ancash, Peru showed no evidence of frost resistance. Canahua and Rea (1980), however, found that quinoa could tolerate frosts if they occurred before flowering, but that after flowering, significant damage occurred. Three accessions from a collection of 144 were resistant to frosts of −3 °C. In Puno, when quinoa varieties were planted in midsummer (15 January), various degrees of frost resistance were observed. The variety Sajama was damaged by the first frosts in March and was killed in April, but Kanccolla and Blanca de Julí were more resistant, whereas Chewecca was damaged by a frost of −15 °C in July, but by this time was already maturing (Canahua and Rea, 1980). At Illpa, Puno, 134 quinoa accessions were found to have some frost resistance, and of these, 12 were resistant to temperatures of −4 °C (Valdivia, 1979). Most of the resistant varieties had red stems.

Frost resistance is an important character for the successful development of quinoa for temperate regions, where frosts are common in early spring. Experiments in Cambridge with the Chilean varieties Baer, Faro, Litu, Valdivia, and Pichaman, and the Peruvian variety Blanca de Junín, sown on 25 March, withstood seven frosts, one of them severe (−5 °C), during their early growth. The Peruvian variety, which was late maturing, was not damaged by autumn frosts (Risi and Galwey, 1984).

Droughts are common on the Altiplano, though they are generally due to bad distribution of rains rather than a deficiency in total rainfall (Aquize, 1977). A serious drought in Bolivia in 1983 caused losses in the potato and barley crops of 66% and 54% respectively, whereas the production of quinoa was barely affected (Cusack, 1984). Quinoa's drought resistance is probably due to its deep, highly ramified root systems, but also to the vesiculate pubescence on the leaves. The vesicles contain crystals of calcium oxalate which are hygroscopic and prevent excessive transpiration. On days of high light intensity, quinoa leaves were found to feel humid when rubbed, whereas leaves of other herbaceous plants were dry, and water stress was conspicuous in these plants (Canahua, 1977). Espíndola (1986) examined the response of quinoa to water

Table 1.5 Responses of quinoa varieties to adverse conditions

Variety	*Lodging*	*Hail*[a]	*Drought*	*Flooding*
Sajama	VS	R	R	VS
Blanca de Julí	S	S	MR	R
Kanccolla	R	S	R	R
Rosada de Puno	R	R	R	R
Ajara	VR	S	VR	VR

VS, very susceptible; S, susceptible; MR, moderately resistant; R, resistant; VR, very resistant.
[a]Conferred by compact inflorescences.

Source: Canahua (1977).

deficit and found that drought reduced yield by 61.5% and 67.5% in two lines when applied at floral initiation but by 37.7% and 42.6% when applied at inflorescence emergence. At both stages transpiration rate was reduced and stomatal resistance was increased. There is considerable genetic variation in the response of quinoa varieties to diverse environmental stresses, as shown in Table 1.5.

The stress tolerance of quinoa has its limits, however. In the cropping season of 1986–87, about 12 000 ha of quinoa were sown in Puno, Peru. Following a long period of drought between October and December, 2500 ha were lost, and another 1000 ha were damaged. In March, three consecutive frosts of -8 °C caused the loss of another 2000 ha (Risi, 1989).

Quinoa's deep tap root enables it to extract nutrients left behind by other crops, and it is often sown after potatoes to take advantage of the residue of phosphate fertilizer applied to the potato crop. In this system quinoa closes the rotation, being followed by the fallow. Its low requirement for fertilizers and other agrochemicals could reduce environmentally damaging inputs on land where it is grown by up to 50%. Some varieties of quinoa show resistance to beet cyst nematode (Doorgeest *et al.*, 1988), and so it would be advantageous to incorporate quinoa in rotations involving sugar beet and rape seed. Quinoa can thus make a contribution to crop diversification and the competitive use of agriculturally marginal land. Its versatility and tolerance to stresses suggest that it could be grown in a wide range of agriculturally marginal locations, such as the foothills of the Rocky Mountains, the Spanish Meseta, the Breckland of eastern England, and perhaps such areas as Ethiopia, the Himalayas and the tropical highlands of Asia and central America (Risi and Galwey, 1984).

1.5 BIOCHEMISTRY

1.5.1 **Biochemical composition of the grain**

Proximate analysis

The chemical composition of quinoa gives it high nutritional value. The proximate composition of the grain is given in Table 1.6 and is compared with that of cereal crops in Table 1.7. It can be seen that quinoa has a higher protein and fat content than the other cereals.

Table 1.6 Proximate composition (%) of quinoa grain

	Romero (1981)			Koziol (1990a)		
Component	N[a]	*Range*	*Average*	N[a]	*Range*	*Average*
Moisture	58	5.4–20.7	12.9	127	6.20–14.09	9.61
Protein	74	9.6–22.1	14.3	127	10.83–21.86	15.72
Carbohydrate	49	46.0–77.4	61.4	69	53.24–67.17	61.70
Fat	54	1.8–8.2	4.6	92	4.26–9.50	7.16
Fibre	38	1.1–5.8	3.0	69	1.22–4.78	2.91
Ash	60	2.4–9.7	3.5	73	1.98–6.13	3.29
Saponin	8	0.4–5.6	2.2	69	0.01–4.65	0.65

[a]Number of determinations.

Table 1.7 Comparison of composition of quinoa with that of cereals

Material	*Composition (%)*					
	Moisture	*Protein*	*Fat*	*Carbohydrate*	*Fibre*	*Ash*
English wheat	15	8.9	2.2	66.8	2.1	1.5
Barley	15	10.0	1.5	66.4	4.5	2.6
Oats, whole	11	10.3	4.7	62.1	9.3	2.6
Rye	10	12.4	1.3	71.7	2.3	2.0
Maize	11	9.4	4.1	72.1	2.0	1.4
Sorghum	11	11.0	3.2	70.9	2.4	1.5
Rice, paddy	12	8.0	1.9	62.7	9.0	6.3
brown	12	9.7	2.4	73.2	1.1	1.6
polished	12	8.6	0.4	78.2	0.3	0.5
Quinoa	12	13.0	5.3	55.7	4.9	3.0

Source: Galwey *et al.* (1990).

Table 1.8 Essential amino acid composition of quinoa compared with that of other grains and with recommended values

Protein source	Amino acid								
	HIS	ILE	LEU	LYS	M+C	P+T	THR	TYR	VAL
Quinoa[a]	3.2	4.9	6.6	6.0	5.3	6.9	3.7	0.9	4.5
Quinoa[b]	3.1	3.8	6.5	6.1	4.2	7.6	3.9	1.3	4.5
Mean	3.2	4.4	6.6	6.1	4.8	7.3	3.8	1.1	4.5
Rice[c]	2.1	4.1	8.2	3.8	3.6	10.5	3.8	1.1	6.1
Maize[d]	2.6	4.0	12.5	2.9	4.0	8.6	3.8	0.7	5.0
Wheat[d]	2.0	4.2	6.8	2.6	3.7	8.2	2.8	1.2	4.4
Recommendations[e]									
Infants	2.6	4.6	9.3	6.6	4.2	7.2	4.3	1.7	5.5
Pre-school	1.9	2.8	6.6	5.8	2.5	6.3	3.4	1.1	3.5
School children	1.9	2.8	4.4	4.4	2.2	2.2	2.8	0.9	2.5
Adults	1.6	1.3	1.9	1.6	1.7	1.9	0.9	0.5	1.3

[a]Means of the values reported by Cardozo and Tapia (1979), Mahoney *et al.* (1975), Risi and Galwey (1984), Romero (1981) and Marroquín (1983).
[b]Values from Latinreco S.A.
[c]Means of values reported by Lásztity (1984) and Romero (1981).
[d]Means of values reported by Lásztity (1984), Risi and Galwey (1984) and Romero (1981).
[e]Recommendations of WHO/FAO/UN.

Abbreviations:
HIS – histidine
ILE – isoleucine
LEU – leucine
LYS – lysine
VAL – valine
M+C – methionine+cysteine
P+T – phenylalanine+tyrosine
THR – threonine
TRY – tryptophan

Source: Koziol (1992).

Protein

The high nutritional value of quinoa grain is due mainly to its high content of good quality protein. The average protein content of the grain is 14.6% but some Ecuadorian varieties have as much as 21.9% protein (Koziol, 1990a). This protein content compares favourably with that of the cereals, and the amino acid balance is superior to that of wheat, barley and soyabean and comparable to that of the casein protein in milk. Values for the amino acid composition of the seed protein obtained in several studies, and corresponding values for cereal proteins, are given in Table 1.8. It is likely that the content and composition of quinoa protein varies with variety and environmental conditions. Variation between studies may be due in part to environmental, agronomic and genetic factors, but may also be due to use of different analytical procedures. Quinoa protein is particularly rich in histidine and lysine (3.2% and 6.1% of protein composition respectively) and also contains more isoleucine and methionine + cysteine than cereals (Koziol, 1992). The amino acid composition of seed proteins is affected by the types of protein which are present, and the high levels of these amino acids may be explained by the composition of the seed protein (Koziol, 1992). Albumin and globulin proteins form the bulk of the seed protein in quinoa, whereas prolamins and glutelins are present only in insignificant amounts (Fairbanks *et al.*, 1990). Scarpati de Briceño and Briceño (1980) report that concentrations of albumins and globulins in quinoa are higher than those found in maize, rice or wheat.

Lipids

The oil content of quinoa grain is about 5.6%, which is higher than that of cereals. Consequently the energy value is comparable with that of cereals, in spite of the higher protein content (Galwey *et al.*, 1990). The composition of quinoa oil is similar to that of soya oil (Table 1.9) and it is a rich source of linoleic and linolenic acids: these essential fatty acids constitute approximately 55–63% of the oil. In spite of the high levels of these fatty acids, which are susceptible to oxidation, quinoa oil is quite stable. This is attributed to its high concentration of vitamin E which is a natural anti-oxidant. Quinoa oil contains from 690–740 p.p.m. of α-tocopherol and 790–930 p.p.m γ-tocopherol. After refining, the concentrations fall to 450 and 230 p.p.m. respectively (Koziol, 1992), but Hudson and Ghavami (1984) report that 100–200 p.p.m. are sufficient for optimal antioxidant activity of these isomers. Koziol (1990a) has calculated that the yield of vegetable oil per hectare which could be extracted from quinoa (80–400 kg/ha) could easily exceed that obtained from maize (20–50 kg/ha) and suggests that, given the high quality of its oil, and the fact that some varieties show lipid contents of up to 9.5%, quinoa is potentially a valuable new oil crop.

Vitamins and minerals

The available evidence indicates that quinoa is a good source of vitamins, notably E, B2, B6, folic acid and biotin, and of minerals, notably calcium, potassium, iron, copper, magnesium, manganese and chloride. However, early studies of the vitamin content (Alvistur *et al.*, 1953; Pulgar-Vidal, 1954; De Bruin, 1964) gave contradictory results, and there is still a need for more reliable studies. Mahoney *et al.* (1975) reported that washed flour, from the cultivar Sajama, contained 0.89 g phosphorus, 26.1 mg calcium, 91.4 mg magnesium, 4.06 mg iron, 3.39 mg zinc and 0.41 mg copper per 100 g. Varriano-Marston and de Francisco (1984), using quantitative X-ray dispersive analysis, have shown that

Table 1.9 Fatty acid composition of quinoa oil and other vegetable oils[1]

Source of oil	*Capric C6:0*	*Caprilic C8:0*	*Lauric C12:0*	*Miristic C14:0*	*Palmitic C16:0*	*Stearic C18:0*
Quinoa variety						
Sajama				0.2	9.8	0.7
Porotoc				0.2	9.1	0.7
Imbaya				0.2	9.8	0.7
Cochasqui				0.3	10.8	1.2
Mean				0.2	9.9	0.8
Quinoa[2]					11.0	0.7
Soya[3]					9.4	4.4
Peanut[3]					9.3	2.0
Olive[3]					9.6	2.8
Palm[3]		4.2	43.9	15.6	8.7	2.9
Coconut[3]	0.7	8.4	48.2	17.7	8.6	2.8

Fatty acid

Souce of oil	*Oleic C18:1*	*Linoleic C18:2*	*Linolenic C18:3*	*Eicosanoic C20*	*Docosanoic C22*	*Tetracosanoic C24*
Quinoa variety						
Sajama	23.8	46.2	9.5	2.8	3.0	0.9
Porotoc	22.2	55.2	4.3	2.6	2.5	0.6
Imbaya	26.8	50.6	3.9	2.4	2.5	0.6
Cochasqui	25.0	48.6	3.9	2.8	2.9	0.6
Mean	24.5	50.2	5.4	2.7	2.7	0.7
Quinoa[2]	22.0	56.0	7.0			
Soya[3]	21.6	55.2	9.4			
Peanut[3]	44.7	35.8		4.2	3.4	1.9
Olive[3]	79.4	7.6	0.6			
Palm[3]	18.1	2.9				
Coconut[3]	5.9	1.0				

Source: [2]Koziol (1990a), [2]Marroquín (1983), [3]Simpson and Osborne (1978).

high levels of magnesium, aluminium, silicon, phosphorus, sulphur and calcium occur in the perianth, but these are not of nutritional significance, as this layer is removed as chaff. The unwashed pericarp showed a similar spectrum, except that the Cl content was much lower. The perisperm contained only trace amounts of potassium, sulphur, chlorine and silicon, and magnesium, sulphur, phosphorus and potassium were present in the embryo. The mineral content of quinoa grain, flour and bran, is summarized in Table 1.10. The mineral composition of quinoa was found to be superior to that of the cereals: for example,

Table 1.10 Mineral composition of quinoa grain, flour and bran

Mineral	*Grain*	*Flour*	*Bran*
Phosphorus (%)	0.47	0.32	0.67
Calcium (%)	0.19	0.15	0.24
Potassium (%)	0.87	0.47	0.16
Magnesium (%)	0.26	0.16	0.46
Sodium (p.p.m.)	115	671	60
Iron (p.p.m.)	205	177	256
Copper (p.p.m.)	67	62	78
Manganese (p.p.m.)	128	121	160
Zinc (p.p.m.)	50	48	77

Source: Ballón *et al.* (1986).

the calcium:phosphorus ratio (1:2.5) was better than that of the cereals (1:10.7), the optimum ratio being 1:1. Sweet varieties tended to have a higher mineral content than bitter ones.

Carbohydrates

Studies of quinoa starch have been carried out by Wolf *et al.* (1950), Scarpati de Briceño and Briceño (1982), Atwell *et al.* (1983), Varriano-Marston and de Francisco (1984) and Lorenz (1990). Various properties of the starch have been determined: these are summarized in Table 1.11, and compared with those of starch from other sources in Table 1.12. The functional properties of starches are affected by their physico-chemical characteristics, the size of the starch granules and their amylose content being important determining factors. Amylose is thought to reinforce the internal micellar network of the starch granule, resulting in close associations between starch polymers. This makes them resistant to enzymolysis. Amylose content is thus an important factor in determining the swelling and water-binding capacities of the starch and resulting viscosity of starch pastes, and in determining the usefulness of the starch in baking processes.

The starch content of quinoa is about 60%, and the starch occurs in perisperm cells as compound granular aggregates. Individual granules have an average particle size of 1–2 μm (Atwell *et al.*, 1983). This is smaller than that of maize (range 1–23 μm) and wheat (2–40 μm) (Wolf *et al.*, 1950; Swinkels, 1985). Quinoa starch has been found to have a much lower amylose content than cereal starches (11% compared to about 28% in wheat). Consequently, its starch paste is more viscous than wheat or barley starch pastes. It has been found to have high swelling and water-binding capacities, and high gelatinization temperatures similar to those of wheat and potato starch. Quinoa starch is rela-

Table 1.11 Characterization of quinoa starch

Analysis	*Range*	
	Quinoa	*Wheat*
Gelatinization	57–64 °C	
Differential scanning calorimetry		
T onset	50.6 ± 1.8 °C	
T complete	87.8 ± 1.4 °C	
dH gelatinization	4.0 ± 0.8 cal/g	
Amylograph viscosities		
Initial	60 °C	84 °C
At 95 °C	570 BU	210 BU
After 60 min at 95 °C	790 BU	180 BU
After cooling to 35 °C	1200 BU	660 BU
X-ray diffraction pattern		
Amylose content	11.0 ± 0.2%	
Average chain length	27.0 ± 2.6%	

[a]BU = Brabender units.

Source: Adapted from Atwell *et al.* (1983).

tively highly branched. The structure of the amylopectin component is thought to be similar to that of amylopectin from other sources (Atwell *et al.*, 1983). The small size of the starch granules could be exploited industrially: for example, a patent application has been made for the manufacture of a carbohydrate-based cream substitute from quinoa starch (European Patent Application No. 89121654.1, 1989). Quinoa starch could also be used for applications in the paper industry, in cosmetic powders, and in composites with other polymers. Lorenz (1990) investigated some potential uses of quinoa in the food industry. Quinoa starch performs well as a thickening agent in fillings, but poorly in cake and bread baking. This can be attributed to its low amylose content; although the starch granules swell, gas retention capacity is poor and structural bonding is weak. Thus the starch cannot withstand the stresses caused by baking and subsequent cooling.

The content and composition of free sugars is important in determining the digestibility of carbohydrates. Gross *et al.* (1989) found sucrose to be the main free sugar in quinoa, whereas González *et al.* (1989) found approximately twice as much glucose as sucrose. This discrepancy may be due to different analytical procedures, or to genetic differences. The concentration of α-galactosides, which are indigestible carbohydrates, is below 1% and poses no problems with regards to digestibility (Gross *et al.*, 1989).

Table 1.12 Comparison of physico-chemical properties of quinoa starch with those of other starches[1]

Starch property	*Wheat*	*Barley*	*Wild rice*	*Amaranth*	*Potato*	*Quinoa*
Amylograph viscosities (BU)						
at 92°C	270	180	970	950	1525	980
after 30 min at 92°C	340	335	800	820	890	895
on cooling to 35°C	510	355	1110	810	720	1690
after 60 min at 35°C	810	480	1190	840	440	1900
Swelling power						
at 60°C	2.03	4.39	4.15	5.76	3.00	10.61
at 70°C	5.39	5.48	6.85	9.71	30.80	12.23
at 80°C	6.89	ND	8.18	ND	62.30	15.75
at 90°C	8.36	7.94	10.00	10.57	90.60	19.73
Solubility (%)						
at 60°C	0.91	0.95	0.70	1.61	3.00	1.14
at 70°C	2.02	2.55	0.86	1.25	17.70	1.20
at 80°C	2.59	ND	0.65	ND	31.00	1.32
at 90°C	3.52	3.74	0.12	0.48	39.30	4.33
Waterbinding (%)	89.1	92.4	113.9	145.6	102.0	118.5
Gelatinization temperature range (°C)						
initial	54.0	55.5	63.0	71.0	60.0	61.0
midpoint	58.0	60.5	67.0	77.5	63.0	64.5
final	61.0	65.0	73.0	82.5	68.0	68.0
Enzyme solubilized (%)	0.44	0.66	2.79	1.67	0.57	3.76

BU=Brabender units
ND=not determined
[1]*Source*: Lorenz (1990)

Saponins

One drawback to the utilization of quinoa is the presence of a class of anti-nutritional compounds called saponins in the pericarp of the grain (Villacorta and Talavera, 1976). These are water-soluble detergent-like molecules which comprise lipophilic sugar chains attached to lipophobic triterpenoid aglycones (sapongenols). Saponins are not unique to quinoa: they are found in many plant species and are thought to have been favoured by natural selection because they provide plants with protection against attack by birds (Risi, 1986). They are generally perceived as having a bitter taste, and have the properties of foaming in water and of lysing cells (Ballón, Tellería and Hutton, 1976). Some saponins are known to distort the intestinal membranes of animals and to reduce the palatability of food. Recently, saponins have been shown to reduce blood cholesterol levels in animals and man (Oakenfull and Siddhu, 1990). Although they are anti-nutritional compounds which detract from the utility of quinoa, they may have uses in their own right as foaming agents or molluscicides, or as raw materials for the chemical or pharmaceutical industries. The level of toxicity depends on the structure of the saponin, the organism and the method of absorption (Koziol, 1992). It may be possible to identify saponins which are toxic to insect pests, but have no toxicity to mammals. Such molecules could be used as potent natural insecticides (Basu and Rastogi, 1967).

There are a number of different methods available for measuring saponin content and composition, differing in their accuracy and specificity:

1. Foam production (Rea and León, 1966; Ríos *et al.*, 1978; Koziol, 1990a, 1990b).
2. Chemical extraction with alcohol, water and diethyl ether (Machicao, 1965; Mizui *et al.*, 1988).
3. Chromatography: Thin layer chromatography (Ballón *et al.*, 1976), gas chromatography (GC) (Elias Peñafiel and Dias Villar, 1988), gas-liquid chromatography (GLC) (Ridout *et al.*, 1991) and high-performance liquid chromatography (HPLC) (Burnouf-Radosevich and Delfel, 1984).
4. Spectrophotometry (Eliaz Peñafiel and Dias Villar, 1988).
5. Haemolysis of red blood cells (Rea, 1948; Aguilar *et al.*, 1979; Burnouf-Radosevich and Paupardin, 1983; Reichert *et al.*, 1986).
6. Fungal growth inhibition, assayed using *Trichoderma viride* (López, reported by Cardozo and Tapia, 1979).
7. Larval growth inhibition, assayed using Red flour beetle, *Tribolium castaneum* (Becker and Hanners, 1990).

Gravimetric and foam assays are believed to over-estimate the saponin content due to their non-specific nature (Galwey *et al.*, 1990). However, Koziol (1990b) has shown that foam production is an adequate method for field testing, as results from this test correlate well with actual saponin contents determined by GLC ($r = 0.992$, $P < 0.001$). It is likely that the structure of individual saponin molecules affects their biological activities. Thus, assays based on anti-fungal or haemolytic properties will probably not give accurate measurements of total saponin content: results will depend on the nature and relative amounts of individual saponin species present (Galwey *et al.*, 1990). Haemolytic assays have been shown to be sensitive to extraction temperature and storage temperature (Reichert *et al.*, 1986). The concentration of saponin

Table 1.13 Saponins in common foodstuffs

Foodstuff	*Saponin content (g/100 g edible portion)*	*Reference*
Chickpeas	5.00	Fenwick and Oakenfull (1983)
Chickpeas	3.47	Jood *et al.* (1986)
Red kidney beans	1.40	Fenwick and Oakenfull (1983)
Peanuts	0.58	Fenwick and Oakenfull (1983)
Spinach	0.55	Fenwick and Oakenfull (1983)
Bean sprouts (mung bean)	0.54	Fenwick and Oakenfull (1983)
Canned baked beans	0.38	Fenwick and Oakenfull (1983)
Brown lentils	0.37	Fenwick and Oakenfull (1983)
Canned broad (Faba) beans	0.31	Fenwick and Oakenfull (1983)
Green peas	0.25	Fenwick and Oakenfull (1983)
Asparagus	0.13	Fenwick and Oakenfull (1983)
Garlic	0.11	Fenwick and Oakenfull (1983)
Garlic	0.10	Smoczkiewicz *et al.* (1982)
Leek	0.10	Smoczkiewicz *et al.* (1982)
Green beans	0.10	Fenwick and Oakenfull (1983)
Onion	0.02	Smoczkiewicz *et al.* (1982)
Quinoa (polished)	< 0.01	Latinreco S.A.

Source: Koziol (1992).

varies with variety (Cardozo and Tapia, 1979) and environmental conditions (Aguilar *et al.*, 1979): estimates range from 0.01–4.65% (dry matter) (Koziol, 1992, estimated by foam production). The composition of saponins also varies with variety (K. Price, unpublished data). Although the presence of saponins in quinoa is a significant problem due to their bitterness, they represent a minimal source of dietary saponin compared to the concentrations present in other foods (Table 1.13).

At least 15 types of saponin molecule have been identified in quinoa; their general molecular structure is shown in Figure 1.3. They are generally derivatives of three main triterpene aglycones, phytoaccagenic acid, hederagenin and oleanolic acid (Burnouf-Radosevich and Delfel, 1984; Mizui *et al.*, 1988; Ridout *et al.*, 1991). Ridout *et al.* (1991) have separated and identified derivatives of the three main aglycones in UK-grown quinoa. Burnouf-Radosevich and Delfel (1984) showed that oleanolic acid and hederagenin are the major aglycones. Seed extracts contained approximately six times more aglycones than callus tissue extracts, but the ratio of hederagenin to oleanolic acid was the same in both materials. Seven other minor silylated pentacyclic triterpenes (β-amyrin, α-amyrin, erythrodiol, ursolic acid, echinocystic acid, gypso-

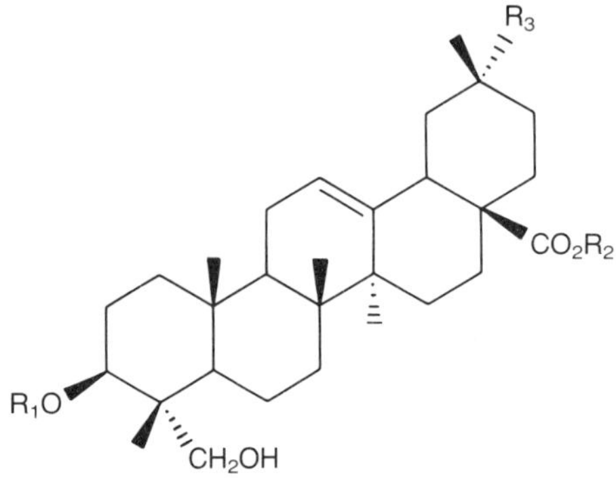

	R_1	R_2	R_3
I	Glu(1→3)Ara–	Glu–	$-CH_3$
II	Glu(1→3)Gal–	Glu–	$-CH_3$
III	Ara–	Glu–	$-CO_2CH_3$
IV	Glu(1→3)Ara–	Glu–	$-CO_2CH_3$
V	Glu(1→3)Gal–	Glu–	$-CO_2CH_3$
VI	Ara–	Glu–	$-CH_3$

Glu: β–D–glucopyranosyl
Ara: α–L–arabinopyranosyl
Gal: β–D–galactopyranosyl

Figure 1.3 General molecular structure of saponin molecule in quinoa.

genin and queretaroic acid) have also been identified (Burnouf-Radosevich, Delfel and England, 1985).

Besides saponins, quinoa contains other classes of anti-nutritional compounds. Trypsin inhibitors in eight varieties of quinoa ranged from 1.36–5.04 Trypsin Inhibitor Units (TIU)/mg of quinoa. This is much less than values reported for common bean (12.9–14.8 TIU/mg) or soya bean (24.5 TIU/mg) (Romero, 1981), and these trypsin inhibitors are inactivated by cooking (Koziol, 1992). Phytic acid is also present in quinoa. This compound can form insoluble complexes with multivalent cations such as Ca^{2+}, Fe^{2+}, Fe^{3+} and Zn^{2+}, reducing their bioavailability (Zemel and Shelef, 1982). The mean concentration in five different varieties of quinoa was 1.18 g/100 g of grain (Tagliaferri, cited by Koziol, 1992). This concentration is higher than that in barley (1.07 g), maize (0.94 g), rice (0.89 g), or wheat (0.99 g) (Souci *et al.*, 1986), but does not affect iron absorption or the incorporation of calcium into bone (Koziol, 1992). Derivatives of the flavonoid aglycone kaempferol are also known to be present in quinoa grain (Harborne, 1984; De Simone *et al.*, 1990). The potential allergenicity of quinoa has been studied and it was found to contain compounds capable of eliciting a hypersensitive reaction at levels less than those in egg white or cow's milk and about equal to those in soya (Pahud and Schwarz, cited by Koziol, 1992).

1.5.2 **Nutritional analysis of the grain**

Experimental approaches

The high protein content and well-balanced amino acid composition of quinoa do not in themselves mean that it has high nutritional value. Saponins, and perhaps other anti-nutritional compounds, may interfere with the digestion and absorption of nutrients. Moreover, the amino acid composition of a protein does not give information on the availability of its individual amino acids to the organism. It is the *in vivo* digestibility of the protein source in the species consuming it which is important. This can be assessed by animal feeding trials, in which characters such as the biological value (BV) and true digestibility (TD) of the protein, net protein utilization (NPU), nitrogen efficiency for growth (NEG), protein efficiency ratio (PER), and weight gain are determined. *In vitro* enzymatic digestion of the protein is also used as an indicator of its *in vivo* digestibility.

In vitro *digestibility studies*

In vitro enzymatic hydrolysis revealed a similar pattern of release of individual amino acids from proteins of raw and washed quinoa, suggesting that saponins do not interfere with the digestion of the protein (Ruales and Nair, 1992). Digestibility values (%D) were 82% for washed quinoa, 78% for raw quinoa and 91% for casein. Coulter and Lorenz (1991a, 1991b) found that adding quinoa to extruded corn products caused a decrease in the *in vitro* digestibility, but improved the nitrogen solubility profile. Nitrogen solubility profiles, over a range of pH values, are a guide to protein functionality. They relate directly to many important properties, such as suitability for use in beverages, emulsification, foaming capacity and gelation (Kinsella, 1976). They also provide an indication of a protein's potential solubility in food systems and degree of denaturation that it may have undergone (Betschart, 1974).

Studies on animals

(a) *Rats*

When quinoa was used as the sole source of protein in otherwise balanced diets fed to rats, it produced rates of growth greater than that obtained with whole dried milk (White *et al.*, 1955). Mahoney *et al.*

(1975) reported that using cooked quinoa (boiled for 30 min) in diets improved the nitrogen efficiency for growth (NEG) by 40%, the weight gain by 100% and the protein efficiency ratio (PER) by 29% relative to the values obtained with uncooked grain. Hurrell (1988, cited by Koziol, 1992) found that cooked quinoa has a PER higher than that of casein and more than three times as high as that of cooked wheat flour. Mahoney *et al.* (1975) found that mixing quinoa with wheat flour in the ratio of 1:4 improved the NEG, weight gain and PER by 43%, 11% and 72% respectively, compared to the values obtained with wheat flour alone. The feeding of cooked quinoa or casein resulted in a tendency towards smaller relative organ weights, lower percentages of carcass nitrogen and ash and higher percentages of carcass lipid in comparison with raw quinoa and with other diets included in the study. Gross *et al.*

Table 1.14 Protein efficiency ratios of quinoa and other foodstuffs

Study	*Protein source*	*PER* (% of casein)
Mahoney *et al.* (1975) (10% dietary protein)	Casein	100
	Wheat flour	32
	Quinoa flour (var. Sajama)	78
	Cooked quinoa	102
	80% wheat flour + 20% quinoa flour	55
Tellería *et al.* (1978) (10% dietary protein)	Casein	100
	Quinoa 'blanca', whole grain flour	44
	Quinoa 'blanca', hot water (87 °C) extraction of saponins	93
	Quinoa 'colorada', whole grain flour	47
	Quinoa 'colorada', hot water (87 °C) extraction of saponins	85
	Sajama (sweet quinoa), washed in water at 50 °C	74
	Sajama, washed in water at 87 °C	85
Hurrell, cited by Koziol (1992) (10% dietary protein)	Casein	100
	Quinoa polished	89
	Cooked quinoa, polished	105
	Wheat flour (type 550)	23
	Cooked wheat flour (type 550)	31
Hurrell, cited by Koziol, 1992. (10% dietary protein)	Casein	100
	Bitter quinoa	69
	Water-washed bitter quinoa	86
	Polished bitter quinoa	90
	Sweet quinoa (var. Sajama)	93

Source: Koziol (1992).

(1989) found that diets based on quinoa were of equivalent value to those based on casein. The PER and true digestibility (TD) values for casein were 3.10 $\pm$ 0.09 and 87.5 $\pm$ 1.60 respectively, whereas those for quinoa were 3.09 $\pm$ 0.09 and 84.1 $\pm$ 2.25 respectively. Interestingly, the serum urea content obtained by feeding quinoa was 11.3 $\pm$ 3.4, whereas the value for casein was 25.5 $\pm$ 3.8, which indicates that quinoa protein is of very high quality. Romero *et al.* (1985) evaluated the effect of extrusion on the protein quality of washed quinoa and found that, whereas cooking increased PER by 30%, expanded and texturized quinoa gave lower increases, of 9% and 22% respectively. Ruales and Nair (1992) found that there was no significant difference in biological value (BV), TD and net protein utilization (NPU) between raw and washed quinoa diets, but that both gave lower digestibility values than a casein diet. However, the NPU values (75.7 for raw and 74 for washed quinoa) were still higher than the NPU value of 70 for mixed diets which are in common use (Passmore and Eastwood, 1986). The protein efficiency ratios (PERs) of raw and washed quinoa are also higher than that of wheat (Table 1.14).

The effects of adding quinoa to diets on the absorption of minerals have also been assessed. Hurrell (1988, cited by Koziol, 1992) found that quinoa, added at a rate of 7% to a milk powder/maize diet, increased the absorption of calcium in rat femurs compared with similar diets to which wheat was added. The bioavailability of the iron is also high (Brown, 1992, cited by Koziol, 1992).

(b) *Chicks*

Negrón *et al.* (1976) showed that quinoa is an acceptable food source for chicks in terms of weight gain, nutrient conversion and mortality over a 56-day experimental period. Weber (1978) described feeding trials in Bolivia in which chicks fed a ration containing cooked quinoa had weight gains equal to those receiving corn or skimmed milk. Raw quinoa, however, caused a depression in weight gain, possibly due to the presence of saponins. In a study of the feeding value for broiler chickens of UK-grown and South American-grown (low saponin) quinoa, it was found that diets containing the South American-grown quinoa produced weight gains and food conversion efficiencies which were at least equal to that of wheat-based control diets. Diets containing UK-grown quinoa, however, had poorer metabolizable energy values and lower amino acid digestibility coefficients than those predicted from their composition. The exact nature of the anti-nutritional factors involved is unknown. Reduction of the level of water-soluble saponins had only a small effect on the nutritional value of the processed material (J. McNabb, unpublished results).

(c) *Pigs and cattle*

Unwashed quinoa depressed the growth of pigs, presumably due to the presence of saponins, whereas cooked quinoa was comparable to powdered milk as a source of protein (Cardozo and Bateman, 1961). Quinoa products, used as a protein supplement for calves, have been reported to produce a greater weight gain than supplements of wheat or faba beans. Animals given quinoa reached commercial weights at a faster rate (Martínez, 1946).

(d) *Humans*

Evaluation of quinoa as human food has been conducted with Peruvian children recovering from severe malnutrition (López de Romana *et al.*, 1978; López de Romana *et al.*, 1981). Better values were obtained with quinoa flour than the whole quinoa grain. Quinoa–wheat diets gave adequate calories and protein for infants whereas quinoa–oat diets gave low absorption of nitrogen and fat. Nitrogen absorption from quinoa diets was lower than that of casein diets, but the BVs of the diets were comparable (43% for whole quinoa, 50.7% for quinoa flour and 45.8% for casein). Atwell *et al.* (1988) evaluated the potential of the α-amylase activity in flour made from germinated quinoa for increasing the palatability and effective calorific density of starch foods used at weaning by the Aymara people living near Lake Titicaca in Bolivia. They concluded that the addition of germinated quinoa flour to the normal diet of potatoes would be of significant nutritional benefit.

Koziol (1992) has calculated the contribution quinoa could make to the human diet on the basis of the recommendations of various statutory bodies. He estimated that a 100 g portion of cooked quinoa (containing 1.11 g of n–6 fatty acids) would provide 7–11% of the daily allowance of the n–6 fatty acids for children aged 1–10 years, and 5–6% of the daily allowance for adolescents and adults. It would also provide 10% of the daily adolescent or adult allowances of vitamin B6, 9–15% of pantothenic acid, 12% of folic acid, 7–24% of biotin, 6% of riboflavin and of niacin equivalents, and 3–6% of the estimated daily intake of choline. Of the recommended daily mineral allowances, 100 g of cooked quinoa would provide 4–6% of calcium, 11–16% of phosphorus, 23–76% of magnesium, 15–19% of potassium, 27–40% of iron, 10–15% of zinc, 47–200% of copper and 1–2% of sodium, depending on age and sex. Quinoa protein would provide 200% of the histidine, 377% of the isoleucine, 347% of the lysine, 312% of the methionine and cysteine, 363% of the phenylalanine and tyrosine, 411% of the threonine, 180% of the tryptophan and 346% of the valine recommended in protein sources for adult nutrition by the FAO/WHO/UN joint expert consultation (1985).

Thus reports indicate that quinoa should be a very good source of dietary protein, providing that anti-nutritional factors can be removed. In many studies quinoa has been found to be equivalent to casein or skimmed milk powder. Further studies on the nutritional value of quinoa in humans and other mammals are required. In particular, the assumption that saponins are the only important anti-nutritional factor in quinoa requires testing. More detailed knowledge of such factors will provide a better focus for breeding programmes.

1.5.3 Biochemical composition of the leaves

Quinoa leaves contain between 2.8% (cv. Sajama) and 4.2% (cv. Blanca Amarga) of protein (Cardozo and Tapia, 1979) and between 1.9% (Blanca Real) and 2.3% (Blanca Amarga) of lipids, on the basis of fresh-weight measurements (Cornejo, 1976). This compares well with spinach which contains 2.8% protein and 0.7% fat (Leung and Flores, 1961). Quinoa leaf protein has more isoleucine and valine and slightly less methionine + cysteine and phenylalanine + tyrosine than that of spinach (Koziol, 1992). Mineral levels in these crops are compared in Table 1.15. Quinoa leaves have been shown to be a rich source of vitamins A and E (Koziol, 1992). The lipid fraction of the leaves consists of 38.9% α-linolenic acid and 22% linolenic acid. No γ-linolenic acid was found in the leaf lipid fraction (Koziol, 1992).

Protein extracted from green crops, known as leaf protein concentrate (LPC), may be an attractive component of fodder for non-ruminants. Experiments to assess the potential of quinoa for LPC production have been conducted in Sweden (Carlsson, 1975, 1980, 1984), southern Poland (Carlsson *et al.*, 1984), Victoria, Australia (Ostrowski *et al.*, 1984), California, USA (Carlsson, 1982), and in Colombia (Amaya and Car-

Table 1.15 Mineral concentrations (mg/100 g fresh weight) in quinoa and spinach leaves (after Koziol (1992))

Mineral	Quinoa[a]	Spinach[b]
Ca	153	126
K	357	633
Na	289	65
P	42	55
Mg	83	58
Fe	0.87	4.10
Cu	0.07	0.12
Zn	0.59	0.50

Sources: [a]Latinreco S.A.
[b]Souci *et al.* (1986)

reño, 1968). Extraction of proteins to produce LPC involves pulping the green material, expressing the juice and drying the protein coagulum. The extracted proteins can be fractionated by heat coagulation into green and non-green LPC at low and high temperatures respectively. Green LPC is a pigment-enriched feed-grade product, whilst non-green LPC is a whitish protein isolate, suitable for direct human consumption (Carlsson *et al.*, 1984). The nutritive value of the white LPC is very high, equivalent to that of casein (Carlsson, 1980). In the production of LPCs, the concentrations of harmful substances such as nitrate and oxalate are reduced, whereas the concentrations of organic nutrients are increased (Carlsson, 1983). Koziol (1992) found that the nitrate concentration of quinoa leaves was double that in spinach (0.43%, compared to 0.2%), a level which would restrict the use of fresh quinoa leaves in infant nutrition.

Quinoa is one of the most promising crops for production of LPCs for feed and human consumption. LPCs are relatively easy to extract from quinoa using methods that require low energy inputs (Carlsson, 1980), and the amino acid composition of quinoa LPCs suggests that they may be a good source of protein for chicks and pigs (Carlsson *et al.*, 1984). In comparative tests carried out in Sweden and California, quinoa out-yielded lucerne (the most commonly used green crop for LPC production) in biomass and protein production. Yields ranged from 8–14 t/ha of biomass and from 1.5–2.0 t/ha of protein (Carlsson, 1982). Growing conditions and extraction conditions affected the quality of the LPC. An optimal product was obtained when plants were grown at a between-plant distance of 25 cm using a nitrogen fertilizer application of 500 kg/ha, and were harvested after 15 weeks of growth. This stage corresponded to the onset of seed formation. About 20% of the extracted protein was in the non-green fraction. The protein fractions had some haemolytic activity (possibly due to the presence of saponins), but this did not interfere with the action of digestive enzymes, and could be removed by washing the protein fractions with ethanol (Carlsson, 1980).

Amaya and Carreño (1968) obtained yields of 60, 84 and 91 t/ha (fresh weight) in Colombia when plants were harvested at heights of 40, 120 and 150 cm respectively. When plants were 150 cm tall, the forage was found to contain 16.88% protein, 11.07% fibre, 2.70% lipids, 15.39% ash and 45.3% nitrogen-free extract. Cornejo (1976) found the composition of leaves from seven varieties to be 17.3–23.7% protein and 19.7–27.1% ash.

Haaber (1991) has investigated the possibility of using quinoa as a green crop for the pelleting industry. Quinoa pellets are believed to be suitable for ruminant animals. The best yields were obtained when plants were harvested 15 weeks after emergence. Leaf tissue was heat-treated in a tumble drier at 200 °C, and rats fed on a diet of heat-treated quinoa for 22 days showed better weight gains than rats fed fresh leaves (4.2 ± 0.2 g and 2.4 ± 0.8 g respectively). This was attributed to a reduction in the saponin content of leaves, though Koziol (1992) has stated that quinoa leaves do not contain saponins. Equivalent diets of alfalfa gave weight gains of 4.5 ± 0.7 g. No alteration of organ tissues due to toxicity was found with diets of heat-treated quinoa.

1.6 AGRONOMY, YIELD AND PRODUCTION

Until recently the predominant agricultural system in which quinoa was grown was subsistence agriculture. This is still the case in many of the Andean regions where quinoa is grown, but recently, mechanized agricultural systems have been used in South American countries to produce grain for commercial uses, both domestic and in export markets. Quinoa usually follows potatoes in the crop rotations of the Altiplano (Rea *et al.*, 1979).

Quinoa requires a level, well-drained seed-bed in order to avoid waterlogging of the small seed, but lack of moisture is often a problem on the Altiplano, and a technique known as dry-season cultivation has been developed to combat this. In December, fallow fields are ploughed so that the rains from December to March can accumulate. Weeds are killed by a further ploughing in early April. This also breaks the soil capillarity so that moisture is conserved. The fields are ploughed a third time and the crop is sown in September (Mujica, 1977).

1.6.1 **Sowing**

A number of sowing techniques are used in the Andes (Tapia *et al.*, 1980).

Broadcast sowing

This has the disadvantage that weed control and any mechanized practices are difficult.

Sowing in rows

This is practised where agriculture is mechanized. Row widths vary from 40–80 cm. Seed is deposited at the bottom of the furrow in dry areas or at the top where rain is more abundant.

Group sowing

This technique is used in the regions of Lipez and Ladislao Cabrera in the Bolivian Salares, where rainfall is low. Four or five seeds are planted in holes about 40 cm in diameter and deep enough to reach humid soil. The holes are 70–120 cm apart (Rea *et al.*, 1979).

Mixed sowing
The field is first cultivated with a ridger and then sown broadcast, so that seeds are deposited both on ridges and in furrows. This assures establishment under either deficiency or excess of moisture.

Transplanting
Small plots of quinoa are sown at high densities and seedlings are transplanted to blank patches in maize fields. This technique is practised in the Urubamba Valley, Cuzco, Peru.

Sowing dates

The sowing of quinoa generally coincides with the onset of the rains in the southern hemisphere summer, in September or October. In Puno, sowings are made as early as August if moisture is available. If irrigation water is available quinoa can be sown earlier (Kenwright, 1989). The late-maturing Altiplano variety Kanccolla can be sown only until mid-September (Mujica, 1977), whereas early-maturing varieties can be sown as late as November (León, 1964). In Concepción, Chile, yields of the local varieties Baer, Litu, Pichamán and Faro were 12% higher from a September sowing than from an October sowing. Yields were sharply reduced by sowing in November, probably due to high temperature (28 °C) and low relative humidity (55%) during flowering in January, which may have affected pollen viability (Etchevers and Avila, 1979).

In Britain, the optimum sowing date has been found to be during the second half of March. Earlier sowings suffer from weed competition. Quinoa is not adversely affected by the frosts which often occur in April (Risi and Galwey, 1991b).

Sowing density

The optimum sowing density is determined by several factors. If the density is too low, the vegetative period will be prolonged, maturity will become uneven, and mechanical harvesting will be difficult (Montenegro, 1976). The period between sowing and closing of the leaf canopy will also be increased, allowing weeds to flourish. On the other hand if the density is too high, competition between plants can adversely affect yield. Their is wide variation in estimates of optimum sowing density. Blanco (1970) reported a value of 4–6 kg of seeds per hectare (1 kg contains about 300 000 seeds) for the Bolivian Altiplano, whereas Canahua (1977) suggested 15–23 kg/ha in Puno. A density of 12 kg/ha has been recommended in Puno for mechanical drilling (Rea *et al.*, 1979). Velásquez (1968) found no significant differences in yield between sowing densities of 15, 20 and 25 kg/ha. However, Ortiz (1974) noted that a combination of high sowing densities (20–30 kg/ha) with high humidity produced small, weak plants with low productivity. Herquinio and Román (1975) obtained higher yields at a sowing density of 8 kg/ha than at 48 or 88 kg/ha, and they reported that the proportion of lodged plants increased at the higher densities. When the Chilean varieties Baer, Faro, Litu and Pichamán were drilled at row widths of 40, 60 and 80 cm, the yield per plant was higher at the wider spacings but the yield per unit area was significantly lower (Etchevers and Avila, 1979). Burgasi *et al.* (1990) recommended a density of 11–13 kg/ha for commercial plots in Ecuador. Risi and Galwey (1991b) found that highest yields in Cambridge were obtained with a row width of 20 cm and sowing density of 20 kg/ha. Plants covered the ground 45 days after sowing, facilitating weed control. At higher sowing rates plants were shorter, a higher proportion was stunted, a lower proportion was branched and maturity was earlier. Increases in within-row density caused greater increases in competition effects than corresponding reductions in row-width, except for the proportion of branched plants, which was more strongly affected by reduction in row width.

Excess quinoa plants are sometimes removed by hand (Mujica, 1977), but use of correct sowing densities, good seed-bed preparation and high-quality seed should make this operation unnecessary.

Sowing depth

The small size of quinoa seed makes it vulnerable to both waterlogging and dehydration, so that particular care must be taken concerning the depth at which it is sown. Bornas (1977) reports an optimum sowing depth of 1.0–1.5 cm, whereas Mujica (1977) recommends a depth of 2–3 cm. Burgasi *et al.* (1990) recommend a depth of 15 mm when the soil is moist. Under dry conditions seed can be sown at a depth of 50 mm. When quinoa seeds were sown in pots at depths of 1, 2 and 3 cm, a lower proportion of seedlings emerged from the deeper sowings (Etchevers and Avila, 1979).

1.6.2 **Fertilizer application**

When quinoa is grown in subsistence agriculture systems, fertilizer is generally not applied. Quinoa usually depends on the residue of fertilizer from preceding crops, sometimes supplemented by an application of nitrogen (Rea *et al.*, 1979), but when quinoa is sown after quinoa an organic fertilizer, such as sheep manure, guano, or peat, is incorporated into the soil. This has been shown to increase the yield (Calzada, 1951). However, quinoa responds well to applications of inorganic fertilizer and to nitrogen in particular. Application rates of 80 kg/ha of N, 40 kg/ha of P_2O_5, and 0 kg/ha of K_2O (80:40:0) have been recommended for Puno (Mujica, 1977; Rea *et al.*, 1979; Condorena, 1979). These proportions reflect the high potassium content of the soils in Puno (Mujica, 1977) rather than any peculiarities in the requirements of the plant. Etchevers and Avila (1979) found that dry-matter production in a pot experiment responded to applications of P_2O_5 or K_2O up to very high levels, equivalent to 450 kg/ha. It is not clear, however, whether the increase in dry-matter production was reflected by any increase in grain yield.

It is advisable that the nitrogen applications should be divided, half of the dose being applied at sowing and the other half about 50 days later (Mujica, 1977), or just before flowering (Alvarez and von Rütte, 1990a). Fertilizer regimes used in Europe are similar to those used for oilseed rape, the only dicotyledonous, non-leguminous species widely grown for grain.

1.6.3 **Weed control**

During the early stages of growth, quinoa is a poor competitor with weeds. Inadequate weed control causes economic losses not only by reducing the quantity and purity of the harvest but also by adding to the weed seed-bank in the soil for subsequent years. Weeds also harbour insect pests and diseases of quinoa.

In subsistence systems, quinoa is weeded by hand, a laborious and time-consuming task, but in mechanized systems it is often desirable to control weeds with chemical herbicides. This is difficult, however, as many herbicides are toxic to *Chenopodium* species, which often occur as weeds. In the following account, commercial names of herbicides are printed with an initial capital letter.

Chaquilla (1976) reported that linuron (Afalon) at a dose of 1 kg/ha gave some control of weeds in Puno as a post-emergence herbicide, but also showed some toxicity to quinoa, causing necrotic spots, smaller inflorescences and later maturity. No pre-emergence herbicide tested was reported to work. Velazco (1977), however, reported that linuron and diuron (Karmex) applied post-emergence at doses of 1.5 kg/ha controlled weeds and did not affect yield significantly in Puno. Cardenas (1949) reported that promising herbicides, both pre- and post-emergence, were available. Quinoa plants treated with these herbicides were given a score of 2–3 on a scale from 0 (no damage) to 10 (death).

Galwey (1989) suggested applications of 2.5 kg/ha of metamitron, pre-emergence, followed by 0.5 kg/ha of propyzamide at the four-true-leaf stage for the control of dicotyledonous weeds. Alternatively, 0.5 kg/ha of propyzamide and 1.9 kg/ha of propachlor could be applied, both pre-emergence. Grasses were controlled by an application of 0.19 l/ha for fluazifop-P-butyl before flowering, and an application of 11.25 kg/ha of sodium monochloroacetate was also found not to produce phytotoxicity. Alvarez (1990) reported that Lazo, a pre-emergence herbicide, applied at a rate of 1.14 kg/ha, reduced the weed population by 95% while causing no phytotoxicity to quinoa. Tribunil, applied pre-emergence at a rate of 1.4 kg/ha, controlled weeds just as effectively but was slightly toxic to quinoa. The use of these herbicides gave higher yields than when only manual control of weeds was practised.

1.6.4 **Earthing-up**

When quinoa plants reached a height of about 40 cm they are often earthed-up, partly to strengthen them and partly to control weeds (Lescano, 1981). There is some debate, however, as to whether this practice is really necessary. Omission of earthing-up in Britain has had no apparent detrimental effects (Risi and Galwey, 1984).

1.6.5 **Harvest**

At physiological maturity, the grains of quinoa become dry and hard, and difficult to break with a fingernail. The rest of the plant dries, in most varieties becoming pale, yellow, and the leaves are shed. The seed-heads remain intact, like those of other grain crops and unlike those of wild *Chenopodium* species, which shatter. Traditionally, the plants are pulled up or cut with a sickle, then left in windrows to dry completely. They are then threshed either on a threshing floor, with sticks, animals or vehicles, followed by winnowing, or else using a stationary thresher. A Triton thresher produced threshed grain at a rate of 600 kg/h (Mujica, 1977; Lescano, 1981).

Combine harvesting is also practised, and is successful provided that the plants are completely dry and provided that the machine is correctly adjusted, so that the grain is not lost nor chaff retained with the seed (Narrea, 1976; Bugasi *et al.*, 1990).

1.6.6 **Yield**

The yield of quinoa is dependent on both variety and growing conditions. On the Altiplano, yields are low due to the harsh climate. Typical values are 0.56, 0.12 and 0.10 t/ha in the northern, central and southern Altiplano respectively (Ordoñez, 1987). Risi (1989) reported that average yields in Puno fluctuated between 0.45–0.48 t/ha from 1980 to 1985. The main reasons given for these low yields are climatic factors (prolonged droughts, poor distribution of rain and frosts), use of sub-

Table 1.16 Experimental yields of quinoa obtained on the Altiplano

Genotype	*Yield (kg/ha)*	*Location*	*Reference*
04–08–186	4708	Tahuaco	Lescano (1984)
03–21–052	4681	Camacani	Lescano (1984)
03–21–051	6591	Puno	Lescano (1984)
03–21–085	5958	Puno	Lescano (1984)
03–02–552	4790	Tahuaco	Mujica *et al.* (1984)
Tahuaco 1	3983	Tahuaco	Mujica *et al.* (1984)
Chewecca	3970	Tahuaco	Mujica *et al.* (1984)
Kanccolla	3941	Tahuaco	Mujica *et al.* (1984)
Blanca de Julí	3021	Tahuaco	Mujica *et al.* (1984)
Sajama	2443	Tahuaco	Mujica *et al.* (1984)

Source: Risi (1989).

optimal agricultural practices and use of low-yielding genotypes. Much higher yields have been obtained experimentally on the Altiplano (Table 1.16), and results obtained with sweet varieties from the experimental station at Patacamaya in farmers' fields in the northern Bolivian Altiplano (Table 1.17) show that higher yields are also obtainable on-farm.

In Ecuador the average yield is 407 kg/ha (MAG, 1984). Experimental yields of a wide range of germplasm accessions at Latinreco ranged from 179–4897 kg/ha (Alvarez *et al.*, 1990), and yields from commercial plots from 1131–4600 kg/ha, values which compared favourably with those obtained from wheat, barley and triticale (Burgasi *et al.*, 1990).

Table 1.17 Yields (kg/ha) of four improved varieties in farmers' fields in Bolivia

Locality	*Variety*	*Yield*	
		With 40 kg N/ha	*Without fertilizer*
Senkata	Chucapaca	715	435
	Huaranga	1465	860
	Kamiri	1225	740
	Sajama	735	500
	Chucapaca	1008	744
	Huaranga	1059	476
	Kamiri	773	530
	Sajama	1312	620

Source: IBTA (1987).

The best experimental yield of quinoa in Colorado was 2556 kg/ha (Johnson and McCamant, 1987). However, D407, the only variety available for commercial production, produced yields of only 545 kg/ha over 2 years of commercial production in the San Luis valley.

Yields of quinoa up to 5142 kg/ha have been obtained in experimental plots in Britain (Risi and Galwey, 1991a), and in Denmark experimental yields of 3.4–4.1 t/ha have been obtained (Denis-Ramirez, cited by Galwey, 1989).

1.6.7 **Diseases and pests**

Effect of plant introduction on pathogens

When a crop is introduced into a new country, there is often a period of freedom from pests and diseases. Pests and pathogens from the region of origin, however, may be inadvertently introduced or may have close relatives already there. In time, pests and pathogens in the new region, possibly already adapted to related plant species, will acquire the ability to use quinoa as a host. In considering the future role of quinoa as a crop both in South America and elsewhere, it is therefore relevant to review both the pests and pathogens which currently attack it there, and those which occur on related species elsewhere. A number of pests and pathogens of quinoa have been identified.

Fungal diseases

(a) *Downy mildew* (Peronospora farinosa)

This is the most important and widespread fungal disease in quinoa. The ideal conditions for the fungus are high relative humidity and moderately low temperatures. It is seed transmitted (Alandia *et al*, 1979). The main symptom of the disease is chlorotic lesions on the upper surfaces of the leaves, with a white or purple mycelium on the lower surfaces. Severe early infections can become systemic, causing the plants to become dwarfed, chlorotic and deformed (Alandia *et al.*, 1979).

Mildew is becoming more of a problem on the Altiplano as the area under quinoa increases (J. Risi, pers. comm.). MACA (1982, cited by Zapata, 1987) recorded losses of 30% in moderately resistant varieties and of 100% in susceptible varieties. The disease was more severe the further north the experimental plots were situated, reflecting the milder weather conditions in the northern Altiplano.

(b) *Brown stalk rot* (Phoma exigua *var.* foveata)

The main symptoms caused by this pathogen are dark-brown lesions with the vitreous edge, from 5–15 cm long, on the stem and inflorescence. Pycnidia are formed on the lesions. The stem is often shrunken, the plant may become chlorotic, and progressive defoliation towards the apex may occur. Artificially inoculated leaves showed a necrotic hypersensitive reaction (Otazu and Salas, 1977). *Phoma exigua* var. *foveata* infects potato foliage (Turkensteen, 1975). It has been isolated from potato tubers (Otazu and Salas, 1977) in Peru, and is the causal agent of potato gangrene, a storage disease in the UK. *Phoma exigua* var. *foveata* isolated in quinoa caused foliar spots on potatoes and tomatoes and gangrene symptoms on potato tubers (Otazu and Salas, 1977). This suggests that potato–quinoa rotations should be avoided, though such rotations are in fact commonly used.

(c) *Stem Gothic spot* (Phoma cava)

This fungus does not require a wound to cause infection (Alandia *et al.*, 1979). The lesions are light grey with brown edges and are surrounded by a vitreous halo. They are 2–3 cm long and are said to have the shape of a Gothic window. They occur mainly on the stem, but also on inflorescence branches and floral peduncles. The lesions produce pycnidia. In severe attacks the lesions can coalesce to girdle the stem, causing leaf and inflorescence branches to collapse (Salas *et al.*, 1977).

(d) *Leaf spot* (Aschochyta hyalospora)

This is a seed-borne disease and was found on between 7.8% and 26.3% of the seed in samples from Bolivia (Boerema *et al.*, 1977). The first symptom is light spots of indefinite area on the leaves. These become well defined, cream-coloured with a light-brown edge, and may coalesce until they cover the whole leaf. At later stages pycnidia can be observed, and the leaves become dry and fall off (Vilca, 1972).

Seed transmission occurs when pycnidia are produced in the inner tissues. The fungus causes mild to severe browning in the roots and hypocotyl of seedlings, and severely infected seedlings die (Boerema *et al.*, 1977). Some degree of resistance has been observed in the varieties Tupiza, Blanca de Chucuito, and Kanccolla (Salas *et al.*, 1977).

(e) *Seed rot and damping-off* (Sclerotium rolfsii)

This disease was observed in California in 1980. In an autumn planting, stem girdling and seedling collapse occurred in patches of the field. The saponin content of the seeds did not influence the development of the disease (Beckman and Finch, 1980).

(f) *Grey mould* (Botrytis cinerea)

This disease was observed on quinoa in Cambridge, England. Grey lesions of irregular shape were observed on the stems and inflorescences of mature plants. Where the central axis of the inflorescence was affected, the secondary axes sometimes collapsed.

Bacterial diseases

Bacterial blight is widely distributed around Puno and causes significant damage, especially when it occurs during grain development. The pathogen has not been fully identified and is described as a *Pseudomonas* sp. (Alandia *et al.*, 1979). The disease is seed-borne, and its dissemination within the field is favoured by rain, damp soil (Salas *et al.*, 1977) and mechanical wounds caused by hail (Alandia *et al.*, 1979).

Viruses

Alandia *et al.* (1979) reported a chlorotic mosaic of quinoa, probably caused by a seed-transmissible virus. Because quinoa is used as an indicator plant for virus testing, its response to a large number of viruses is known. Of those tested, many produce symptoms, but probably few would cause significant damage in the field.

Insect pests

A range of insects pests can attack quinoa during germination, vegetative growth, reproductive growth and seed storage (Zanabria and Mujica, 1977). Losses due to pests can be great: in the salares of Bolivia, insects constitute an important limitation on quinoa production, causing yield losses of up to 40% in the Ladislao Cabrera region (Quispe, 1976, cited by Ortiz and Zanabria, 1979). In the Peruvian Altiplano, insects cause losses of about 8% (Ortiz and Zanabria, 1979). When no control measures were used, total loss of the crop occurred at the experimental station of Los Lipez in the southern Altiplano (MACA, 1981, cited by Saravia and Calle, 1984; MACA, 1986a).

Quinoa insect pests can be classified as cutworms, mining and grain-destroying insects, and piercing and sucking insects (Zanabria and Mujica, 1977).

(a) *Cutworms*

This group includes several polyphagous species of the family Noctuidae, order Lepidoptera, including *Feltia experta, Spodoptera frugiperda, Spodoptera eridania, Copitarsia turbata* and *Agrotis ypsilon*. The adult moths have a wingspan of 35–40 mm and have grey or beige forewings with dark stripes characteristic of the species. The hind wings are generally white, cream or transparent. The caterpillars reach a maximum length of 35 mm and vary in colour from yellow–brown to grey along their dorsal surfaces (Zanabria and Mujica, 1977). The caterpillars do the damage, cutting through the stems of seedlings, and sometimes make it necessary to re-sow patches. They also attack the leaves of older plants, leaving only the veins (Lescano, 1981).

(b) *Mining and grain-destroying insects*

Leaf sticker (Panicle destroyer)
This is probably the most important pest of quinoa on the Altiplano. It was formerly classified as *Scrobipalpula* sp. but has recently been identified as *Eurisaca melanocompta*, family Gelechiidae, order Lepidoptera (P. Delgado, cited by Galwey, 1989). The adult moths have a wingspan of 15–16 mm and are brown–grey or yellow–grey. The caterpillars reach a maximum length of 12 mm and vary in colour from yellow–green to dark–brown, with diffuse brown or pink spots on their dorsal surface (Zanabria and Mujica, 1977).

This pest is favoured by warm, dry weather. The larvae mine the inflorescences of young plants and stick the young leaves together to shelter their cocoons. Mature inflorescences and seeds are also attacked, leaving a white dust on the plants (Lescano, 1981). This gives the insect its Quechua name of 'kcona kcona', meaning 'grain grinder' (Ortiz and Zanabria, 1979). The pest can also attack stored grain.

A moth probably belonging to the genus *Scrobipalpula*, and certainly belonging to the family Gelechiidae, has been observed on quinoa in Cambridge.

Quinoa moth and inflorescence caterpillar
These two similar pests have been identified as *Pachyzancla bipunctalis* and *Hymenia recurvalis* respectively, both of the family Pyralidae, order Lepidoptera (Ortiz and Zanabria, 1979). The adult moths of *P. bipunctalis* have a wingspan of about 25 mm and are pale yellow, whereas those of *H. recurvalis* are slightly smaller and are dark brown with a yellow spot and a darker transverse band on the forewing. The caterpillars of *P. bipunctalis* reach about 20 mm in length, those of *H. recurvalis*, 17 mm. The two are similar in form and colour, ranging from light to dark green with a dark-brown head (Zanabria and Mujica, 1977; Ortiz and Zanabria, 1979). The caterpillars destroy leaves and inflorescences and stick them together at the apex of the plant (Ortiz and Zanabria, 1979).

Leaf miner
This pest has been identified as *Lyriomiza brasiliensis*, family Agromyzidae, order Diptera. The adult fly is about 2.5 mm long and is black with yellow markings. The larva is vermiform, reaching about 3 mm in length, and is pale cream in colour (Ortiz and Zanabria, 1979). The larvae produce serpentine mines on the leaves and occasionally on stems. They may coalesce into blotches. The pest is favoured by warm, dry weather (Ortiz and Zanabria, 1979). An insect of the same family has been observed in quinoa in Cambridge.

Looper
The adults of this species (*Perisoma sordescens*) are dark-grey moths with a wingspan of 30 mm. The caterpillars reach about 20 mm in length and vary in colour from pale green to yellow–cream. They destroy leaves and immature and mature seeds, and during severe attacks empty inflorescences are found (Zanabria and Mujica, 1977).

(c) *Biting and defoliating insects*

This group includes three species of Coleoptera, *Epicauta latitarsis*, *Epicauta willei* and *Epitrix subscrinata*, which are also potato pests. The adult beetles of both *Epicauta* species are about 15 mm long and have cylindrical bodies, but whereas *E. latitarsis* is black and hairless, *E. willei* is dark grey and is covered with yellowish-grey velvety hairs. *Epitrix subscrinita* is a flea beetle about 2 mm long and is dark brown or shiny black (Zanabria and Mujica, 1977).

In all three species, the damage is caused by the adults. *Epicauta latitarsis* and *E. willei* attack leaves and inflorescences. They are favoured by dry, warm weather, and a severe attack can destroy entire fields in a few days. *Epitrix subscrinata* makes small perforations in the leaves and its attack is most serious at early plant growth stages (Ortiz and Zanabria, 1979).

(d) *Piercing and sucking insects*

This group comprises aphids, including *Mysus persicae* and *Macrosiphum* spp., leafhoppers, including *Bergallia* spp., and thrips, including *Franklinella tuberosi*. Feeding by these polyphagous species weakens the plant, but the damage that they cause by transmitting viruses is probably more important. All three groups of pests are favoured by warm, dry weather (Ortiz and Zanabria, 1979). Black bean aphids (*Aphis fabae*) have been observed on quinoa in Cambridge.

Nematodes

The false nodule nematode (*Nacobbus* sp.) and the cyst nematode (*Heterodera* spp.) have been reported on quinoa. *Nacobbus* sp. is a common pest, the females forming nodules on the roots (Alandia *et al.*, 1979). *Heterodera* is an occasional pest, the adult females forming yellow, white or brown cysts on the roots. Solano (1976) and Muñoz (1979) evaluated the resistance of quinoa varieties to *Nacobbus* sp. The varieties Chewecca and Real were highly resistant, whereas Sajama was susceptible. Resistance to *Heterodera* spp. has been reported among germplasm accessions evaluated in Holland (Doorgeest *et al.*, 1988).

Birds

Birds attack quinoa seedlings and mature inflorescences. In Cambridge about 40% of seedlings were attacked by birds, which destroyed the cotyledons. If the apex of the seedling was not damaged, the plant recovered (Risi and Galwey, 1984). On the Altiplano, losses at maturity of up to 40% occur in sweet varieties, especially around Lake Titicaca where pigeons are abundant (Mujica, 1977). Losses to birds are lower in bitter varieties (Aguilar, 1980).

Pests and diseases found on *Chenopodium album*

Chenopodium album is the closest European relative of quinoa and has been carried as a weed in European crops to North America, Australasia and Africa, as well as to South America (Williams, 1963). It is therefore likely to be a source of pests and diseases for quinoa cultivated outside South America. Fungi found on *C. album* include Peronospora farinosa, *Ligniera verrucosa, Olpidium brassicae, Cercospora dubia, Cercospora beticola, Cercospora chenopodii, Verticillium albo-atrum, Septoria chenopodii* and *Phoma lavendulae* (Williams, 1963). The insects include various species of collembola, thrips, aphids, Lepidoptera, and Diptera, and the nematode species *Aphelenchoides ritzema-bosi, Heterodera schactii* and *Ditylenchus dipsaci* have also been found (Williams, 1963; Basset and Crompton, 1978).

1.6.8 **Control of pests and diseases**

In subsistence agriculture, pests and diseases are traditionally controlled by eliminating diseased plants, and caterpillars or adult insects, by hand, and by the burning of sulphur or old tyres to deter the insects. Kcona kcona are repelled by putting branches of the aromatic plant muña (*Hedeoma mandosiana*) in the quinoa inflorescences (Ortiz and

Zanabria, 1979). The fungicides Polyram-Combi, Cupravit OB 21, Manzate D and Lanacol at a dose of 1.5 kg/ha were found to inhibit the growth of *Peronospora farinosa* (Laura, 1977).

In the Andes it has been found that the severity of insect attacks depends principally on the ecology of the area in which quinoa is grown (Quispe and Savaria, 1987; MACA 1986b; Saravia and Calle, 1984). Quispe and Savaria (1987) found that pests were more common in the southern part of the Bolivian Altiplano where it is both cooler and drier than in the northern part. The importance of regular rains in the control of insect proliferation is clear from the comparison of two surveys carried out in the region of Salinas de Garci Mendoza in the southern Bolivian Altiplano, one in 1983, a particularly dry season (Saravia and Calle, 1984), the other in 1985–86 (MACA, 1986b), a wet season. In the dry season survey, serious infestations of thrips and aphids were found throughout the season. Maximum populations of 24 kcona kcona per plant and nine ticona per plant were recorded at maturity and mid-cycle respectively. In the wet-season survey, however, no pests were found between October and December, a period of frequent precipitation. During January and February the number of kcona kcona observed per plant increased gradually from 2.3 to 18.1. A week later the numbers had decreased dramatically to 0.7 and continued declining until none were detected in mid-April. Ticona were present from the end of January until the beginning of March. Their numbers fluctuated from 0–0.4 per plant.

Frosts have also been shown to act as a natural pest control. Infestations of kcona kcona were reported to decline from 46.4 to 4.6 insects per plant in response to a 7 day period of frost in the Salinas de Garci Mendoza region of the southern Bolivian Altiplano. Frosts maintained the population of kcona kcona between 0.3 and 1.1 per plant until mid-February; thereafter, the population increased from 0.6 per plant on 16 February to 19.6 on 23 February (Germán and Saravia, 1987).

Chemical control of insects is commonly practised in traditional production zones, but there is little technical assistance (Saravia and Calle, 1984) and it is therefore likely that dosages and timings of applications are often inappropriate. Farmers tend to use large doses (Saravia and Calle, 1984; Germán and Saravia, 1987). As a number of quinoa pests also use the potato as a host, regimes for the control of potato pests may reduce pest populations on subsequent crops of quinoa. Table 1.18 gives details of commercial products which have been used in pest control.

In traditional production zones the natural enemies of quinoa pests are important in their control (Saravia and Calle, 1984; Ruales, 1987; Peralta, 1986; Ruales and Fiallo, 1990). Ruales and Fiallo report several beneficial insects associated with the quinoa crop. The insects parasitize the larvae of the major quinoa pests, as shown in Table 1.19. Peralta

Table 1.18 Commercial products and dosages used in the chemical control of pests of quinoa in Peru and Bolivia and of potato in Ecudor

Pest	*Commercial product*	*Dosage*	*Reference*
Agroti ypsilon	Dipterex 2.5	20 kg/ha	Zanabria (1981)
	Sevin 5	20 kg/ha	Zanabria (1981)
	Sevin 85%	3 kg/ha	Muñoz and Cruz (1978)
Copitarsia sp.	Sevin 85%	2.7 kg/ha	Merino and Vazquez (1978)
	Monitor 600	1.6 l/ha	Merino and Vazquez (1978)
	Thiodan methyl 20%	2.2 l/ha	Merino and Vazquez (1978)
	Orthene 50%	1.0 kg/ha	Merino and Vazquez (1978)
	Lannate 90%	0.5 kg/ha	Merino and Vazquez (1978)
C. turbata	Azodrin 60	0.30%	Zanabria (1981)
Eurisaca	Lannate	0.40%	Zanabria (1981)
melanocampta	Azodrin 60	0.20%	Zanabria (1981)
	Lannate	0.30%	Zanabria (1981)
	Ambush 50	0.05%	Zanabria (1981)
	Decis 2.5	0.01%	Zanabria (1981)
	Ripcord 10	0.01%	Zanabria (1981)
Naupactus sp.	Thiodan	1.5 kg/ha	Merino (1986)
Ephestia sp.	Phostoxin	4–6 tablets/t grain	Bayer (1970)

Source: Ruales and Fiallo (1990).

Table 1.19 Quinoa pests and their natural predators

Pest	*Beneficial insect*
Agrotis spp.	*Incamyia* sp. (Diptera, Tachinidae)
Copitarsia spp.	*Incamyia* sp. (Diptera, Tachinidae) Diptera, Tachinidae A Diptera, Tachinidae B Hymenoptera, Icheumonidae A Hymenoptera, Braconidae A
Eurisaca melanocampta	*Elfia* sp. (Diptera, Tachinidae) Hymenoptera, Icheumonidae A

Source: Ruales and Fiallo (1990).

(1986) considered that good cultural practices may be sufficient to control most pests. Pest populations were found to be maintained at a low level by their natural predators, and remaining insects could be removed by hand. When fields were neglected, however, one or several pest species might proliferate so that it became necessary to treat this crop with an insecticide. Another potential method for controlling some pests is cultivation of the land to destroy the hibernating pupae of Noctuidae, Gelechiidae and Pyralidae before the adult moths emerge.

1.7 GENETICS AND BREEDING

1.7.1 **The breeding system and its implications for breeding strategies**

Range of breeding systems

The breeding system of a crop has important implications for the genetic structure of its populations and for the breeding strategies which can be used in its improvement. In the past it was assumed that quinoa is predominantly self-pollinated; however, there is evidence which suggests that substantial levels of outcrossing can occur. This has important implications for breeding methods and, since inbreeding depression and heterosis can be expected to occur, for the types of varieties which should be produced. Methods used to maintain the distinct-

Table 1.20 Breeding systems observed in quinoa

1. Plants with hermaphrodite flowers only, perianth not reflexed at anthesis. Homogamous, self-compatible.

2. Plants with hermaphrodite flowers only, perianth fully reflexed at anthesis. Homogamous, self-compatible.

3. (a) Plants with hermaphrodite flowers only, perianth fully reflexed at anthesis. Partially dichogamous, self-compatible. (b) Plants with female and hermaphrodite flowers (gynomonoecious). Self-compatible.

4. Plants of two kinds, bearing functionaly female flowers only or bearing female and hermaphrodite flowers (gynodioecious). Self-compatible.

5. Plants with hermaphrodite flowers only, but with well developed dichogamy (protogynous; protandry probably does not occur). Stigmatic branches strongly exerted (up to 2 mm in length). Pollen not shed. Thus plant is effectively female. Obligate outbreeder.

6. Plants with female and hermaphrodite flowers (gynomonoecious). Self-compatible. Obligate outbreeder.

Source: Adapted from Nelson (1968).

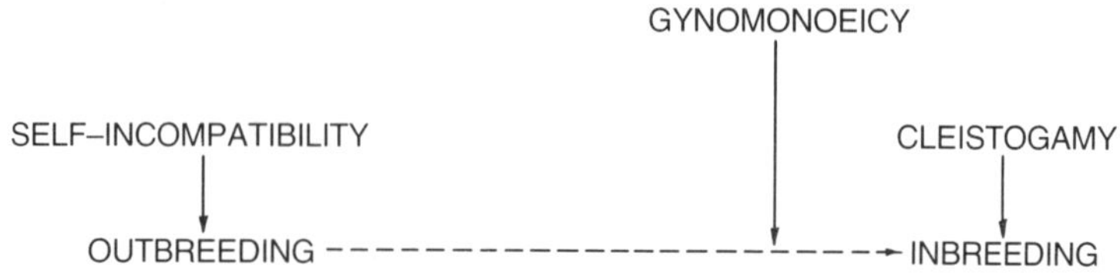

Figure 1.4 Breeding systems in quinoa.

ness and purity of germplasm collections and breeding lines will also be affected.

Quinoa exhibits quite a range of reproductive mechanisms, from complete self-incompatibility to virtual cleistogamy. Table 1.20 gives details of the systems observed by Nelson (1968), in order of their potential for outbreeding.

Gynomonoeicy

The predominant breeding system in quinoa is gynomonoeicy (Figure 1.4). Within this system, the ratio of hermaphrodite to female flowers is under both genetic and environmental control (Rea, 1969). In extreme cases only the apical flower of a group is hermaphrodite (León, 1964). The male fertility of hermaphrodite flowers also varies with variety and environment, thus effectively increasing the proportion of female flowers (Simmonds, 1965; Gandarillas, 1969). Aguilar (1980) reported that low temperatures inhibited the expression of male fertility.

Hunziker (1943) observed that hermaphrodite flowers generally have five perianth lobes, five anthers, and a superior ovary from which two, or rarely three, stigmatic branches emerge. Many flowers in experiments conducted in Cambridge had six or seven anthers and perianth lobes. Three stigmatic branches have also been quite common.

Self-incompatibility

Self-incompatibility can be assessed by enclosing an inflorescence in a bag during anthesis and observing the effect on seed-set. However, partial self-incompatibility will not be detected by this method as it will produce a similar low seed-set to that caused by damage to the flowers during the bagging process, or to heat, humidity, insects or reduced pollination in still air (Nelson, 1968). Nelson showed that the variety Rosada de Junín is self-incompatible. It set no seed in 18 selfing attempts, although its pollen is viable, and it did set seed when pollinated by other accessions. Rea (1969) observed that neither Rosada de

Junín nor Blanca de Junín bred true in trials, but that each appeared to have hybridized extensively with the other.

Male sterility

Male sterility, due to the anthers being non-functional in all the hermaphrodite flowers, has been reported by Gandarillas (1969), Simmonds (1971) and Aguilar (1980). Studies of the inheritance of this character suggested that male sterility is controlled by a single recessive gene, designated ms (Gandarillas, 1969; Simmonds, 1971). Plants homozygous for this gene, however, do not breed true; they seem to have an erratically expressed or transmitted cytoplasmic factor for male sterility (Simmonds, 1971). Aguilar (1980) reported the possible existence of restorer genes. UNTA 292 (a male sterile line) produced only male sterile plants in the F_1 generation and in two successive backcrosses to three male fertile lines. When it was crossed with the variety Sajama, however, its progeny were male fertile, and in the F_2, fertile and sterile plants were reported to occur in the ratio 3:1. Sajama seems, therefore, to possess a dominant gene for the restoration of male fertility in UNTA 292 cytoplasm. Reliable cytoplasmic male sterility has recently been reported in Colorado in selections from the cultivar Apelawa, and has been transferred into four additional genetic backgrounds (Ward, 1991). Male sterility could be of considerable use for the production of F_1 hybrid varieties, or even for pedigree breeding, in view of the difficulties associated with manual emasculation. However, restorer genes for the Apelawa male sterile cytoplasm have yet to be found.

Rate of outcrossing

The rate of outcrossing can be affected by the environment in several ways: wind speed will determine the distance over which pollen can travel, and other environmental factors will affect the ratio of hermaphrodite to female flowers. Temperatures at anthesis will affect pollen production, and the occurrence of rain will influence pollen transmission. It is necessary, therefore, to obtain an estimate for outcrossing in each area where quinoa is grown. As environmental factors vary from year to year, and as genotypes exhibit a range of breeding systems and interact with the environment in different ways, estimates of the rates of outcrossing will not be widely applicable to many genotypes and locations. Kenwright (1989) reported circumstantial evidence which suggested that rates of outcrossing in her experiments at Cambridge were unexpectedly high. There were greater levels of dominance than expected in the F_2 and F_3 generations and there was evidence of introgression of 'exotic' characters (such as purple leaf pubescence) into Chilean germplasm. Outcrossing was thought to occur mainly between near neighbours, and rarely over great distances. Subsequent experiments have confirmed that a significant amount of outcrossing occurs under field conditions in Cambridge.

The rate of outcrossing in quinoa in the conditions of the Bolivian Altiplano has been estimated by Gandarillas (1979b). He found that outcrossing was most frequent over short distances (9.9% at 1 m), but could also occur over considerable distances: at 20 m, for example, 1.5% outcrossing occurred. Lescano (1980) estimated rates of outcrossing in eight landraces and five commercial varieties of quinoa and obtained an average value of 5.8%. Studies of outcrossing in Colorado under field conditions suggest rates exceeding 10% (D. Johnson, unpublished). However, controlled wind-pollination studies have detected lower levels of outcrossing; 1–2% at distances of up to 36 cm, falling to 0.1% at 400 cm, with no further outcrossing beyond this distance (S. Ward, unpublished).

1.7.2 **Genetic analysis**

Inheritance of Mendelian characters

Plant colour

Gandarillas (1974) identified three basic patterns of plant coloration, namely red plants with pigment in all parts, purple plants with pigment only in the apical leaves and inflorescence (which may turn yellow at maturity), and green plants with no pigmentation. The intensity of colour may vary: Simmonds (1971) reported that when plants are grown under low light intensities, pigmented plants may be misclassified as green plants. Plant colour is determined by a series of alleles at one locus, designated R for red, rp for purple, and r for green. R is dominant to both rp and r, and rp is dominant to r (Gandarillas, 1968, 1974; Simmonds, 1971). This allelomorphic series also controls inflorescence colour (Gandarillas, 1974), the environment exerting a similar influence on colour intensity (Simmonds, 1971).

Axial pigmentation

Axial pigmentation is controlled by a single gene designated Ax, with pigmented axils being dominant to non-pigmented (Simmonds, 1971). This character cannot be observed in red plants.

Inflorescence type

The two types of inflorescence (glomerulate and amaranthiform) are controlled by a single gene, designated G, with glomerulate being dominant to amaranthiform (Gandarillas, 1974).

Grain colour

The ancestral colour of the pericarp is black (Heiser and Nelson, 1974). In quinoa, red and yellow grain are both dominant to white, and black is dominant to yellow (Paca, 1970; Gandarillas, 1974). Gandarillas (1979b) suggested that grain colour depends on two series of alleles, situated on different chromosomes, and that it is the combination of these alleles which determines the colour of the grain. However, large segregating populations from a wide range of carefully planned crosses must be studied in order to confirm the features of such a complex genetic system, and this scheme should therefore be regarded as provisional.

Alvarez and von Rütte (1990b) have pointed out that Mendelian ratios of seed colours are not found in natural field populations, but such ratios are to be expected only in the progeny of crosses between genetically pure stocks, and this observation therefore requires no special explanation.

Saponin content

Gandarillas (1974) concluded that saponin content was controlled by a single dominant gene, designated *D*, high saponin content being dominant to low saponin content. However, semi-sweet varieties such as Kanccolla and Blanca de Julí exist, and so it is more likely that saponin content is polygenically controlled. Alvarez and von Rütte (1990b) have used the natural variation found in this character to select sweet varieties (LR-01201 and LR-012-2) from a bitter variety (LR-012). They concluded that the genes controlling this character are not linked to a particular pattern of plant pigmentation or phenology: high- and low-saponin types are found in red, green and purple plants, in plants with and without axil pigmentation and in early- and late-maturing plants. Grain colour may, however, be linked to saponin content: all the known sweet varieties have opaque white seeds. Sweet varieties also produced grain of higher density than bitter varieties, the average values for bitter and sweet varieties being 64.6 and 68.9 kg/hl respectively. Alvarez and von Rütte have also shown that the presence of saponins is not strongly linked with any nutritional constituent: in particular, the correlation between saponin content and protein content was low ($r = 0.22$). They found no difference in the tolerance of pests and disease between the two types. However, this result should be treated with caution, since other investigators have suggested that saponins confer resistance to diseases and pests (Torres and Minaya, 1980). A positive correlation ($r = 0.72$, $P < 0.01$) has been found between seed weight and saponin content (Reichert *et al.*, 1986). However, large-seeded, low-saponin cultivars such as Sajama and small-seeded, high-saponin cultivars such as Baer do exist.

Other seed characters

Most varieties of quinoa have an opaque endosperm, but some, known as *chullpi* varieties, have translucent endosperms. This character is reported to be controlled by a single gene designated Su, opaque endosperm being dominant (Gandarillas, 1974). However, other results suggest that this character varies quantitatively. Opaque (floury) endosperms are more common further north (Wilson, 1990). In some varieties of quinoa from the Titicaca basin, the pericarp can be easily separated from the endosperm. This character is called koytu (Gandarillas, 1974) or ccoito (Tapia, 1979) and is commonly found in low-saponin varieties. It is reported to be determined by a recessive gene (Gandarillas, 1974).

Male sterility

Simmonds (1971) found a weak linkage between a recessive gene determining male sterility, ms, and the loci determining plant pigmentation (R) and axil pigmentation (Ax), with the following recombination frequencies:

Ax——8%——R——————35%——————Ms

However, Aguilar (1980) found recombination frequencies for these characters that did not differ significantly from the 50% expected for unlinked genes.

Inheritance of quantitative characters

Risi (1986) and Kenwright (1989) analysed the inheritance of a number of continuous characters in the progeny of an 8×8 diallel cross involving two parents of each of four ecotypes, namely Valley, Altiplano, Salares and Sea-level. Risi investigated the assumptions of the diallel cross-analysis, namely that there are no cytoplasmic effects or epistatic effects, and that there is independent action of non-allelic genes in the

Table 1.21 Broad- and narrow-sense heritabilities of quantitatively varying characters

Character	*Heritability (%)*		
	Narrow sense[a]	*Broad sense*[b]	*Narrow sense*[b]
Height	34.3	63.39	39.25
Leaf protrusions	89.7	30.31	24.63
Inflorescence dimensions			
length	44.9	NR	NR
diameter	37.7	NR	NR
density	76.7	87.29	61.63
Number of days to reach growth stages			
flower initiation	84.4	NR	NR
anthesis	83.7	73.03	52.01
maturity	93.7	87.2	72.76
Saponin content	87.0	62.96	58.50

NR = not reported.

Sources: [a]Risi (1986).
[b]Kenwright (1989).

F_1 generation. He concluded that for inflorescence length and diameter these assumptions were not met, but that long, wide inflorescences were nevertheless probably dominant. In the case of plant height, tallness was shown to be overdominant. For the periods between germination and floral initiation, anthesis and maturity, there appear to be both dominant and recessive genes which affect earliness. The behaviour of parents and crosses was different for each period, indicating that there are different genes regulating the length of each of these sub-periods. Thus it should be possible to manipulate the development pattern through breeding. Saponin content was shown to be a continuously variable characteristic, with high saponin being generally dominant to low saponin content. Kenwright (1989) studied the F_2 generation of the same diallel cross. In both generations, the narrow sense heritabilities of most of the characters studied were found to be high (Table 1.21).

Kenwright found that there was close genetic control of flowering date, maturity date, height and saponin content, although there was a fairly strong environmental component in the variation for flowering date and saponin content. Stem diameter, and the length and diameter of the inflorescence, were found to be greatly influenced by the environment. In contrast to Risi's finding that the number of leaf protrusions was under close genetic control, Kenwright found this variable to be subject to environmental variation. However, it was likely that the Chilean varieties used (Baer and Faro) contained recessive alleles producing a high number of leaf protrusions. Directional dominance was found for inflorescence compactness, plant height, maturation date and saponin content, but not for flowering date. Characteristics typical of cultivated types tended to be produced by recessive alleles.

1.7.3 **Breeding**

Breeding objectives and methods in South America

Objectives

Scientific improvement of quinoa in South America began in 1965 at Patacamaya in Bolivia (Gandarillas, 1979c). Other breeding programmes have since been established. Germplasm collections have been formed at the Bolivian Institute of Andean Crops (Gandarillas, 1979c), at the Universidad Nacional Agraria, La Molina, Lima, Peru (Salhuana *et al.*, 1977), at the Universidad Nacional Técnica del Altiplano, Puno, Peru (Lescano, 1981), and at the University of Concepción, Chile (Risi and Galwey, 1984). Breeding and research work is being conducted at Nestle's South American company, Latinreco S.A., in Ecuador.

Breeder's objectives have mostly been those required to adapt the crop for modern agricultural and processing systems. Short, compact genotypes with uniform maturity, which is responsive to increasing levels of inputs and to agricultural mechanization, are desirable. Frost resistance and drought tolerance are important objectives for cultivation on the Altiplano. Lescano and Tapia (1977) pointed out the needs of producers for higher yields and resistance to pests, diseases and other environmental factors, of processors for low-saponin varieties, and of consumers for high protein quality. These objectives are not always compatible: for example, breeding for reduction in saponin content may result in reduced yields due to increased attack by birds and other pests. On the other hand, bird damage may not be an important factor in large commercial plots. Inevitably, the objectives used in breeding for mechanized agriculture may conflict with the needs of peasant farmers practising subsistence agriculture, where reliable yields are more important than high yields.

The main methods employed by breeders have been mass selection in genetically heterogeneous landraces, and hybridization followed by various selection schemes.

Mass selection

The improvement of local landraces by elimination of defective genotypes has been successfully used in many crops when variable landraces are common (Allard, 1960). This is still the case for quinoa, as the practice of subsistence agriculture has conserved the genetic diversity of the crop. Peasant farmers commonly cultivate their own or the local landrace, which is generally a mixture of genotypes, and they use this method to maintain the purity of a variety for the characteristics that they regard as important (Lescano, 1977). The farmers of the southern Bolivian Altiplano sow mixtures of a number of varieties and retain the best seed-heads for sowing in the following year (Kenwright, 1989). The varieties produced by mass selection include Blanca de Junín, Amarilla de Maranganí and Dulce de Quitopamba, belonging to the Valley ecotype; Kanccolla, Chewecca, Blanca de Julí and Pasankalla, belonging to the Altiplano ecotype; Real, belonging to the Salar ecotype; and Baer, belonging to the Sea-level ecotype. These varieties have been described in more detail by Risi and Galwey (1984). The Ecuadorian varieties LR-012 and LR-013, produced by Latinreco S.A., and INIAP-Cochasquí and INIAP-Imbaya, produced by the Instituto Nacional de Investigaciones Agropecuarias, were also obtained by mass selection (Nieto *et al.*, 1986; Alvarez and von Rütte, 1990b). These Ecuadorian varieties belong to the Valley ecotype. Other Bolivian varieties include Huaranga, Chucapaca, Kamiri, Sajama Amarantiforme and Real Blanca (also known as Calcha). These are low-saponin varieties belonging to the Altiplano ecotype (Nina and Saravia, 1987). Sajama Amarantiforme was produced by reselection out of the variety Sajama, but the breeding method by which the others were produced is not stated.

Hybridization

Hybridization offers the possibility of making improvements with more precision than mass selection, since desired characters or combinations of characters can be incorporated into the same genotype from varieties of diverse origin.

Once suitable genotypes have been identified, they must be crossed. Quinoa produces a large number of very small flowers, which generally produce abundant pollen. Manual emasculation of the flowers is possible, but very difficult and tedious. A number of methods have been described. Rea (1948) suggested that the five or six most apical flowers of a chosen glomerulus or group of flowers should be emasculated by removing the anthers with a needle before they shed their pollen, without damaging the stigmas. The rest of the hermaphrodite flowers should be removed with a needle or pair of forceps. Female flowers can be left in place. Pollen from the other genotype should immediately be placed on the stigmas with a fine brush, and the glomerulus should then be covered with a grease-proof paper bag to exclude alien pollen. Pollen can be collected by shaking an inflorescence of the pollinator genotype into a paper bag. The glomerulus should be examined after a period of 5 or 6 days and any pollen-producing flowers that remain should be removed. Gandarillas (1967) suggested modifications to the method. All the glomeruli except two or three at the base of the inflorescence can be removed. The newly opened hermaphrodite flowers in these glomeruli are emasculated or removed every morning. The female and emasculated flowers are pollinated and each glomerulus is covered with a grease-proof paper bag. This process is repeated every day throughout anthesis, thus reducing the opportunities for self-pollination. Lescano and Palomino (1976) describe a technique for emasculating the whole inflorescence. Before anthesis occurs in any flower, all the small and poorly deformed glomeruli are removed. Hermaphrodite flowers are removed or emasculated and the whole inflorescence is covered with a grease-proof bag. Receptive stigmas, which can be recognized by their shiny white appearance, are pollinated and all the flowers with dull stigmas are removed.

These techniques have a number of limitations. They rely on the ability to recognize and remove hermaphrodite flowers before their anthers have dehisced. Due to the large number of flowers in one inflorescence, their small size and abundant pollen produced, this is quite difficult. The emasculation technique and the use of paper bags can easily damage the stigmas and thus prevent fertilization. Anthesis occurs over a period of several days and so daily attention must be given to the female parents. If the male parent does not reach anthesis at the same time as the female parent, the pollen must be stored under conditions in which it will remain viable.

These problems make it quite likely that there will be a mixture of selfs and hybrids in the progeny. It is important, therefore, to use genetic markers, such as axil pigmentation or plant colour, to distinguish between these two types of progeny. As the breeding of quinoa advances there will be a desire to cross more closely related varieties. A wider range of marker genes, morphological and/or molecular, will thus be needed.

Hybridization would be much easier if it were possible to male-sterilize inflorescences by other means such as the use of chemicals. Aguilar (1980) investigated the use of hot water to sterilize flowers. Flowers showed only slight scalding when submerged for 20 min in hot water at 41–45 °C, though higher temperatures caused excessive damage. Soilleux (1986) found that immersion in water at temperatures low enough not to kill the inflorescence reduced pollen viability by only

45%. Kenwright (1989) reported that research was underway at the Universidad Nacional Agraria, Lima, to develop alternative methods for emasculation. The traditional method was being compared with the use of a suction pump to remove the anthers and with immersion in water or alcohol to inactivate the pollen. The most reliable means for hybrid production will be the use of genetically male-sterile lines, but these are not readily available for routine use. Experience in Cambridge has indicated that if the parent lines are sufficiently genetically distinct, it is not necessary to attempt emasculation of the flowers. The inflorescences of the parents can be shaken together at anthesis, pollen from the male parent being transferred to the stigmas of the female parent. The inflorescence heads are then covered with grease-proof paper bags to prevent contamination from other plants. If this process is repeated each day throughout anthesis a substantial number of hybrids can be generated. The success of this method is influenced by the breeding system in the parent varieties: it is particularly successful when the female parent shows protogyny.

Varieties obtained by selection from hybrids include Sajama, bred at Patacamaya for production on the Altiplano, and UNC-20 and UNC-76, bred by the Universidad Nacional del Centro, Huancayo, Peru, for production in the Mantaro Valley. These varieties have been described in more detail by Risi and Galwey (1984).

Use of improved varieties

Some improved varieties are widely accepted into traditional systems and have displaced the local landraces. This has occurred most widely in the northern and central Bolivian Altiplano, where the experimental station of the Instituto Boliviano de Tecnología Agropecuaria at Patacamaya is influential (Kenwright, 1989). At Patacamaya, in the central Altiplano, 47% of the farmers sow improved varieties from the experimental station (MACA, 1987).

Breeding methods and objectives outside South America

North America

Breeding and research on quinoa in Colorado began in 1983, when Sierra Blanca Associates and the Colorado State University, in collaboration with the Instituto Boliviano de Tecnología Agropecuaria (IBTA), began to develop quinoa as a crop for the higher altitude valleys and the foothill regions of the Rocky Mountains (Cusack, 1983; Ballón *et al.*, 1986). The shortness of the growing season restricts the areas in which quinoa can be grown in this region. The average temperatures in the State of Colorado vary from −12 °C to 32 °C. Rainfall in summer months is minimal, though in the spring there is an accumulation of meltwater from the Rocky Mountains (Ballón *et al.*, 1986). Because of these restrictions, cultivation is limited to locations at altitudes between 2000 and 3500 m (Lorenz and Coulter, 1991).

A germplasm collection of 160 accessions was evaluated at five locations between these altitudes in Colorado in 1984, and the best-adapted varieties were selected and grown out in subsequent years. Peasant varieties were found to perform better than the 'improved' varieties, and those with higher saponin contents performed better than those with lower saponin contents (McCamant, 1985). Irrigation and fertilizer trials, and small-scale and commercial-scale yield trials, were performed. In 1986, the highest yield obtained was 2556 kg/ha (Johnson and McCamant, 1987). The maximum temperatures that quinoa could tolerate were found to be between 29 °C and 34 °C. It was found to withstand temperatures as low as −6 °C: this is lower than the minimum of −4 °C reported from the Andes (McCamant, 1985). Only the named variety, D407, is currently available for experimental production in Colorado. It has early maturity, a semi-dwarf growth habit, yellow, compact heads and medium-small seeds, and has produced 1200 lbs/ha (530 kg/ha) over 2 years in the San Luis Valley (Lorenz and Coulter, 1991).

Europe

The potential for using quinoa as a break crop in cereal rotations in eastern Britain and other similar European regions was also recognized in the early 1980s. The development of quinoa in Britain began with the assembling of a germplasm collection of nearly 300 accessions, obtained from the Andes between 2 °N and 22 °S and also from low altitudes in

Table 1.22 Characteristics of the quinoa ideotype

Structural characteristics	*Responses to the environment*
Short stature	Responsive to nitrogen fertilization
Thick stem	in terms of yield
Little or no branching	Tolerant to nitrogen fertilization in
Compact inflorescence	terms of vegetative growth
Even, early maturity	Insensitive to daylength
High harvest index	Frost resistant
White or light-coloured seed	Drought resistant

Source: Risi and Galwey (1984).

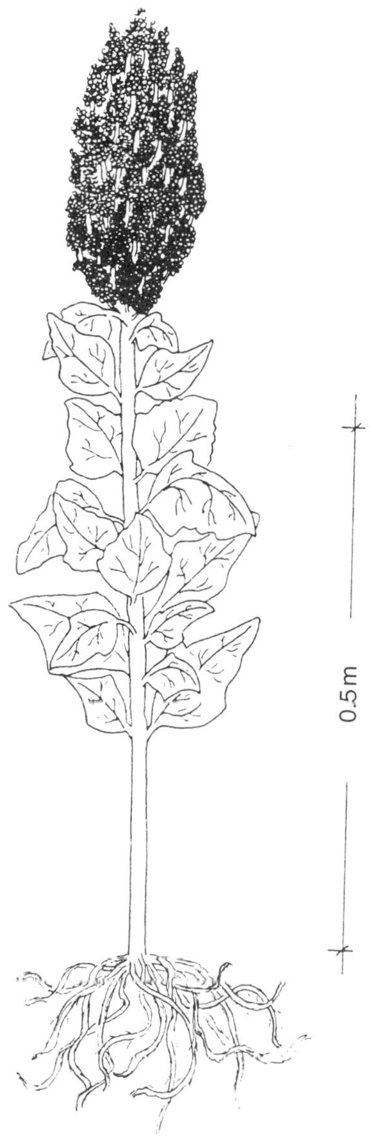

Figure 1.5 Quinoa ideotype.

central and southern Chile (35–40 °S). After establishment of appropriate agronomic practices, the collection was grown out and the accessions best adapted to English growing conditions were selected. An ideotype for Britain was drawn up to provide a basis for evaluation and selection (Table 1.22; Figure 1.5). The germplasm evaluation revealed that there is wide diversity for many characters and combinations of characters in quinoa and that there is great scope for the manipulation of the plant's developmental sequence through breeding (Risi and Galwey, 1989a). The accessions from Chile were chosen as a basis for the breeding programme, as they mature within the British growing season and have many good agronomic characters. The grain from these accessions, however, is small and buff-coloured, with a relatively high saponin content.

A breeding programme was initiated, two strategies being used:

1. mass selection within the Chilean variety Baer;
2. pedigree selection from a diallel cross between two Chilean varieties, two Valley varieties, two Altiplano varieties and two Salar varieties.

In addition, selections were made from new accessions. After two cycles of selection significant advances had been made relative to the diallel cross material: for example, maturation dates were on average 50 days earlier. A considerable number of selections closely resembled the ideotype for all agronomic characters, but seed characters required further improvement. Subsequent selection cycles were carried out and the resulting lines have been grown and harvested successfully on a commercial scale. An additional crossing cycle has been performed between these lines and varieties with large white seeds with low-saponin content, in an attempt to improve the seed characteristics of the British-adapted lines.

Breeding work has also been conducted in other European countries. In Denmark, Jacobsen *et al.* (1994) have conducted evaluations of quinoa genotypes, selected genotypes suited to the Danish climate and growing conditions, and determined appropriate agronomic practices. Haaber (1991) has evaluated genotypes suited for green-fodder production. Quinoa was first introduced into the Netherlands in 1986 (L. van Soest, pers. comm.). A germplasm collection of 40 accessions was evaluated and methods for disease resistance screening were developed. Resistance to beet-cyst nematode was identified.

1.8 POST-HARVEST TECHNOLOGY

1.8.1 **Removal of saponins**

In the Andean region quinoa is processed for household use by washing. Traditionally, the grain is washed and rubbed in cold water and then either cooked immediately or left to dry in the sun. If it is soaked for too long, the saponins may enter the internal parts of the grain

(Vargas *et al.*, 1977), and it must dry in a matter of hours to prevent germination or fermentation. It can then be ground into flour or used whole. Saponins can also be removed by toasting the grain and rubbing it on stone mortars to remove the outer layers (Torres and Minaya, 1980). With the resurgence of interest in the crop, however, these processes have had to be converted to an industrial scale so that quinoa products can be made available to the urban consumer. The major processing problem is the removal of saponins from the grain. Although low-saponin varieties do exist, high-saponin varieties are often preferred because of their great resistance to bird attack, and possibly also to pests and diseases (Torres and Minaya, 1980).

Converting these processes to an industrial scale poses problems. The large amount of heat required to dry the grain makes the process expensive, and disposing of the bitter foaming waste water is also difficult. Hernández (1987) reported successful prototypes of machines for both wet and dry methods of saponin removal. Machines used for dry methods of saponin removal are similar to those used to remove wheat bran. A machine designed at the Instituto de Investigaciones Agroindustriales, Lima, Peru, has been successfully used to remove most of the saponin from grain samples. With the machine retaining 94.6% of the grain, and producing processed grain at a rate of 850 kg/h, the saponin content was reduced to less than 0.05 mg/kg seed (Torres and Minaya, 1990). Dry scouring of the seed in a laboratory rice scourer (70% flour extraction) has been investigated (Junge, 1973), as has roller-milling (Amaya-Farfán *et al.*, 1978). Abrasive dehullers have been developed by Reichert *et al.* (1984) and Reichert *et al.* (1986), and have been tested on a number of quinoa varieties. Abrasive dehulling resulted in saponin extraction ranging from 85.2–98.8%, and reduced the saponin content to that found in sweet varieties. In general, the higher the initial saponin content of the grain, the greater the loss of bran required to remove the saponins. Methods based on abrasion probably carry the penalty of reducing the content of minerals and vitamins (Koziol, 1992), as these are concentrated in the outer layers of the seed (Varriano-Marston and de Francisco, 1984).

1.8.2 **Food uses**

Traditionally, quinoa grain is used in a variety of ways. It can be toasted, ground into flour, or boiled and added to soups, or made into breakfast foods or pasta. The flour has been used to make a coarse bread called 'kispina' (Weber, 1978). It can also be fermented to make beer (Simmonds, 1965). Quinoa flour, flakes, tortillas, pancakes and puffed grains are commercially available in Bolivia and Peru. The flour can be used as a partial substitute for wheat flour in bread production.

Flour production

It has been argued that the efficiency of white flour extraction from quinoa should be much higher than from cereals, as its outer layers are softer than those of the cereal grains. Thus after removal of saponins, all parts of the grain can be used for flour production (Llerena, 1973; Ballón and Coca, 1984). Potentially, therefore, the flour yield in sweet varieties is 100%, whereas the yield from wheat is 70.8–76.8%. Ballón and Coca (1984) found that the time taken to mill a 100 g sample of grain varied from 1.0–2.9 min for soft grain to 5.97–7.0 min for very hard grain. Grain from the southern Bolivian Altiplano was very hard, whereas grain from the central Bolivian Altiplano was semi-soft and that from the Peruvian and northern Bolivian Altiplano was soft. This probably reflects the distinction between floury and transparent endosperms, the latter being more common further south.

Food processing

The development of a market for quinoa outside South America has involved investigation of novel ways in which it can be incorporated into food products. The size of starch granules in a flour influences the stability of doughs used for baking and the manufacture of pastas. Small starch granules have been shown to destabilize wheat doughs (Kulp, 1973), and the low amylose content of quinoa starch will also tend to reduce its quality in baking applications. However, Weber (1978) reported the successful use of quinoa flour as a partial substitute for wheat flours in breads, cookies and pasta, and Luna (1957) reported that when a mixture of 90% wheat flour and 10% quinoa flour was used in bread-making, the weight and volume of the loaf were not affected. At substitution rates of 20%, 30% and 40% the weight and volume of the loaf were reduced and at rates of 30% and 40% the loaf became brown. In biscuit-making, substitution of quinoa flour in wheat mixtures by up to 70% increased the nutritional value of the product, but at the 70% substitution level the quality of the biscuits was adversely affected (Luna and Querzola, 1958). In noodles, substitution rates of 30% and 40% were acceptable, but at a rate of 50% the cooking quality, appearance and taste were adversely affected (Luna and Marchetti, 1957). Bean (1981) found that addition of quinoa flour at 5% and 10% to wheat flour caused a decrease in loaf volume, but this could be counteracted by the addition of 20 or 40 p.p.m. sodium bromate. Lorenz and Coulter (1991) found that breads containing 5% and 10% quinoa flour were quite acceptable. At levels higher than this the structure became more open and the texture slightly harsh, and using 30% quinoa flour gave the bread an unpleasant aftertaste. The crumb colour of the bread became darker at higher quinoa contents. In cakes and cookies, substitution of 5% or 10% quinoa flour improved the taste, but higher levels of quinoa flour (20% or 30%) resulted in a bitter after-taste. Vela and Cabrera (1984) report that pre-cooked quinoa flour can substitute for up to 15% of pre-cooked maize flour used in the production of 'arepas', or for wheat semolina in the production of pastas, without a reduction in the quality of the products.

Coulter and Lorenz (1991a, 1991b) have investigated the production of a snack-type product through extrusion processing, which pre-cooks, shapes, textures and restructures food constituents. They extruded blends of corn grits, which are low in protein, with quinoa to produce products with improved protein content and amino acid composition. Analysis of various physical and nutritional properties showed that increasing the proportion of quinoa in the blend caused an increase in the ash, ether extract, crude protein and fibre contents of the extruded products. Adding 30% quinoa to the corn grits improved the lysine content by 50% and also raised the threonine, valine and isoleucine content. Frying of the extruded products decreased the lysine content, as lysine is lost at high temperatures. Sensory evaluation of the flavour, texture and appearance showed the products to be moderately acceptable. They were described as having a crunchy texture and a unique shape and flavour, described as nutty or wheaty. Increasing the proportion of quinoa in the blend caused an increase in the extrusion rate, and the degree of expansion of product and its shear strength decreased. The high protein and lipid content and the lower amylose content of quinoa are thought to be the causes of these detrimental changes in product quality. Romero *et al.* (1985) reported that the use of extrusion processing to produce puffed or textured snack or breakfast type foods gave good organoleptic profiles and did not affect the nutritional quality compared to that of the whole grain.

Peñaloza *et al.* (1992) have investigated the use of quinoa as a substrate for tempe production. Tempe is a solid-substrate fermentation product of pulses and legumes produced by the fungus *Rhizopus oligosporus* Saito. Quinoa yielded a good quality tempe and was considered to be an ideal raw material as it was easier to prepare than soya bean and other substrates investigated.

1.8.3 **Industrial uses**

As mentioned earlier, various fractions of the crop have potential industrial uses. Research is currently underway in several European countries to develop uses for quinoa starch as a filler for biodegradable plastics, and in anti-offset and dusting powders, cosmetics and hair preparations, and paper coatings. Saponin molecules may find uses as potent natural pesticides. The stalks could be used for the paper industry (De Braeckelaer, 1992).

1.9 MARKET DEVELOPMENT

Quinoa is used for subsistence wherever it is grown in South America, and this use probably accounts for the majority of production. However, certain patterns of trade in the crop can be identified, though very little quantitative information is available concerning the scale of these movements. Several urban areas in the Andes which have experienced immigration from the rural highlands provide a market for quinoa. Thus the grain is transported from the Bolivian Altiplano to La Paz, and from the Mantaro Valley to Lima. Market traders in Lima have also reported that quinoa is imported from Bolivia. It has been estimated that the destinations of quinoa produced on the Bolivian Altiplano are as follows:

Consumption by the producer	45–48%
Processing plants	30%
Internal market	18%
Contraband export to Peru and Chile	2–5%

(T. Ordoñez, pers. comm.). However, the values for consumption by the producer are probably overestimates, and those for contraband export probably underestimates, as farmers can be expected to be secretive about the latter.

Since the mid 1980s, quinoa has been exported from the Andes to Europe and North America. Packets on sale in Britain state that the grain is produced in Peru, but it seems likely that grain produced in Bolivia may also find its way into consignments shipped from Peru. In the late 1980s a new monoculture production area was developed around Quito in Ecuador, for export to North America (Wahli, 1990). The North American market is also being supplied from local production in Colorado. In England locally grown quinoa was recently available in health food shops and some supermarkets. However, it has now been withdrawn due to changes in guidelines from the Ministry of Agriculture, Fisheries and Food for the introduction of novel foods. A fuller investigation of the toxicology of locally produced grain is required. At present, the grain imported from South America is white-seeded, with floury endosperm, whereas that produced in northern temperate regions is similar to the Chilean *quingua* type, with darker seeds and translucent endosperm. However, plant breeding programmes are underway, aimed at producing varieties with improved seed types adapted to temperate latitudes. Alongside development of production technology, post-harvest technology and food processing methods have been developed, and a wide range of quinoa products is now commercially available in the USA.

Little information is available on the size of these markets. The area of quinoa cultivated in Peru reached a low point of about 15 000 ha in 1975, but increased to about 25 000 ha by 1981 (Risi and Galwey, 1984). Thereafter about 20 000 ha have been cultivated annually (Risi, 1989). The area cultivated in Bolivia was about 12 000 ha in 1979 (Risi and Galwey, 1984), and had risen to about 40 000 ha by 1991 (J. Risi, pers. comm.). In the USA, the market for quinoa expanded rapidly in the late 1980s, and in 1987 the projected demand was 147 000 kg (Johnson and McCamant, 1987). In England, the first field-scale production of quinoa occurred in 1986, and in 1988, 20 t was combine-harvested. In 1991, quinoa was grown on an area of 1300 ha, though mainly as a cover crop for game birds.

Traditionally, farmers have perceived quinoa as a crop for marginal land, tolerant of adverse conditions. The corollary of quinoa's tolerance to adversity is that it has been regarded as a poor man's food. In Bolivia, where indigenous traditions are strong, eating quinoa is not a matter for shame, though it is a second-preference food in comparison with rice. However, further north, in Ecuador and Colombia, farmers sometimes describe quinoa as chicken food and deny eating it. Such attitudes may be changed by the development of markets for quinoa in developed countries and urban areas. The high content and quality of quinoa protein may be reflected in reports from Andean consumers that it is a 'strong' food, able to substitute for meat in the diet.

1.10 FUTURE POTENTIAL

The renewed cultivation of quinoa, both in the Andes and in temperate regions, has stimulated not only plant breeding but also research into other aspects of the crop. As reported in earlier sections, work has been undertaken on the characterization of the starch and saponin components of the grain, the nutritional evaluation of the grain in poultry and other animal diets, and the evaluation of the green matter as fodder. Such a multidisciplinary approach is essential if the development of quinoa is to continue. It may be possible to utilize the high protein content, protein quality and energy value of quinoa grain in animal diets. It may also be possible to utilize the starch, with its unusually small grains and high viscosity, for specialized industrial applications (Galwey *et al.*, 1990). If quinoa can be used for such purposes, its cultivation could provide a valuable alternative for land use in areas currently devoted to cereal production, notably in western Europe. In addition to reducing cereal surpluses, the cultivation of quinoa would facilitate the control of the weeds and diseases of cereal crops, since continuous cultivation of cereals allows pathogen populations to build up, and precludes the application of some herbicides for the control of grass weeds. The ability of quinoa to extract fertilizer residues left in the soil by previous crops may mean that its cultivation will help to restrain the accumulation of nitrates in rivers and in the water table. It has been estimated that quinoa could be grown on an acreage of 500 000 ha within the EEC, of which 20% could be set-aside land cultivated for non-food uses (De Braeckelaer, 1992). In addition to utilization of the grain, it is possible that the green fodder can be economically fed to ruminants in a pelletized form. Quinoa can also be used as a cover crop for game birds.

The utilization of quinoa grain for purposes other than human nutrition will depend on raising grain yields from the values of 0.5–1.0 t/ha which are typical of, and acceptable in, subsistence agriculture, to at least 5 t/ha in favourable environments. Yields approaching this value have sometimes been achieved in field experiments (Galwey, 1989; Galwey *et al.*, 1990), and with further breeding work may be reliably attainable.

There is also much scope for further adaptation of the crop to specific environments, agricultural systems and markets through breeding. It should be possible to combine the earliness and compact plant habit of the *quingua* form with the large white seed, low-saponin content and floury endosperm of some Andean types. Such varieties would permit the production of high-quality grain in temperate latitudes. So far, the improvement of quinoa has been based on fairly inbred populations and pure lines, but there may be scope for the development of hybrid varieties. Male sterility has often been reported (Risi and Galwey, 1984), though its genetic basis is uncertain.

If current efforts to develop quinoa for affluent markets are successful, this should raise the status of the crop in the Andes, leading to increased production and consumption, and research input, in that region. However, in adapting quinoa for more intensive production systems, it will be important not to undermine its ability to fulfil its traditional role, in which hardiness and resistance to stresses – including bird resistance based on saponins – will continue to be paramount. It is likely that many Andean landraces will in future be displaced by more uniform cultivars, and it is important that these genetic resources should be conserved. Quinoa has a unique ability to produce high-protein grain under ecologically extreme conditions (Wilson, 1988a), and the long-term aim should be to exploit this ability more widely, both in the Andes and beyond.

REFERENCES

Aellen, P. (1929) Bietrag zur systematik der *Chenopodium*. Arten Amerikas. *Fedde. Rep. Spec. Nov. Regn. Veg.*, **26**, 31–64.

Aellen, P. (1960) *Chenopodium*, in *Illustrierte Flora von Mitteleuropa*, 2nd edn, vol. 3C, (ed. G. Hegi), Hanser, Munich, pp. 533–657.

Aguilar, P.C. (1980) Identificación de Mecanismos de Androesterílidad, Componentes de Rendimiento y Contenido Protéico en Quinua (*Chenopodium* quinoa Willd.). M.S. thesis, Universidad Nacional Agraria, La Molina, Lima, Peru.

Aguilar, R.H., Guevara, L. and Alvarez, J.O. (1979) A new procedure for the quantitative determination of saponins and its application to several types of Peruvian quinoa. *Acta. Cient. Venez*, **30**, 167–74.

Aguirre, L. and Tapia, M.E. (1982) Estudio sobre quinuas de valle, in *Tercer Congreso Internacional sobre Cultivos Andinos*, Instituto Boliviano de Technología Agropecuaria, Instituto Interamericano de Ciencias Agrícolas, La Paz, Bolivia, pp. 55–61.

Alandia, S., Otazu, V. and Salas, B. (1979) Enfermedades, in *Quinua y Kañiwa. Cultivos Andinos*, (ed. M.E. Tapia), Serie Libros y Materiales Educativos No. 49, Instituto Interamericano de Ciencias Agrícolas, Bogotá, Colombia, pp. 137–40.

Allard, R.W. (1960) *Principles of Plant Breeding*. Wiley, New York.

Alvarez, M. (1990) Herbicidas, in *Quinua: Hacia su Cultivo Comercial*, (ed. Ch. Wahli), Latinreco S.A., Casilla 17–110–6053, Quito, Ecuador, pp. 61–8.

Alvarez, M., Pavón, J. and von Rütte, S. (1990) Caracterización, in *Quinua: Hacia su Cultivo Comercial*, (ed. Ch. Wahli), Latinreco S.A., Casilla 17–110–6053, Quito, Ecuador, pp. 5–32.

Alvarez, M. and von Rütte, S. (1990a) Fertilización, in *Quinua: Hacia su Cultivo Comercial*, (ed. Ch. Wahli), Latinreco S.A., Casilla 17–110–6053, Quito, Ecuador, pp. 95–104.

Alvarez, M. and von Rütte, S. (1990b) Genética, in *Quinua: Hacia su Cultivo Comercial*, (ed. Ch. Wahli), Latinreco S.A., Casilla 17–110–6053, Quito, Ecuador, pp. 35–58.

Alvístur, J.E., White, P.L. and Chiriboga, C.C. (1953) El valor biológico de la quinua. *Soc. Quim. Perú Lima Bol.*, **19**, 197–209.

Amaya, M. and Carreño, L. (1968) La quinua forrejera en alimentación animal. Ingeniero Agrónomo thesis, Facultad de Agronomía, UPTC, Tunja.

Amaya-Farfán, J., Gacco, C.F., Agapito, F.J.L. and Augusto-Ruiz, W. (1978) *Removal of saponins from quinoa* (Chenopodium quinoa *Willd.) grain by milling*. (Abstract) 5th International Congress on Food Science and Technology, Kyoto, Japan.

Anonymous (1930) *Chenopodium nuttalliae*, a Mexican cereal. *Kew Bulletin 1930*, 332–3.

Aquize, J.E. (1977) Climatología del cultivo de la quinua, in *Curso de Quinua*, Fondo Simon Bolívar, Ministerio de Alimentación, Instituto Interamericano de Ciencias Agrícolas, Universidad Nacional Técnica del Altiplano, Puno, Peru, pp. 119–26.

Atwell, W.A., Hyldon, R.G., Godfrey, D.D., Galle, E.L., Sperber, W.H., Pedersen, D.C., Evans, W.D. and Rabe, G.O. (1988) Germinated quinoa flour to reduce the viscosity of starchy foods. *Cereal Chem.*, **65**, 508–9.

Atwell, W.A., Patrick, B.M., Johnson, L.A. and Glass, R.W. (1983) Characterisation of quinoa starch. *Cereal Chem.*, **60**, 9–11.

Ballón, E., Tellería, W. and Hutton, J. (1976) Aproximación a la determinación de saponinas por cromatografía de capa fina, in *Segunda Convención Internacional de Quenopodiaceas*. Universidad Boliviana Tomas Frías, Comité Departamental de Obras Públicas de Potosí, Instituto interamericano de Ciencias Agrícolas, Potosí, Bolivia, pp. 89–94.

Ballón, E. and Coca, A. (1984) Extracción de harineas en variedades de quinua, in *Cuarto Congreso Internacional de Cultivos Andinos*, Instituto Colombiano Agropecuario, Centro Internacional de Investigaciones parpa el Desarrollo, Programa de Cultivos Multiples, Pato, Colombia, pp. 299–307.

Ballón, E., Johnson, D. and McCamant, J.F. (1986) Adaptación de la quinua en el estado de Colorado, in *Quinto Congreso Internacional de Sistemas Agropecuarios Andinos*, pp. 135–42.

Basset, I.J. and Crompton, C.W. (1978) The biology of Canadian weeds. 32. *Chenopodium album. Can. J. Plant Sci.*, **58**, 1061–72.

Basu, N. and Rastogi, R.P. (1967) Triperpenoid saponins and their sapogenins. *Phytochemistry*, **6**, 1249–70.

Bayer (1970) *Phostoxin. Información Técnica.* Juan H. Kruger, Quito.

Bean, M.M. (1981) Composite flours for breadmaking in Bolivia: Technical aspects, in *Final Report: Improving the Nutritional Value of Wheat Foods*, Appendix, B1–B8. Agency for International Development 231–11–76/USDA Western Regional Research Center, Albany, California.

Becker, R. and Hanners, G.D. (1990) Compositional and nutritional evaluation of quinoa whole grain flour and mill fractions. *Lebensm.-Wiss. u.-Technol.*, **23**, 441–4.

Beckman, P.M. and Finch, H.C. (1980) Seed rot and damping off of *Chenopodium quinoa* caused by *Sclerotium rolfsii. Plant Disease*, **64**, 497–8.

Betschart, A.A. (1974) Nitrogen solubility of alfalfa protein concentrate as influenced by various factors. *J. Food Sci.*, **39**, 1110–12.

Blanco, C. (1970) *La Quinua. Como se Debe Cultivar.* Universidad Técnica de Oruro, Oruro, Bolivia.

Boerema, G.H., Mathur, S.B. and Neergaard, P. (1977) *Ascochyta hyalospora* (Cook and Ell.) *comb. nov.* in seeds of *Chenopodium quinoa. Neth. J. Plant Pathol.*, **83**, 143–59.

Bornas, E.A. (1977) Repuesta de la Quinua (*Chenopodium quinoa* (Willd.) variedades Sajama y Kanccolla a la Profundidad de Siembra en Cuatro Clases Texturales de Suelo. Ingeniero Agrónomo thesis, Universidad Nacional Técnica del Altiplano, Puno, Peru.

Burgasi, G., Pavón, J. and von Rütte, S. (1990) Cultivo Comercial, in *Quinua: Hacia su Cultivo Comercial*, (ed. Ch. Wahli), Latinreco S.A., Casilla 17–110–6053, Quito, Ecuador, pp. 107–34.

Burnouf-Radosevich, M. and Delfel, N.E. (1984) High performance liquid chromatography of oleanane-type triterpenes. *J. Chromatogr.*, **292**, 403–9.

Burnouf-Radosevich, M., Delfel, N.E. and England, R.E. (1985) Gas chromatography – mass spectrometry of oleanane- and ursane-type triterpenes – application to *Chenopodium quinoa* triterpenes. *Phytochemistry*, **24**, 2063–6.

Burnouf-Radosevich, M. and Paupardin, C. (1983) Triterpene saponins from plant organs and *in vitro* tissue cultures of *Chenopodium quinoa* Willd. *C.R. Acad. Sci. Paris*, **296**, 429–32.

Calzada, J. (1951) *Variedades de Quinua Recomendadas para los Sembríos de la Sierra.* Bol. Inf. No. 30., Dirección General de Agricultura, Ministerio de Agricultura, Lima, Peru.

Canahua, A. (1977) Observaciones del compartamiento de quinua a la sequía, in *Primer Congreso Internacional sobre Cultivos Andinos*, Universidad Nacional San

Cristobal de Huamanga, Instituto Interamericano de Ciencias Agrícolas, Ayacucho, Peru, pp. 390–9.
Canahua, M.A. and Rea, J. (1980) Quinuas resistentes a heladas avances a la investigación, in *Segundo Congreso Internacional sobre Cultivos Andinos*, Facultad de Ingenería Agronómica, Escuela Superior Politécnica de Chimborazo, Instituto Interamericano de Ciencias Agrícolas, Riobamba, Ecuador, pp. 143–9.
Cardenas, M. (1949) Plantas alimenticias de los Andes de Bolivia. *Folia Univ.*, **3**, 109–13.
Cardozo, A. and Bateman, J.V. (1961) La quinua en la alimentación animal. *Turrialba*, **11**, 72–7.
Cardozo, A. and Tapia, M.E. (1979) Valor nutrivo, in *Quinua y Kañiwa. Cultivos Andinos*, (ed. M.E. Tapia), Serie Libros y Materiales Educativos No. 49, Instituto Interamericano de Ciencias Agrícolas, Bogotá, Colombia, pp. 149–92.
Carlsson, R. (1975) Selection of *Centrospermae* and other species for production of leaf protein concentrates. Ph.D. thesis, University of Lund, Sweden.
Carlsson, R. (1980) Quantity and quality of leaf protein concentrates from *Atriplex hortensis* L., *Chenopodium quinoa* Willd. and *Amaranthus caudatus* L., grown in southern Sweden. *Acta Agric. Scand.*, **30**, 418–26.
Carlsson, R. (1982) Trends for future applications of wet-fractionation of green crops. Proceedings E.E.C. Green Crops Meeting, Forage Protein Conservation and Utilisation, Dublin, 13–16 September 1982, pp. 57–81.
Carlsson, R. (1983) Nitrate and oxalate contents in leaf and stem of wild and cultivated leafy vegetables – leaf nutrient concentrate production as a means for detoxification. Proceedings of 6th World Congress of Food Science and Technology, Dublin, September 1983, Vol I, pp. 37–8.
Carlsson, R. (1984) Effect of species, plant physiological development stages, growth conditions, and processing conditions on the nutritive value of leaf protein concentrates. *Progress in Leaf Protein Research*, (ed. N. Singh), *Current Trends in Life Sciences*, **Vol XI**, pp. 201–8.
Carlsson, R., Hanczakowski, P. and Kaptur, T. (1984) The quality of green fraction of leaf protein concentrate from *Chenopodium quinoa* Willd. grown at different levels of fertilizer nitrogen. *Anim. Feed Sci. Technol.*, **11**, 239–45.
Carmen, M.L. (1984) Acclimatization of quinoa and cañihua to Finland. *Ann. Agric. Fenn.*, **23**, 135–44.
Chaquilla, O. (1976) Exploración de herbicidas en el cultivo de la quinua (*Chenopodium quinoa* Willd.), in *Segunda Convención Internacional de Quenopodiaceas*, Universidad Boliviana Tomas Frías, Comité Departmental de Obras Públicas de Potosí, Instituto Interamericano de Ciencias Agrícolas, Potosí, Bolivia, pp. 127–9.
Condorena, J.B. (1979) Efecto de Tres Formulaciones N–P–K en Ocho Cultivares de Quinua (*Chenopodium quinoa* Willd.). Ingeniero Agrónomo thesis, Universidad Nacional Técnica del Altiplano, Puno, Peru.
Cook, O.F. (1925) Peru as a centre of domestication. Tracing the origin of civilisation through the domesticated plants. *J. Hered.*, **16**, 33–46, 95–110.
Cornejo, de Z.G. (1976) Hojas de quinua (*Chenopodium quinoa* Willd.) fuente de proteína, in *Segunda Convención Internacional de Quenopodiaceas*, Universidad Boliviana Tomas Frías, Comité Departamental de Obras Públicas de Potosí, Instituto Interamericano de Ciencias Agrícolas, Potosí, Bolivia, pp. 177–80.
Coulter, L.A. and Lorenz, K. (1991a) Extruded corn grits – quinoa blends: I Proximate composition, nutritional properties and sensory evaluation. *J. Food Proc. and Pres.*, **15**, 231–42.
Coulter, L.A. and Lorenz, K. (1991b) Extruded corn grits – quinoa blends: II

Physical characteristics of extruded products. *J. Food Proc. and Pres.*, **15**, 243–59.

Cumming, B.G. (1959) Extreme sensitivity of germination and photoperiod reaction in the genus *Chenopodium* (Tourn.) L. *Nature*, **184**, 1044–5.

Cusack, D.F. (1983) *The 1983 Sierra Blanca Colorado Quinoa Project: Conclusions, Recommendations and Plans for 1984*. Sierra Blanca Associates, Colorado, USA.

Cusack, D.F. (1984) Quinoa: grain of the Incas. *Ecologist*, **14**, 21–31.

De Braeckelaer, P. (co-ordinator) (1992) Quinoa, a Multi-Purpose Crop for E.C.'s Agricultural Diversification. Research proposal to the Commission of the European Communities' agricultural and agro-industry programme 1991–1994.

De Bruin, A. (1964) Investigation of the food value of quinoa and cañihua seed. *J. Food Sci.*, **26**, 872–6.

De Simone, F., Dini, A., Pizza, C., Saturnino, P. and Schettino, O. (1990) Two flavonol glycosides from *Chenopodium quinoa*. *Phytochemistry*, **29**, 3690–2.

Doorgeest, M., Harrewijn, J.L. and Wijkstra, A. (1988) Reactions of new crops to *Meloidogyne hapla* and *Heterodera schachtii*. Proceedings of Nematology Symposium, Uppsala, Sweden.

Elias Peñafiel, C.C. and Dias Villar, L. (1988) Determinación espectrofotométrica de ácido oleanólico y saponinas de quinua (*Chenopodium quinoa* Willd., variedad Kanccolla). *Arch. Latinoamer. Nutr.*, **38**, 113–31.

Elmer, L.A. (1942) *Chenopodium quinoa*. *E. Afr. Agric. J.*, **8**, 21–3.

Espíndola, G. (1986) Repuestas Fisiológicas, Morfológicas y Agronómicas de la Quinua (*Chenopodium quinoa* W.) al Deficit Hídrico. MS thesis, Colegio de Postgraduados, Institución de Enseñanza e Investigaciones en Ciencias Agrícolas, Chapingo, Mexico.

Etchevers, B.J. and Avila, T.P. (1979) Factores que afectan el crecimiento de quinua (*Chenopodium quinoa*) en el centro-sur de Chile. 10th Latin American Meeting Agricultural Science.

European Patent Application No. 89121654.1 (1989) *Carbohydrate Cream Substitute*. Applicant: The Nutrasweet Corporation. Publication No. 0 403 696. European Patent Office.

Fairbanks, D.J., Burgener, K.W., Robinson, L.R., Andersen, W.R. and Ballón, E. (1990) Electrophoretic characterisation of quinoa seed proteins. *Plant Breeding*, **104**, 190–5.

FAO/WHO/UN (1985) *Energy and Protein Requirements*. World Health Organisation, Geneva.

Fenwick, D.E. and Oakenfull, D. (1983) Saponin content of food plants and some prepared foods. *J. Sci. Food Agric.*, **34**, 186–91.

Flores, F.G. (1977) Estudio Preliminar de la Fenología de la Quinua (*Chenopodium quinoa* Willd.). Ingeniero Agrónomo thesis, Universidad Nacional Técnica del Altiplano, Puno, Peru.

Fuller, H.J. (1949) Photoperiodic response of *Chenopodium quinoa* Willd. and *Amaranthus caudatus*. *L. Am. J. Bot.*, **36**, 175–80.

Galwey, N.W. (1989) Exploited plants. Quinoa. *Biologist*, **36**, 267–74.

Galwey, N.W., Leakey, C.L.A., Price, K.R. and Fenwick, G.R. (1990) Chemical composition and nutritional characteristics of quinoa. *Food Sciences and Nutrition*, **42**, 245–61.

Gandarillas, H. (1967) Observaciones sobre la biología reproductiva de las quinoa. *Sayana*, **5**, 26–9.

Gandarillas, H. (1968), *Razas de Quinua*. Bol. No. 34, Instituto Boliviano de Cultivos Andinos, División de Investigaciones Agrícolas, Ministerio de Agricultura, La Paz, Bolivia.

Gandarillas, H. (1969) Esterilidad genética y citoplásmica de la quinua (*Chenopodium quinoa*). *Turrialba*, **19**, 429–30.

Gandarillas, H. (1974) *Genética y Origen de la Quinua*. Bol. No. 9, Instituto Nacional del Trigo, Ministerio de Asuntos Campesinos y Agropecuarios, La Paz, Bolivia.

Gandarillas, H. (1979a) Botánica, in *Quinua y Kañiwa. Cultivos Andinos*, (ed. M.E. Tapia), Serie Libros y Materiales Educativos No. 49, Instituto Interamericano de Ciencias Agrícolas, Bogotá, Colombia, pp. 20–44.

Gandarillas, H. (1979b) Genetica y origen, in *Quinua y Kañiwa. Cultivos Andinos*, (ed. M.E. Tapia), Serie Libros y Materiales Educativos No. 49, Instituto Interamericano de Ciencias Agrícolas, Bogotá, Colombia, pp. 45–64.

Gandarillas, H. (1979c) Mejoramiento genetico, in *Quinua y Kañiwa. Cultivos Andinos*, (ed. M.E. Tapia), Serie Libros y Materiales Educativos No. 49, Instituto Interamericano de Ciencias Agrícolas, Bogotá, Colombia, pp. 65–82.

Germán, M. and Saravia, R. (1987) Estudio de las fluctaciones poblacionales de insectos en Salinas de Garcí Mendoza, in *Informe Anual 1986–1987*, Instituto Boliviano de Tecnología Agropecuaria, Ministerio de Asuntos Campesinos y Agropecuarios, Bolivia, pp. 66–72.

Giusti, L. (1970) El genero *Chenopodium* en Argentina. I. Número de cromosomas. *Darwiniana*, **16**, 98–105.

González, J.A., Roldán, A., Gallardo, T., Escudero, T. and Prado, F.E. (1989) Quantitative determinations of chemical compounds with nutritional value from Inca crops: *Chenopodium quinoa* ('quinoa'). *Plant Foods for Human Nutrition*, **39**, 331–7.

Gross, R., Koch, F., Malaga, I., de Miranda, A.F., Schoenberger, H. and Trugo, L.C. (1989) Chemical composition and protein quality of some local Andean food sources. *Food Chemistry*, **34**, 25–34.

Haaber, J. (1991) *Chenopodium quinoa* Willd. as a green crop for the pelleting industry – the effect of heat treatment on the palatability in green pellets made of quinoa. First European Symposium on Industrial Crops and Products, Maastricht, The Netherlands (Abstracts, p. 19).

Harborne, J.B. (ed.) (1984) Phytochemical Methods. A Guide to Modern Techniques of Plant Analysis. 2nd edn. Chapman & Hall, London.

Heiser, C.B. and Nelson, D.C. (1974) On the origin of the cultivated Chenopods. *Genetics*, **78**, 503–5.

Hernández, C. (1987) Experiencias preliminares sobre procesamiento de semilla de quinua, in *Reunión Nacional sobre Producción, Uso y Comercialización del Cultivo del Quinua. Memorias*, Instituto Nacional de Investigaciones Agropecuarios, Programa de Cultivos Andinos, Estación Experimental 'Santa Catalina', Quito, Ecuador, pp. 45–7.

Herquinio, F. and Román, F. (1975) Estudio de la Densidad de Siembra en el Cultivo de la Quinua en Tres Campañas Agrícolas. Universidad Nacional del Centro, Huancayo, Peru.

Hudson, B.J.F. and Ghavami, M. (1984) Stabilising factors in soybean oil – natural components with antioxidant activity. *Lebensm. Wiss U.-Technol.*, **72**, 82–5.

Hunziker, A.T. (1943) Los especies alimenticas de *Amaranthus y Chenopodium* cultivadas por los Indios de America. *Rev. Argent. Agron.*, **30**, 297–353.

IBTA (1987) Sistemas de Producción de Quinua en el Altiplano Boliviano. Instituto Boliviano de Tecnología Agraria, La Paz, Bolivia.

Jacobsen, S.E., Jorgensen, I. and Stolen, O. (1984) Cultivation of quinoa (*Chenopodium quinoa* Willd.) under temperate climatic conditions in Denmark *J. Agr. Sci.*, **122**, 47–52.

Johnson, D. and McCamant, J.F. (1987) *Quinoa Research and Development 1986.* Sierra Blanca Associates, Colorado, USA.

Jood, S., Chauhan, B.M. and Kapoor, A.C. (1986) Saponin content of chickpea and black gram: varietal differences and effects of processing and cooking methods. *J. Sci. Food Agric.*, **37**, 1121–4.

Junge, I. (1973) *Lupin and Quinoa, Research and Development in Chile.* An. Escuela Ind. Publ. No. 1, University of Concepción, Chile.

Kenwright, P.A. (1989) Breeding the Andean Grain Crop Quinoa (*Chenopodium quinoa*) for Cultivation in Britain. Ph.D. thesis, University of Cambridge, England.

Kinsella, J.E. (1976) Functional properties of proteins in foods: a survey. *Crit. Rev. Food Sci. Nutr.*, 219–80.

Koziol, M.J. (1990a) Composición quimica, in *Quinua: Hacia su Cultivo Comercial*, (ed. Ch. Wahli), Latinreco S.A., Casilla 17–110–6053, Quito, Ecuador, pp. 137–59.

Koziol, M.J. (1990b) Afrosimetric estimation of threshold saponin concentration for bitterness in quinoa (*Chenopodium quinoa* Willd.). *J. Sci. Food Agric.*, **54**, 211–19.

Koziol, M.J. (1992) Chemical composition and nutritional evaluation of quinoa (*Chenopodium quinoa* Willd.). *J. Food Composition and Analysis*, **5**, 35–68.

Kulp, K. (1973) Characteristics of small granule starch of flour and wheat. *Cereal Chem.*, **50**, 666–79.

Lásztity, R. (1984) *The Chemistry of Cereal Proteins.* CRC Press, Boca Raton, Florida, USA.

Laura, R. (1977) Control Química del Mildiú (*Peronospora effusa* Grev.) en Quinua. Ingeniero Agrónomo thesis, Universidad Nacional Técnica del Altiplano, Puno, Peru.

León, J. (1964) Plantas Alimenticias Andinas. Bol. Tec. No. 6, Instituto Interamericano de Ciencias Agrícolas – Zona Andina, Lima, Peru.

Lescano, J.L. (1977) Métodos de mejoramiento, in *Curso de Quinua*, Fondo Simon Bolivar, Ministerio de Alimentación, Instituto Interamericano de Ciencias Agrícolas, Universidad Nacional Técnica del Altiplano, Puno, Peru, pp. 63–5.

Lescano, J.L. (1980) Avances en la genética de la quinua, in *Primera Reunión de Genética y Fitomejoramiento de la Quinua*, Universidad Nacional Técnica del Altiplano, Instituto Boliviano de Technología Agropecuaria, Instituto Interamericano de Ciencias Agrícolas, Centro Internacional de Investigaciones para el Desarrollo, Puno, Peru, pp. 81–9.

Lescano, J.L. (1981) *Cultivo de la Quinua.* Universidad Nacional Técnica del Altiplano, Centro de Investigaciones en Cultivos Andinos, Puno, Peru.

Lescano, J.L. (1984) Obtención y comparativo de diez y seis lines de alto rendimiento en Puno, in *Cuarto Congreso Internacional de Cultivos Andinos*, Instituto Colombiano Agropecuaria Pasto, Colombia, pp. 259–61.

Lescano, J.L. and Palomino, C. (1976) Metodología del cruzamiento en quinua, in *Segunda Convención Internacional de Quenopodiaceas*, Universidad Boliviania Tomas Frías, Comité Departamental de Obras Públicas de Potosí, Instituto Interamericano de Ciencias Agrícolas, Potosí, Bolivia, pp. 78–80.

Lescano, J.L. and Tapia, M.E. (1977) Investigación en el mejoramiento de la quinua, in *Curso de Cultivos Andinos*, C1–C13. Instituto Interamericano de Ciencias Agrícolas – Zona Andina, Instituto Boliviano de Technología Agropecuaria, La Paz, Bolivia, C1–C13.

Leung, W.-T.W. and Flores, M. (1961) *Food Composition Tables for use in Latin*

America. Instituto de Nutrición de Centro America y Panamá, Ciudad de Guatemala, Guatemala.

Llerena, H. (1973) Ensayo de Panificación con Harina de dos variedades de Quinua (*Chenopodium quinoa* Willd.) y Cañihua (*Chenopodium pallidicaule* Aellen). Ingeniero Agrónomo thesis, Universidad Nacional Técnica del Altiplano, Puno, Peru.

López de Romano, G., Creed, A.M. and Graham, G. (1978) Alimentos comunes Peruanos, tolerancia y digestibilidad en infantes desnutridos. *Arch. Latinoamer. Nutr.*, **28**, 419–33.

López de Romano, G., Graham, G., Rojas, M. and McLean, Jr. W.C. (1981) Digestibilidad y calidad proteinica de la quinua: estudio comparativo en niños entre semilla y harina de quinua. *Arch. Latinoamer. Nutr.*, **31**, 485–97.

Lorenz, K. (1990) Quinoa (*Chenopodium quinoa*) starch – physico-chemical properties and functional characteristics. *Starch/Stärke*, **42**, 81–6.

Lorenz, K. and Coulter, L.A. (1991) Quinoa flour in baked products. *Plant Foods for Human Nutrition*, **41**, 213–23.

Luna, R. (1957) Ensayo de panificación con mezclas de harina de trigo y quinua. *Inf. Mens. Estac. Exp. Agric. 'La Molina', Lima*, **359**, 6–8.

Luna, R. and Marchetti, I. (1957) Ensayo de elaboración de fideos con mezclas de harinas de trigo y quinua. *Inf. Mens. Estac. Exp. Agric. 'La Molina', Lima*, **364**, 1–7.

Luna, R. and Querzola, I. (1958) Ensayo de elaboración de galletas con mezclas de harina de trigo y quinua. *Inf. Mens. Estac. Exp. Agric. 'La Molina', Lima*, **374**, 1–6.

Ma, W.W., Heinstein, P.F. and McLaughlin, J.L. (1989) Additional toxic, bitter saponins from the seeds of *Chenopodium quinoa*. *J. Natural Products-Lloydia*, **52**, 1132–5.

MACA (1986a) Comparativo de rendimientos de líneas tolerantes a helados vs. susceptibles, in *Informe Anual 1985–1986*, Instituto Boliviano de Tecnología Agropecuaria, Ministerio de Asuntos Campesinos y Agropecuarios. Proyecto Quinua, Convenio CIID Canada, pp. 20–4.

MACA (1986b) Fluctuación poblacional de insectos, in *Informe Anual 1985–1986*, Instituto Boliviano de Tecnología Agropecuaria, Ministerio de Asuntos Campesinos y Agropecuarios, La Paz, Bolivia. Proyecto Quinua, Convenio CIID Canada, pp. 35–7.

MACA (1987) *Sistemas de Producción de Quinua en el Altiplano Boliviano*. Instituto Boliviano de Tecnología Agropecuaria, Ministerio de Asuntos Campesinos y Agropecuarios. Proyecto Quinua, Convenio CIID Canada.

Machicao, E. (1965) Las saponins de la quinua. *Sayana*, **4**, 24–5.

MAG (1984) *Estimación de la Superficie Coshechada y la Producción Agrícola del Ecuador*. Ministerio de Agricultura y Ganadería, Dirección General de Informática, Quito, Ecuador.

Mahoney, A.W., López, J.G. and Hendricks, D.G. (1975) An evaluation of the protein quality of quinoa. *J. Agric. Food Chem.*, **23**, 190–3.

Marroquín, A.S. (1983) Dos cultivos olvidados, de importancia agroindustrial. *Arch. Latinoamer. Nutr.*, **23**, 11–32.

Martínez, C. (1946) *La Quinua. Divulgación Botánica, Agrícola y de Utilización Industrial y Culinaria*. Dirección de Agricultura, Ministerio de Agricultura, Lima, Peru.

McCamant, J.F. (1985) *Report on the 1984 Colorado Quinoa Project*. Sierra Blanca Associates, Colorado, USA.

Merino, G. (1986) Observaciones preliminares sobre *Naupactus* sp. (Coleoptera, Curculionidae) un insecto potencialmente peligroso. *Sanidad Vegetal (Quito)*, **1**, 75–7.

Merino, G. and Vázquez, V. (1978) Control del Gusano de la Hoja de la Papa. Instituto de Investigaciones Agropecuarias, Quito, Ecuador.

Mizui, F., Kasai, R., Ohtani, K. and Tanaka, O. (1988) Saponins from brans of quinoa, *Chenopodium quinoa* Willd. *Internat. Chem. Pharm. Bull.*, **38**, 375–7.

Montenegro, B.C. (1976) Investigación sobre quinua Dulce de Quitopamba, in *Primera Reunión Binacional Sobre la Planificación de la Producción de la Quinua*, Comité Interinstitucional Colombiano de la Quinua, Ministerio de Salud Pública, Instituto Colombiano de Bienestar Familiar, Pasto, Colombia, pp. 47–56.

Mujica, S.A. (1977) Technología de Cultivo de Quinua. *Bol. Tecn. – Ser. Quinua No. 2*, Fondo Simon Bolivar, Ministerio de Alimentación Zona XII-Puno, Instituto Interamericano de Ciencias Agrícolas, Puno, Peru.

Mujica, S.A., Quispe, B. and Valdivia, F.R. (1984) Comparativo de variedadas de quinua, in *Cuarto Congreso Internacional de Cultivos Andinos*, Instituto Colombiano Agropecuaria, Pasto, Colombia, pp. 259–61.

Muñoz, D.H. (1979) Evalación de 100 Ecotipos de Quinua (*Chenopodium quinoa*) al Aataque de Nacobbus sp. Ingeniero Agrónomo thesis, Universidad Nacional Técnica del Altiplano, Puno, Peru.

Muñoz, F. and Cruz, L. (1978) Manual del Cultivo de Papa. Instituto Nacional de Investigaciones Agropecuarias, Quito, Ecuador.

Narrea, A. (1976) La producción de quinua en el Perú, in *Segunda Convención Internacional de Quenopodiaceas*, Universidad Boliviana Tomas Frías, Comité Departamental de Obras Públicas de Potosí, Instituto Interamericano de Ciencias Agrícolas, Potosí, Bolivia, pp. 170–6.

Narrea, A. (1977) El cultivo de la quinua y perspectivas de su incremento en el Perú, in *Primer Congreso Internacional sobre Cultivo Andinos*, Universidad Nacional San Cristobal de Huamanga, Instituto Interamericano de Ciencias Agrícolas, Ayacucho, Peru.

National Research Council (1989) *Lost Crops of the Incas: Little Known Plants of the Andes with Promise for Worldwide Cultivation*, National Academy Press, Washington D.C.

Negrón, A., Alvarez, G.E. and Calmet, U.E. (1976) La quinua y la cañihua en raciones de pollos parrilleros en Puno, Perú, in *Segunda Convención Internacional de Quenopodiaceas*, Universidad Boliviana Tomas Frías, Comité Departamental de Obras Públicas de Potosí, Instituto Interamericano de Ciencias Agrícolas, Potosí, Bolivia, pp. 170–6.

Nelson, D. (1968) Taxonomy and Origins of *Chenopodium quinoa* and *Chenopodium nuttaliae*. Ph.D. thesis, University of Indiana.

Nieto, C., Peralta, E. and Castillo, R. (1986) *'INIAP-Imbaya' e 'INIAP-Cochasquí', Primeras Variedades de Quinua para le Sierra Ecuatoriana*. Boletín Divulgación No. 187, Estación Experimental 'Santa Catalina', Instituto Nacional de Investigaciones Agropecuarias, Quito, Ecuador.

Nina, M.G. and Saravia, R. (1987) Comparativo de 5 variedades dulces de quinua mejorada (Salinas de Garci Mendoza), in *Informe Anual 1986–1987*, Instituto Boliviano de Tecnología Agropecuaria, Ministerio de Asuntos Campesinos y Agropecuarios, Proyecto Quinua, Convenio CIID Canada, pp. 102–12.

Oakenfull, D. and Siddhu, G.S. (1990) Could saponins be a useful treatment for hypercholesterolaemia? *Eur. J. Clin. Nutr.*, **44**, 79–88.

Ordoñez, T. (1987) *Sistemas de Producción de Quinua en el Altiplano Boliviano.*

Ministerio de Asuntos Campesinos y Agropecuarios, Instituto Boliviano de Tecnología Agropecuaria, Junta del Acuerdo de Cartagena, Comunidad Económica Europea, La Paz, Bolivia.

Ortiz, V. (1974) Densidades de siembra en quinua. *Av. Invest.*, **3**, 3–4.

Ortiz, V. and Zanabria, E. (1979) Plagas, in *Quinua y Kañiwa. Cultivos Andinos*, (ed. M.E. Tapia), Serie Libros y Materiales Educativos No. 49, Instituto Interamericano de Ciencias Agrícolas, Bogotá, Colombia, pp. 111–30.

Ostrowski-Meissner, H.T., Carlsson, R. and McKenzie, D.R. (1984) Recovery and evaluation of chloroplast-free protein concentrates from leaves of *Chenopodium quinoa* and *Atriplex hortensis* grown in Sweden and Australia. *J. Food Quality*, **7**, 27–41.

Otazu, V. and Salas, B. (1977) La podredumbre marrón del tallo de la quinua (*Chenopodium quinoa*) causada por *Phoma exigua* var *foveata*. *Fitopatología*, **12**, 54–8.

Paca, F.R. (1970) Herencia de Colores de Quinua (*Chenopodium quinoa* Willd.) y Cañihua (*Chenopodium pallidicaule* Aellen). Ingeniero Agrónomo thesis, Universidad Nacional Técnica del Altiplano, Puno, Peru.

Partap, T.E.J. and Kapoor, P. (1987) The Himalayan grain chenopods. III. An under-exploited food plant with promising potential. *Agric. and Ecosystems*, **19**, 71–9.

Passmore, R. and Eastwood, M.A. (1986) Proteins, in *Human Nutrition and Dietetics*, 8th edn, Churchill Livingston, pp. 40–53.

Peñaloza, W., Davey, C.L., Hedger, J.N. and Kell, D.B. (1992) Physiological studies on the solid-state quinoa tempe fermentation, using on-line measurements of fungal biomass production. *J. Sci. Food Agric.*, **59**, 227–35.

Peralta, T. (1986) Evaluación de insectos de quinua y metodos campesinos de control de las plagas, in *Quinto Congreso Internacional de Sistemas Agropecuarios Andinos*, pp. 120–1.

Pulgar-Vidal, J. (1954) *La Quinua o Suba en Colombia*. Publ. No. 3, Fichero Científico Agropecuario, Ministerio de Agricultura, Lima, Peru.

Quispe, E. and Savaria, R. (1987) Dinámica poblaciónal de insectos en el cultivo de la quinua, in *Informe Anual 1986–1987*, Instituto Boliviano de Technología Agropecuaria, La Paz, Bolivia, pp. 53–65.

Rea, J. (1948) *Observaciones sobre Biología Floral y Estudio de Saponinas en* Chenopodium quinoa *Willd.* Ser. Tec. No. 3, Ministerio de Agricultura, La Paz, Bolivia.

Rea, J. (1969) *Biologia Floral de la Quinua*. Instituto Interamericano de Ciencias Agrícolas, Lima, Peru.

Rea, J. and León, J. (1966) Determinación práctica del contenido de saponinas en quinua, in *Informe Técnico. No. 6*, Instituto Interamericano de Ciencias Agrícolas, San José, Costa Rica, p. 190.

Rea, J., Tapia, M.E. and Mujica, S.A. (1979) Practicas agronómicas, in *Quinua y Kañiwa. Cultivos Andinos*, (ed. M.E. Tapia), Serie Libros y Materiales Educativos No. 49, Instituto Interamericano de Ciencias Agrícolas, Bogotá, Colombia, pp. 83–120.

Reichert, R.D., Oomah, B.D. and Youngs, C.G. (1984) Factors affecting the efficiency of abrasive-type dehulling of grain legumes investigated with a new intermediate-sized batch dehuller. *J. Food Sci.*, **49**, 267–72.

Reichert, R.D., Tatarynovich, J.T. and Tyler, R.T. (1986) Abrasive dehulling of quinoa (*Chenopodium quinoa*): effect on saponin content as determined by an adapted haemolytic assay. *Cereal Chem.*, **63**, 471–5.

Ridout, C.L., Price, K.R., Dupont, M.S., Parker, M.L. and Fenwick, G.R. (1991)

Quinoa saponins – analysis and preliminary investigations into the effects of reduction by processing. *J. Sci. Food Agric.*, **54**, 165–76.

Ríos, M.L.T., Sgarbieri, V.C. and Amaya, J. (1978) Chemical and biological evaluation of quinoa (*Chenopodium quinoa* Willd.). Effect of extracting the saponins by heat treatment. *Arch. Latinamer. Nutr.*, **28**, 253–63.

Risi, J. (1986) Adaptation of the Andean Grain Crop Quinoa (*Chenopodium quinoa* Willd.) for Cultivation in Britain. Ph.D. thesis, University of Cambridge, England.

Risi, J. (1989) *La Investigación de la Quinua en Puno.* Bogotá, Colombia.

Risi, J. and Galwey, N.W. (1984) The *Chenopodium* grains of the Andes: Inca crops for modern agriculture. *Adv. in Appl. Biol.*, **10**, 145–216.

Risi, J. and Galwey, N.W. (1989a) The pattern of genetic diversity in the Andean grain crop quinoa. I. Associations between characteristics. *Euphytica*, **41**, 147–62.

Risi, J. and Galwey, N.W. (1989b) The pattern of genetic diversity in the Andean grain crop quinoa. II. Multivariate methods. *Euphytica*, **41**, 135–45.

Risi, J. and Galwey, N.W. (1991a) Genotype X environment interaction in the Andean grain crop quinoa in temperate environments. *Plant Breeding*, **107**, 141–7.

Risi, J. and Galwey, N.W. (1991b) Effects of sowing date and sowing rate on plant development and grain yield of quinoa (*Chenopodium quinoa*) in a temperate environment. *J. Agric. Sci.*, Cambridge, **117**, 325–32.

Romero, A., Bacigalupo, A. and Bressani, R. (1985) Efecto de la extrusión sobre las caracteristicas funcionales y la calidad proteínica de la quinua (*Chenopodium quinoa* Willd.). *Arch. Latinoam. Nutr.*, **35**, 148–62.

Romero, J.A. (1981) Evaluación de las Características Físicas, Químicas y Biológicas de ocho variedades de Quinua (*Chenopodium quinoa* Willd.). Masters thesis, Instituto de Nutrición de Centro America y Panamá, Facultad de Ciencias Químicas y Farmacia, Universidad de San Carlos de Guatemala, Guatemala.

Romero, M., Gómez, L. and Mont, R. (1977) Ampliación de la frontera agrícola en el Perú, in *Primer Congreso Internacional sobre Cultivos Andinos*, Universidad Nacional San Cristobal de Huamanga, Instituto Interamericano de Ciencias Agrícolas, Ayacucho, Peru, pp. 111–18.

Ruales, C. (1987) Insectos asociados al cultivo de quinua, in *Reunión Nacional sobre Producción, Uso y Comercialización del Cultivo del Quinua. Memorias*, Instituto Nacional de Investigaciones Agropecuarias, Programa de Cultivos Andinos, Estación Experimental 'Santa Catalina', Quito, Ecuador.

Ruales, C. and Fiallo, E. (1990) Plagas, in *Quinua: Hacia su Cultivo Comercial*, (ed. Ch. Wahli), Latinreco S.A., Casilla 17–110–6053, Quito, Ecuador, pp. 91–2.

Ruales, J. and Nair, B.M. (1992) Nutritional quality of the protein in quinoa (*Chenopodium quinoa* Willd.) seeds. *Plant Foods for Human Nutrition*, **42**, 1–11.

Salas, B., Otazu, V. and Vilca, A. (1977) Enfermedades de la quinua, in *Curso de Quinua*, Fondo Simon Bolivar, Ministerio de Alimentación, Instituto Interamericano de Ciencias Agrícolas, Universidad Nacional Técnica del Altiplano, Puno, Peru, pp. 143–9.

Salhuana, W., Gómez, L. and Romero, M. (1977) Selección de colleciones similares en el banco de germplasma de quinua de la UNA, in *Primer Congreso Internacional sobre Cultivos Andinos*, Universidad Nacional San Cristobal de Huamanga, Instituto Interamericano de Ciencias Agrícolas, Ayacucho, Peru, pp. 282–9.

Saravia, R. and Calle, E. (1984) Fluctuaciones poblacionales de insectos en el cultivo de la quinua, in *Cuarto Congreso Internacional de Cultivos Andinos*, Instituto Colombiano Agropecuaria, Centro Internacional de Investigaciones para el Desarrollo, Programa de Cultivos Multiples, Pasto, Colombia, pp. 228–32.

Scarpati de Briceño, Z. and Briceño, P.O. (1980) Evaluación de la composición química y nutricional de algunas entradas de quinua (*Chenopodium quinoa* Willd.) del banco de germoplasma de la Universidad Técnica del Altiplano. *An. Cient. (Universidad Nacional Agraria, La Molina, Lima, Peru)*, **18**, 125–43.

Scarpati de Briceño, Z. and Briceño, P.O. (1982) Evaluación de la composición química y nutricional de algunas entradas de quinua del banco de germoplasma de la Universidad Técnica del Altiplano, in *Tercer Congreso Internacional de Cultivos Andinos, 8–12 Febrero 1982*. Ministerio de Asuntos Campesinos y Agropecuarios, La Paz, Bolivia, pp. 67–77.

Simmonds, N.W. (1965) The grain chenopods of the tropical American highlands. *Econ. Bot.*, **19**, 223–35.

Simmonds, N.W. (1971) The breeding system of *Chenopodium quinoa*. I. Male sterility. *Heredity*, **27**, 73–82.

Simmonds, N.W. (1976) Quinoa and relatives, in *Evolution of Crop Plants*, (ed. N.W. Simmonds), Longman, London, pp. 29–30.

Simpson, B.W. and Osborne, W.J. (1978) A rapid method for estimation of fatty acid levels in vegetable oils. *Lab. Pract.*, **27**, 642–3.

Singh, H. (1961) *Grain Amaranths, Buckwheat and Chenopods*. Indian Council of Agricultural Research, New Dehli, India.

Smoczckiewicz, M.A., Nitschke, D. and Wieladek, H. (1982) Microdetermination of steroid and triterpene saponin glycosides in various plant materials. I. *Allium* species. *Mikrochim. Acta*, **2**, 43–53.

Soilleux, A. (1986) Intra- and Inter-Specific Hybridisation in Wild and Cultivated *Chenopodium* Species. *M. Phil. thesis*, University of Cambridge, England.

Solano L., M.A. (1976) Evaluación de 30 ecotipos de quinua (*Chenopodium quinoa* Willd.) al ataque de *Naccobus* sp. Ingeniero Agrónomo thesis, Universidad Nacional Técnica del Altiplano, Puno, Peru.

Souci, S.W., Fachman, W. and Kraut, H. (1986) Food Composition and Nutrition Tables 1986/87. Wissenschaftliche Verlagsgesellschaft mbH, Stuttgart.

Swinkels, J.J.M. (1985) Sources of starch, its chemistry and physics, in *Starch Conversion Technology*, (eds. G.M.A. van Beynum and J.A. Roels), Dekker, New York, pp. 15–46.

Tapia, M.E. (1979) Historia y distribución geográfica, in *Quinua y Kañiwa.* Cultivos Andinos, (ed. M.E. Tapia), Serie Libros y Materiales Educativos No 49, Instituto Interamericano de Ciencias Agrícolas, Bogotá, Colombia, pp. 11–15.

Tapia, M.E. (1982) *El Medio, los Cultivos y los Sistemas Agrícolas en los Andes del Sur del Perú*. Proyecto de Investigación de los Sistemas Agrícolas Andinos, Instituto Interamericano de Ciencias Agrícolas, Centro de Investigación Internacional para el Desarrollo, Cuzco, Perú.

Tapia, M.E., Mujica, A. and Canahua, A. (1980) Origen distribución geográfica y sistemas de producción en quinua, in *Primer Reunión sobre Genética y Fitomejoramiento de la Quinua*, A1–A8, Universidad Nacional Técnica del Altiplano, Instituto Boliviano de Technología Agropecuaria, Instituto Interamericano de Ciencias Agrícolas, Centro de Investigación Internacional para el Desarrolla, Puno, Peru.

Tellería Ríos, M.L., Sgarbieri, V.C. and Amaya, J. (1978) Evaluación química y biológica de la quinua (*Chenopodium quinoa* Willd.). Influencia de la extracción de las saponinas por tratamiento térmico. *Arch. Latinoam. Nutr.*, **28**, 253–63.

Torres, H.A. and Minaya, I. (1980) *Escarificadora de quinua. Diseño y construcción.* Publicaciones Miscelaneas No. 243, Instituto Interamericano de Ciencias Agrícolas, Oficina en Perú, pp. 1–27.

Turkensteen, L. (1975) Hongos picnidiformantes asociados con el tizón foliar de la papa. *Fitopatología*, **10**, 83–4.

Valdivia, F. (1979) *Investigación en el Cultivo de la Quinua.* Bol. Tec. No. 6, Estación Experimental de Puno, Instituto Nacional de Investigaciones Agrarias, Puno, Perú.

Vargas, R., Begazo, V., Moreyra, P., Rojas, H., Cochella, T., Castro, C., Soriano, G. and Vega, C. (1977) *La Quinua como Sucedanae y Fortificante de la Harino de Trigo en Panificación Fideería y Galletería.* Programa Nacional de Alimentación Popular. Instituto de Investigaciones Agro-Industriales, La Molina, Lima, Perú.

Varriano-Marston, E. and de Francisco, A. (1984) Ultrastructure of quinoa fruit. (*Chenopodium quinoa* Willd.). *Food Microstruct.*, **3**, 165–73.

Vela, N.P. and Cabrera, L., J.A. (1984) Utilización de la semilla de quinua el la alimentación humana. *Technología*, **25**, 7–27.

Velásquez, F. (1968) Efectos del Sistema de Siembra, Densidad y Estado de Maduración sobre la Producción de Quinua (*Chenopodium quinoa* var. *Kanccolla*). Ingeniero Agrónomo thesis, Universidad Nacional Técnica del Altiplano, Puno, Peru.

Velazco, M.S. (1977) Control Químico de Malezas en Quinua (*Chenopodium quinoa* Willd.) var. *Kanccolla*. Ingeniero Agrónomo thesis, Universidad Nacional Técnica del Altiplano, Puno, Peru.

Vilca, A. (1972) Estudio de la Mancha Foliar en Quinua. Ingeniero Agrónomo thesis, Universidad Nacional Técnica del Altiplano, Puno, Peru.

Villacorta, L. and Talavera, V. (1976) Anatomía del grano de quinua. *An. Cient.*, **14**, 39–45.

Wahli, Ch. (ed) (1990) *Quinua: Hacia su Cultivo Comercial.* Latinreco S.A., Casilla 17-110-6053, Quito, Ecuador.

Ward, S. (1991) Male sterility in quinoa (*Chenopodium quinoa* Willd.). M.S. thesis, Colorado State University, Fort Collins.

Weber, E.J. (1978) The Inca's ancient answer to food shortage. *Nature*, **272**, 486.

White, P.L., Alvistur, J.E., Dias, C., Vinas, E., White, H.S. and Collazos, C. (1955) Nutrient content and protein quality of quinoa and cañihua, edible seed products of the Andes mountains. *J. Agric. Food Chem.*, **3**, 531–4.

Williams, J.T. (1963) Biological flora of the British Isles: *Chenopodium album* L. *J. Ecol.*, **51**, 711–25.

Wilson, H.D. (1976) Genetic control and distribution of leucine aminopeptidase in the cultivated chenopods and related weed taxa. *Biochemical Genetics*, **14**, 913–19.

Wilson, H.D. (1980) Artificial hybridization among species of *Chenopodium*. *Systematic Botany*, **5**, 253–63.

Wilson, H.D. (1981) Genetic variation among South American populations of tetraploid *Chenopodium* subsect. *Cellulata*. *Systematic Botany*, **6**, 380–98.

Wilson, H.D. (1985) *Chenopodium quinoa* Willd.: variation and relationships in southern South America. *Nat. Geog. Soc. Res. Repts.*, **19**, 711–21.

Wilson, H.D. (1988a) Allozyme variation and morphological relationships of *Chenopodium hircinum*. *Systematic Botany*, **13**, 215–28.

Wilson, H.D. (1988b) Quinua biosystematics I.: Domesticated populations. *Econ. Bot.*, **42**, 416–77.

Wilson, H.D. (1988c) Quinua biosystematics II.: Free-living populations. *Econ. Bot.*, **42**, 478–94.

Wilson, H.D. (1990) Quinua and relatives. *Econ. Bot.*, **44**, 92–110.
Wilson, H.D. and Heiser Jr., C.B. (1979) The origin and evolutionary relationships of 'Hauzontle' (*Chenopodium nuttalliae* Safford), domesticated chenopod of Mexico. *Amer. J. Bot.*, **66**, 198–206.
Wilson, H.D., Barber, S.C. and Walters, T.W. (1983) Loss of duplicate gene expression in tetraploid *Chenopodium. Biochem. Syst. Ecology*, **11**, 7–13.
Wolf, M.J., MacMasters, M.M. and Rist, C.E. (1950) Some characteristics of the starches of three South American seeds used for food. *Cereal Chem.*, **27**, 219–22.
Zanabria, E. (1981) Plagas de la quinua en el Altiplano y las perspectivas de control integrado, in *Segundo Curso Intensivo de Control Integrado de Plagas y Enfermedades Agrícolas, 1–7*, Fasciculo 56, Universidad Nacional Agraria, 'La Molina', Lima, Peru.
Zanabria, E. and Mujica, S. (1977) Plagas de las quinua, in *Curso de Quinua*, Fondo Simon Bolívar, Ministerio de Alimentación, Instituto Interamericano de Ciencias Agrícolas, Universidad Nacional Técnica del Altiplano, Puno, Peru, pp. 129–42.
Zapata, O. (1987) Distribución y epifitiología del mildiú en el Altiplano Norte y Centro en el cultivo de la quinua, in *Informe Anual 1986–1987*, Instituto Boliviano de Tecnología Agropecuaria, La Paz, Bolivia, pp. 73–80.
Zemel, M.B. and Shelef, L.A. (1982) Phytic acid hydrolysis and soluble zinc and iron in whole wheat bread as affected by calcium containing additives. *J. Food Sci.*, **47**, 535–7.

2

Buckwheat (*Fagopyrum esculentum*)

B.D. Joshi and R.S. Rana

2.1 HISTORY AND USES

Common or sweet buckwheat (*Fagopyrum esculentum* Moench.) is a pseudocereal cultivated on a limited scale but holding promise for more profitable cultivation. The grains have excellent protein quality in terms of amino acid composition. In a number of regions, particularly the Himalayas, it is a staple food crop and a mainstay of subsistence agriculture.

From its ancestral home in temperate Asia (northern India and China), buckwheat was introduced to Europe during the Middle Ages. In Asia cultivation is significant in India, Nepal, Bhutan, China, Mongolia, North Korea, far eastern Russia and Japan. It is less significant in Pakistan, Afghanistan and Iran. In Europe there has been a degree of production in the UK, France, Germany, Switzerland, Austria, Italy, Yugoslavia, Bulgaria, Hungary, Romania, Poland and Russia. In the New World, cultivation is found in the USA, Canada and Brazil. Production has recently fallen in Russia and France, once major producers. Canada's production also declined until the 1960s but has since recovered as a result of export demands. Cultivation has also spread to Australia and South Africa.

A second species, (*F. tataricum* Gaertn.), tartary or bitter buckwheat, is cultivated in India and China in cooler and harsher conditions. Both species are adapted to cool, temperate climates, but will tolerate tropical climates at higher altitudes.

Buckwheat was introduced into Europe in the Middle Ages, probably

Cereals and Pseudocereals. Edited by J.T. Williams. Published in 1995 by Chapman & Hall, 2–6 Boundary Row, London SE1 8HN. ISBN 0 412 46570 1.

from Siberia, reaching Germany early in the fifteenth century. Darlington (1956) mentioned that in Germany tartary buckwheat grows as a weed in crops of common buckwheat. In Russia it was apparently not cultivated to any appreciable extent until the fifteenth century (Krotov, 1963) and this was the time when it was introduced (Tsvetoukhine, 1952; Giacomini, 1955). In Poland buckwheat was brought from Asia by Mongol people and the first record of buckwheat being grown in Poland also dates to the fifteenth century (Ruszkowski and Noworolnik, 1983). Seeds and fossil buckwheat grains have been discovered in Poland (Krol, 1986). In Yugoslavia earliest records suggest that it was not widely spread in this area until the sixteenth and seventeenth centuries.

It was presumed to be introduced into Kashmir after 1200 AD (Gohil, 1984). According to de Candolle (1883), the common buckwheat was not cultivated in ancient times in India. There is no name for any type of buckwheat in Sanskrit literature. In the nineteenth century, Roxburgh had not seen it in northern India. The first descriptive record of buckwheat cultivation in Kashmir is found in the writings of the nineteenth century (Lawrence, 1895). However, buckwheat almost certainly has been cultivated in the Himalayas for centuries especially between 500 m and 4200 m above sea level from Pakistan to Myanmar. The southern boundaries of this area are contiguous with the Chinese centre of crop diversity and its western fringe with the Central Asia and West Asian centres.

2.1.1 **Origins of the crop**

The crop is believed to have been domesticated in China but probably in an area of temperate central Asia spreading from China to India (Elagin, 1959). The genus *Fagopyrum* has its range of wild species centred in central Asia spreading to Manchuria and Siberia (Morris, 1947; Tsvetoukhine, 1952) and the progenitor of the two major cultivated species is indigenous to the Himalayan region of western China and northern India. The progenitor is perennial and domestication involved the selection for annual growth form, non-shattering seeds and low seed dormancy. It is thought that *F. tataricum* is less selected than *F. esculentum* since it has more shattering and more seed dormancy, even if there is some evidence that *F. tartaricum* is phylogenetically the more ancient (Kovarenko, 1986). Almost certainly the two cultigens have been derived by divergent evolution from the same progenitor, *F. cymosum* Meissen.

Buckwheat is probably not an ancient crop. Although known to have been cultivated in China for at least 1000 years there is reference to buckwheat production in the fifth and sixth centuries AD (Krotov, 1963; Yurui and Zhongging, 1984).

Buckwheat is believed to have been introduced into Japan along with common millet (*Panicum miliaceum*), foxtail millet (*Setaria italica*) and rice (*Oryza sativa*). The first descriptive record of buckwheat in Japan dates to 722 AD. It has been inferred that buckwheat was introduced into Japan via the Korean peninsula from China (Nagatomo, 1983).

The diversity seen is partitioned among local landraces which have undergone local selection pressures for grain quality, and other uses such as vegetable, fodder and medicinal properties. The genepool is well adapted to the traditional agricultural practices (Joshi and Paroda, 1991; Singh and Thomas, 1978).

2.1.2 **Uses**

Flour from the grain is a staple food in the hills of India, Bhutan and Nepal. In India flour is used for making 'chillare', an unleavened bread fired with ghee and also for crisp 'pakora'. The bitter form (*F. tataricum*) requires the seed treating with boiling water to remove the bitterness.

Flour is also used for making sweet preparations and is sometimes mixed with wheat and barley flour for making 'chapattis'.

Occasionally whole grains are used. In Kulu Valley a variety of bitter buckwheat (var. *himalianum*, locally called 'gangri') is easily husked and grain is used and cooked like rice. In Japan flour is used to make noodles, called 'soba'.

In Europe and North America the flour is mixed with flour of wheat, rice or maize for a range of cooked products ranging from cakes to noodles and breakfast cereals. In Russia and Poland the flour is used in soups and porridges.

The flour and the groats must be used fresh since there is a high fat content and they soon become rancid.

1. **Source of honey** Honey produced from sweet buckwheat is prized in India. One acre of the buckwheat crop can produce 60 kg of honey; it is dark-coloured and strong in flavour. Analysis shows average contents of 78–58% invert sugar, sucrose 0–3%, dextrose 1–22% as well as formic acid 0–21%, ash 0–9% and water 18–56% (Wilson, 1948).
2. **Source of medicines** Some crops are harvested for medicinal use. Some plants contain a glucoside, rutin (see section 6.2). Rutin has the property of reducing increased capillary fragility associated with hypertension. Although a promising source of rutin the crop has not been commercialized for this purpose.
3. **Animal feed** Grain is mostly used for animal feed in the USA and Russia but not in India. Buckwheat flour is mixed with that of other

cereals and skimmed milk and salt. Its most extensive use is for poultry. The use of the plant as fodder can cause fagopyrism.

4. **Use as green manure and for soil conservation** The crop is very suitable as a green manure. It is also excellent for soil improvement, e.g. by cultivating rye with buckwheat followed by pasture (Ladefoged, 1952).

Due to its temperate range international attention was not focused on buckwheat through the US National Academy of Sciences report of 1976. However a series of events has promoted collaborative work on the crop. First, interest in the developed world in alternative crops has been stimulated as a result of surpluses (Taylor, 1989; Kephart *et al.*, 1990). Second, interest in the developing world in new plant sources for food and industry resulted in a major co-ordinated research programme in India (Paroda and Mal, 1989), and international meetings on life support species of the Commonwealth Science Council has also highlighted buckwheat. Third, international research has developed strong links between scientists through the International Buckwheat Research Association which issues a journal and organizes international symposia every 3 years. Finally, regional attention to genetic resources of buckwheat was promoted through regional workshops of the International Board for Plant Genetic Resources (IBPGR) in 1980 and 1989 and a specific workshop on buckwheat genetic resources held in 1991.

Although the main product is the grain used for flour and porridge and in a wide range of cooked products such as biscuits, noodles and pancakes other parts of the plants are also useful. The tender shoots are used as leafy vegetables. The flowers and green leaves are used for extraction of rutin (see above). Rutin has been reported to afford protection against harmful effect of radiation on high mountain areas and X-rays, indicating that it may be of use to persons exposed to atomic radiation. It has also been used with success as a prophylactic against gangrene due to frost bite. Buckwheat flowers help in producing honey of good quality and this use can be expanded. The crop is therefore multipurpose and frequently under-exploited.

Expansion of cultivation has been driven by the need for export earnings, e.g. in China, and by deficits in supplying demand, e.g. Japan only produces 20% of domestic demand. The nutritious attributes of the grain have led to countries such as India expanding production into areas where it was not a local subsistence crop. Such developments have been boosted by attention to breeding in several developed countries.

2.2 TAXONOMY AND ECOLOGY

The genus *Fagopyrum* is in the herbaceous subdivision of the subfamily Polygonoideae of the family Polygonaceae. It contains 15 species distributed in temperate Eurasia. The taxonomy of the cultivated species is still somewhat confused (Gohil *et al.*, 1983). Some minor cultivated species have been described, e.g. coarse buckwheat, *F. sagittatum* Gilib, and Kashmir buckwheat *F. kashmirianum* Munshi, (Munshi, 1986) but their morphological separation is probably not justified in terms of the buckwheat cultivated genepool. *F. kashmirianum* is allied to *F. tataricum* and *F. sagittaricum* to *F. esculentum*. A variety of sweet buckwheat var. *emarginatum* is called the notch-seeded buckwheat and a morphological variety of bitter buckwheat is recognized as var. *himalianum*.

Farooq and Tahir (1987) carried out a comparative study on growth forms of the species and found that *F. tataricum* behaved separately from the other three, *F. esculentum*, *F. kashmirianum* and *F. sagittatum*.

In general these various taxa are all recognized as conspecific with the two cultivated species.

The parental wild type is the perennial *F. cymosum* Meissn. and its putative ancestral status has been supported by Ohnishi (1983) who compared the electrophoretic band pattern of seven enzymes among this species and the two cultigens. He suggested *F. esculentum* and *F. tataricum* are derived from *F. cymosum* independently. Although *F. tataricum* has more morphological variability than *F. esculentum*, there is limited allozyme variability within the species (Ohnishi, 1986).

2.2.1 **Description of species**

Morphological details of the species have been provided by Bailey (1958), Marshall (1969), McGregor (1976) and Munshi (1982).

F. cymosum

This is a tall branched perennial which shows vigorous growth and is found commonly in the temperate Himalayas from Kashmir to Sikkim, at an altitude of 1800–2800 m. Plants are tall and stout. The leaves are large, long-stalked and triangular, angle acute or obtuse, upper ones usually narrow and amplexicaule; petiole long and slender. The pedicelled white flowers are borne on long recurved branches of peduncled terminal and axillary cymes. The nut (achene) is ovate and triangular, more than twice as long as the persistent perianth, larger than that of *F. esculentum* and *F. tataricum*, 7–8 mm long and 6–7 mm wide. In the North West hills of India, this species is locally called 'phaphru' or 'khurfa'. In Assam it is known as 'jarain'.

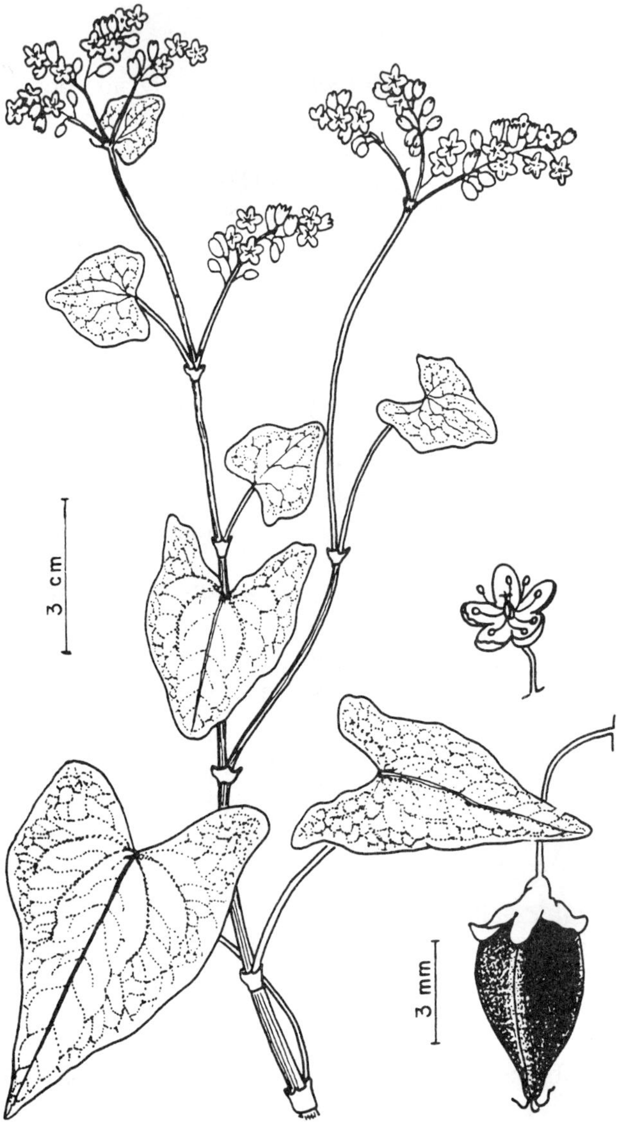

Figure 2.1 Common buckwheat (*Fagopyrum esculentum*). (Courtesy of New York State Museum.)

F. esculentum (Figure 2.1)

An erect annual attaining a height of 60–80 cm. The stem is hollow, angular with swollen nodes with red, pink and green coloration. The leaves are alternate, triangular, blades hastate or cordate, acute 5–10 cm long. The upper leaves are almost sessile, lower with petioles up to 10 cm long; ochrea short truncate. The inflorescence is axillary and terminal cyme with more or less densely clustered flowers. Each flower consists of five-partite persistent perianths, white, pink or red in colour. There are eight stamens alternating at the base with eight glands secreting honey. The ovary is one-celled and contains a single erect ovule; the style is tripartite, each with a knob-like stigma. The flower is dimorphic, with long- and short-styled flowers. On each plant there is only one type of flower. The fruit is a dry, one-seeded, three-sided nut or achene, measuring 6 mm × 3 mm. Some varieties have distinctly winged seeds. The seed colour is brown or grey, marked with dark spots and lines and triangular in shape. This species is locally called 'bharesha', 'ogla' and 'fullen'.

Notch-seeded buckwheat (*F. esculentum* var. *emarginatum* Moench.) is shorter in height and has small dehiscent fruits, the seeds of which are exposed when ripe. It is easier to separate the seeds of this variety. Its grain differs from tartary and common buckwheat by having the angles of edges of the hull extended into wide, rounded margins or wings. The seed colour is brown or blackish.

F. tataricum (Figure 2.2)

Plants are annual, and are taller and coarser than *F. esculentum*. The stems have shorter internodes which are stouter and less fragile than those of *F. esculentum*. The leaves are narrow and arrow-shaped with a long triangular membraneous ochrea. The flowers are borne in axillary racemes and are small with inconspicuous light-green sepals that make them distinguishable from the large white- or red-flowered varieties of common buckwheat. The flowers are all of one type. The fruit is ovoid, conical, with more or less wavy outline, brownish, grey or black in colour, with dull irregular faces on each of which is a deep furrow. The angles of the fruit are rounded except near the tips where they are slightly keeled. Some varieties have top-shaped, and others tooth-shaped, seeds. This species is locally called 'phaphara', 'kathu' and 'trumba'.

2.2.2 **Seeds and cotyledons**

Seed measurements differentiate the two cultigens and the wild *F. cymosum*. Materials from Iran and Iraq collected in 1955 by Kyoto University showed the following length/breadth ratios of seeds: *F. esculentum* 1.63, *F. tataricum* 2.0 and *F. cymosum* 1.21. Cotyledons of *F. esculentum* are oblong, the other species rounded. The ratio length of hypocotyl and petiole of cotyledons is diagnostic: *F. esculentum* 5.95, *F. tataricum* 3.22 and *F. cymosum* 0.24 (Nagatomo, 1965).

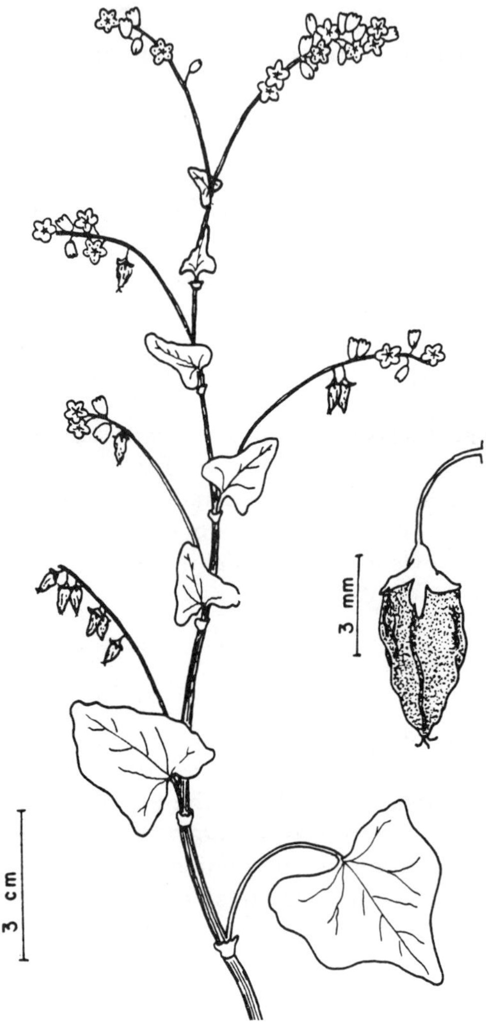

Figure 2.2 Bitter buckwheat (*Fagopyrum tataricum*). (Courtesy of New York State Museum.)

2.2.3 **Cytology**

The three species of buckwheat described above have been examined cytologically by various workers and were found to have diploid chromosome numbers. Jarotzky (1928) reported that *F. esculentum* has 2n = 16 and *F. esculentum* var. *emarginatum* has 2n = 16. Quisenberrg (1972) and Sando (1935) observed that *F. tataricum* has 2n = 16. For the wild perennial species (*F. cymosum*), Jarotzky (1928) reported 2n = 16. The chromosome numbers are thus 2n = 2x = 16, and the karyotype consisting of four metacentrics and four submetacentrics. Monsurova (1948), while studying the cytology, observed that a translocation involving chromosomes V and VI may be responsible for an additional satellite in *F. esculentum* var. *emarginatum*.

2.2.4 **Ecology**

Very little ecological information is available except that buckwheat can withstand poor, infertile and acidic soils and buckwheat is a potential food plant in cold deserts. It does not have potential in semi-arid or saline areas. It thrives best on sandy, well-drained soils but can be grown in dry regions with poor soils and drainage.

The frequency of the wild perennial and the wide diversity seen in landraces of cultivated buckwheat with adaptations to altitude and growing seasons suggest a wide ecological tolerance and opportunities to readily expand cultivation into new areas.

Buckwheat plants are very sensitive to freezing. Young plants can be affected by spring frosts when sown early in the spring. Ripening plants can be exposed to early autumn frost and ripening buckwheat is particularly sensitive.

Buckwheat is a short duration crop (3–4 months) and requires moist, cool climates. Hence it is admirably suited for short summer seasons at higher altitudes. Bitter buckwheat withstands more cool temperatures than common buckwheat and can be grown at higher altitudes.

2.3 MORPHOLOGY AND PHYSIOLOGY

2.3.1 **Growth development**

Distribution of photosynthates to plant parts relevant to buckwheat yield have been reviewed by Kroft (1986). The modelling of plant development is relatively easy for determinate species, i.e. plants with terminal inflorescences, because such plants have a fixed pattern of development (de Vries and Van Laar, 1992). The development of indeterminate plants such as buckwheat is more complex (Kreft, 1986). They continue to grow and produce flowers while the earliest formed seeds are ripening, so growth of vegetative organs continues parallel with formation of grain. The development is not only genetically programmed, but depends on environmental conditions, so buckwheat has specific peculiarities in the distribution of assimilates between organs (Fesenko, 1968).

Temperature affects metabolic processes in plants by its influence on the reaction kinetics of chemical events and on the effectiveness of the various enzymes involved. The temperature dependence of net photosynthesis results from the difference between the rate of photosynthetic CO_2 incorporation and the rate of repiration prevailing at a given temperature.

The temperature dependency of net photosynthesis, namely limiting temperatures (minimum, maximum) and temperature optimum, was measured and analysed by Bogataj and Gaberscik (1986) for two varieties of buckwheat, for a diploid cv. Siva and a tetraploid cv. Bednja 4. Results showed that the net photosynthesis response curves of Siva and Bednja varieties are not very different, although the values of net photosynthesis were a little higher for Siva. The lowest temperature limit for net photosynthesis can be set at 0 °C, which is common for plants of temperate zones. The net photosynthesis lies within a wide interval between 10 °C and 30 °C (typical for C3 plants), and optimum temperature is obtained between 15 °C and 25 °C. At high temperatures, above 40 °C, the value of net photosynthesis falls off sharply, most likely due to an intensified rate of respiration which released a larger amount of CO_2. The highest temperature limit for net photosynthesis for both varieties of buckwheat tested seems to be about 42 °C.

The dry matter accumulation of various plant parts over time is shown in Figure 2.3 and is partitioned into five stages (Bogataj and

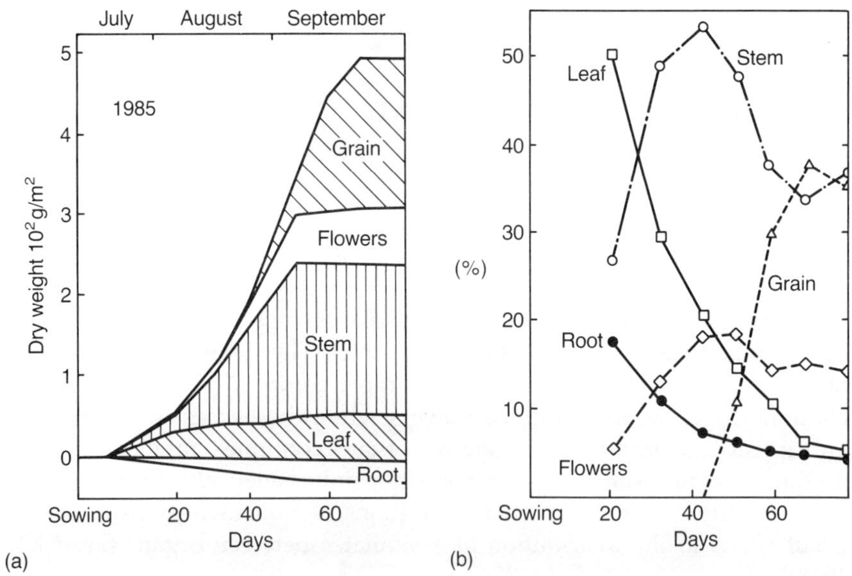

Figure 2.3 Dry matter accumulation of various plant parts of buckwheat over time.

Knava, 1985). Leaves reach a maximum weight before the end of the flowering period, after which most of the photosynthates were probably used for grain production. The weight of stems increases throughout the growth period but showed a stagnation between stages 3 and 4 due to bad weather conditions, primarily high precipitation. The same was observed for the total dry weight. The weight of flowers reaches a maximum in stage 3, the full flowering stage. The grains began to develop between stages 2 and 3 which is about 35 days after sowing. The yield in the year of the experiment was below normal. Leaf area and leaf area index (LAI) varies through the growing season and reaches the maximum value in the middle of the flowering period. Approximately 45 days after the leaves yellow, LAI gradually diminishes.

Bogataj (1987) reported dry weights of different plant organs at

Figure 2.4 Pollination of buckwheat by bees. (Courtesy of Rodale Press Inc.)

Table 2.1 Dry weights over time expressed as g/plant part

Days after leaf germination	*Leaf*	*Stem*	*Root*	*Flowers*	*Grains*	*Total*
20	30.8	16.4	10.6	2.8	–	60.6
31	40.7	67.2	15.6	13.8	–	137.3
42	54.6	145.6	21.0	49.9	–	271.1
50	57.8	187.0	26.0	71.3	43.8	386.1
58	53.7	179.2	25.9	69.1	141.7	469.6
76	28.2	183.7	27.7	73.2	179.2	492.2

various stages of growth development after germination, above ground (Table 2.1).

Exponential growth did not begin until day 30 when all the photosynthetic products were invested in the formation of new roots, leaves and stems. From 30 to 50 days, growth rate remained constant with photosynthetic products being invested more in maintenance of the plant and less for expansion of, for instance, leaf area. After day 50,

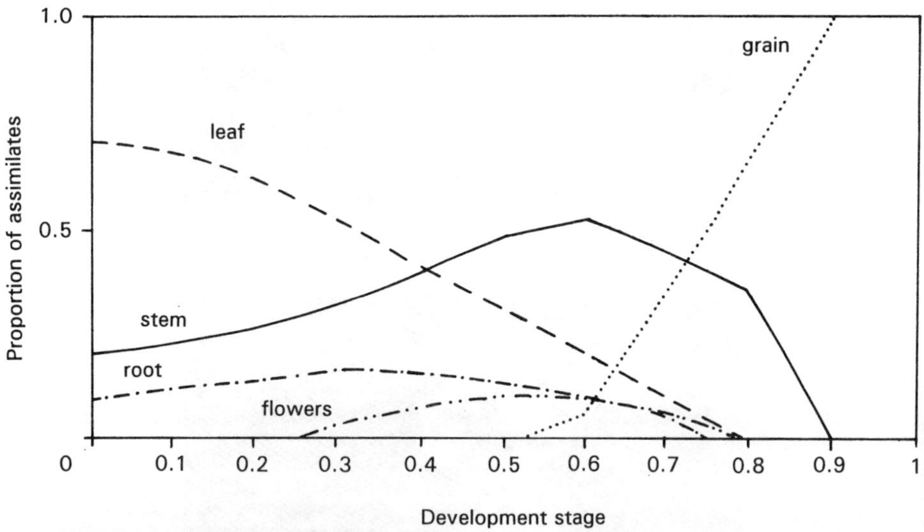

Figure 2.5 Distribution of assimilates during development. Development stage 0 corresponds to germination and 1.0 maturity.
Source: Bogataj, (1988).

plants changed from the vegetative phase into the reproductive phase. The quantity of leaves decreased, the quantity of stems and roots stagnated but flowers and grain began further growth.

Bogataj (1988) was able to model the growth of buckwheat on a development scale (DS) of 0 = germination and 1.0 = maturity. At DS 0.3 more than half the assimilates are invested in leaf formation but by DS 0.8 this ceases. Investment in stems is largely between DS 4 and 7, in flowers 0.25–0.55 stopping at 0.8, and in grains 0.5 to 0.9. In general, the reproductive structures seem to be the strongest sink after DS reaches 0.90 (Figure 2.5). Earlier the assimilates seem to be stored mainly in stems for later use to support grain growth. It can be assumed that where water or nutrients limit growth this model would change, for instance towards increasing the fraction going towards roots and decrease of fractions of shoot organs.

Farooq and Tahir (1987) compared the growth attributes of the four cultivated buckwheats of Kashmir and found that *F. esculentum* attains maximum dry weight, leaf area and height earlier than *F. sagittatum*, *F. kashmirianum* and *F. tataricum*. A higher productivity potential of *F. tataricum* was attributed to its longer growth span, higher biomass duration and also leaf area duration.

2.3.2 Photoperiodicity

De Jong (1972) considers buckwheat to be photoperiodically indeterminate, but this is far from the case. For instance buckwheat ecotypes in Japan are classed into three agro-ecotypes: summer ones not very sensitive to day length, autumn ones which are highly photosensitive and intermediate ones which show moderate photosensitivity (Komeichi *et al.*, 1992).

2.3.3 Germination

Cultivated forms produce non-dormant seeds (Tahir and Farooq, 1983) but occasionally viability of a seed lot may be low. Dormancy is enforced by high temperatures at harvest, e.g. at 25 °C and lasts for 60–70 days; and also by low temperatures at harvest, e.g. at 2–3 °C which might last for 6 months. In this respect both cultigens behave similarly. Singh and Mall (1977) have reported viviparous germination of both ripe and unripe seeds of *F. esculentum*.

Growth and metabolic changes have been studied by Tahir and Farooq (1983) during seed germination in both cultivars (including *F. sagittatum*, *F. tataricum* and *F. kashmirianum*). Germination, early seedling growth and mobilization of reserves into utilizable substrates is rapid in *F. esculentum* and slower in *F. tataricum*. The accumulation of total sugars was most pronounced in seedlings of *F. esculentum* and less so in *F. tataricum* compared to *F. sagittatum* and *F. kashmirianum*. In each of the four taxa, almost 60% of starch was depleted during 72 h of germination. The soluble protein content increased progressively during germination. Seedlings of *F. esculentum* maintained a higher total carbohydrate/total protein ratio. However, the ratio decreased in the seedlings of each of the four taxa during germination.

2.3.4 **Floral biology**

Buckwheat is an entomophilous and allogamous plant due to heterostylic self-incompatibility and seeds can only develop when compatible pollen grains are deposited by reciprocal legitimate pollination between the pin and the thrum-flower plants. The former have long-styled flowers with small pollen grains and the latter have short-styled flowers with big pollen grains.

Ren and Liu (1986) studied the flying distance of various insect pollinators in large buckwheat fields, applying dye spraying. They released about 10 000 dyed pollinators and then caught visiting pollinators at six different flying distances from 500–5000 m. Among the pollinators, bumblebee had the greatest range reaching 4500 m followed by honeybees that could reach 4000 m. The flying ability of other pollinators was poor. They also studied pollen spreading distance by wind: 500–600 m when the velocity of the wind was less than 3 m/s, but up to 1000 m when higher than 6 m/s. It was, therefore, concluded that a safe isolation distance is about 4000 m in areas where there are a lot of bumblebees and honeybees and about 1000 m where such bees are absent.

In traditional agriculture, floral abortion is not uncommon. The causative factors are not fully known. Kusiorska and Koszykowska (1981) reported that the decay of pollinated flowers and failure of seed-set is a limiting factor in grain yield. Factors which contribute to the low seed-set in buckwheat include incompatibility due to heterostyly, defectiveness of embryosac development and failure of fertilization (Adachi *et al.*, 1983). Studies carried out by Morton (1966) suggested that low grain yield is not attributable to lack of pollination, inviable pollen or to a self-incompatibility mechanism. Tatebe (1958) found that self-compatibility was not enhanced by spraying self-incompatible plants with boron. Sugawara (1960) believes that translocation of carbohydrates to the developing ovaries could be a limiting factor in grain set.

2.4 BIOCHEMICAL COMPOSITION

Buckwheat grain is highly nutritious, being a rich source of protein (Coe, 1931). The nutritive value of buckwheat is superior as compared

Table 2.2 Average composition of buckwheat grains kg/100 g compared with some other grains

Name of food grains	*Food energy*	*Moisture (%)*	*Protein (g)*	*Fat (g)*	*Total Carbo-hydrates*	*Calcium (mg)*	*Iron (mg)*	*Phosphorus (mg)*
Buckwheat	335	11.0	13.5	7.4	72.9	114	13.2	282
Amaranth	391	9.3	15.3	7.1	63.1	490	22.4	453
Maize	355	12.0	9.2	3.9	73.7	20	3.5	256
Rye	334	11.0	12.1	1.7	73.4	38	–	376
Wheat	333	12.0	13.3	2.0	71.0	41	10.5	372

Source: USDA data.

Table 2.3 Average composition of buckwheat products (%)

Product	*Dry matter*	*Digestible protein*	*Total digestible nutrient*	*Protein*	*Fat*	*Fibre*	*Minerals*
Grain	90.4	8.9	64.4	11.9	2.4	10.3	2.9
Flour	87.5	7.9	86.1	8.6	1.7	0.7	1.2
Millings	88.7	25.8	75.7	29.7	7.3	7.4	4.9
Feed (good grade)	89.1	14.7	59.1	18.6	5.0	18.3	4.1
Hulls	90.0	0.1	28.0	4.3	0.9	44.4	2.1
Straw	90.1	1.2	32.3	5.2	1.3	43.0	5.5
Green fodder	36.6	2.9	21.7	4.6	0.9	8.0	3.6

Source: Morrison (1945).

to millet or even the cereals like rice and wheat. Apart from protein it is a rich source of minerals and fat. The composition of buckwheat grains is shown in Table 2.2.

The average composition of buckwheat and its products is shown in Table 2.3. The minerals includes calcium, iron, zinc, selenium, manganese, copper and phosphorous.

Protein contents and net protein utilization of buckwheat in relation to other foods are shown in Table 2.4.

Pomeranz *et al.* (1975) determined protein content in whole grains and the germ of 10 genetically diverse buckwheats (*F. esculentum*). The hull was found to contain up to 4%, the groat up to 16.4% and the whole grain up to 13.8% protein, suggesting that the groat is a good protein supplement. Nicholson *et al.* (1976) reported that tartary buckwheat contains about 10.2% protein.

Table 2.4 Protein contents (PC %) and net protein utilization scores (NPU %) of buckwheat flour and other foods

Food	*PC (%)*	*NPU (%)*
Buckwheat flour	12	74
Wheat flour	10	47
Maize	10	42
Brown rice	7	70
Chicken egg	12	94
Beef	20	67
Chicken	18	65

Source: Citations in Namai (1992).

2.4.1 **Amino acid composition**

The amino acid composition (Table 2.5) shows excellent amino acid proportions and in particular high amounts of lysine which is often limiting in plant proteins.

Sokolov and Semihov (1968) fractioned grain proteins of diploid and tetraploid buckwheats. Globulins comprised the main protein fraction in both. Buckwheat proteins contain relatively low concentrations of glutamic acid and proline, but are rich in arginine and aspartic acid. No significant correlation was found between amino acid composition and protein content, indicating that increasing protein content could not impair the excellent amino acid pattern of buckwheat proteins. Eggum *et al.* (1981) reported that buckwheat protein quality is high due to high

Table 2.5 Amino acid composition g/100 g proteins

Amino acid	*Buckwheat flour*	*Wheat flour*	*Chicken egg*
Lysine	6.2	3.2	7.0
Arginine	10.1	4.2	6.6
Tyrosine	1.5	1.0	1.5
Methionine	1.8	1.9	4.0
Valine	5.1	4.6	7.2
Isoleucine	3.7	4.2	7.7
Cysteine	1.6	1.6	2.4
Leucine	6.2	7.4	9.2
Threonine	3.7	3.5	4.3
Phenylalanine	4.2	4.4	6.3
Histidine	2.2	2.1	2.4

Source: Citation in Namai (1992).

concentration of most essential amino acid especially lysine, threonine, tryptophan and the sulphur containing amino acids. However, due to the high content of crude fibre the true digestibility is below 80%. Ono *et al.* (1978) observed that the albumin fraction consisted of 12 proteins but the molecular weights of these proteins were close to one another, ranging between 7000 and 8000 daltons in presence of 2-mercaptoethanol. Albumins and globulins comprise the main protein fraction in buckwheat grains (Javornik *et al.*, 1981). They concluded that low prolamine content in buckwheat corresponds to a favourable amino acid composition compared with other cereals.

Yang Keli (1992) cites values of amino acids in buckwheat in China and showed both common cultigens are almost identical in amino acid contents but there are big differences between local cultivars. Some are high quality with total amino acid content higher than 14%.

2.4.2 **Other food compounds**

Buckwheat grains have been found to contain a high carbohydrate content, chiefly starch (Javornik *et al.*, 1981). Sando (1956) has recorded a high fat content in buckwheat flour. Belova *et al.* (1970) found palmitic, oleic linoleic and linolenic acids in buckwheat grains.

2.4.3 **Vitamins**

Buckwheat contains the important vitamin P and contains the valuable flavonoid rutin. Rutin is known to keep the capillaries and arteries strong and flexible, and also protects the blood vessels from rupturing

Table 2.6 Vitamin B1 and B2 in buckwheat flour and some other foods mg/100 g edible portion

Food	*Vitamin B1*	*Vitamin B2*
Buckwheat flour	0.46	0.11
Boiled soda noodles	0.05	0.02
Wheat flour	0.20	0.04
White bread	0.07	0.07
Brown rice	0.54	0.06
Cooked white rice	0.03	0.01
Maize	0.03	0.14
Boiled maize	0.16	0.30
Chicken egg	0.08	0.48

Source: Science and Technology Agency, Japan (1982).

and forming clots. Moreover, rutin given together with vitamin C is used for reducing high blood pressure. Buckwheat flour contains about 6.5 mg of rutin per 100 g which decreases to 1.2 mg by boiling. Tartary buckwheat contains more vitamin P than common buckwheat: 3.42 mg/100 mg versus 3.11 mg/100 mg (Yang Kets, 1992).

Buckwheat also contains considerable amounts of vitamins B1 (thiamine) and B2 (riboflavin) (see Table 2.6).

The average vitamin E content is 0.4–0.5 mg/100 g higher in common buckwheat (range in China 0.09–8.15 mg/100 g but commonly in the range 0.63–2.20) than in tartary buckwheat (normal range in China 0.45–1.52 mg/100 mg but as low as 0.21 and as high as 4.58) (Yang Keti, 1992).

2.4.4 **Other compounds**

Buckwheat grains have been found to contain several novel compounds such as unique amino acid amides (Koyama *et al.*, 1973): fagomine 1,3,4 dihydroxy-2-piperdine, methanol, glucosides salicylamide and a new amino acid L-2(2 furogl) alanine.

A proteinaceous inhibitor of trypsin and chymotrypsin has been isolated from buckwheat seeds. The inhibitor shows polymorphicity, existing in three forms, i.e. I, II and III, of which I and II are thermostable and III is thermolabile (Ikeda *et al.*, 1983).

A good deal of work has been carried out with buckwheat on the phenolic constituents and their biosynthesis. The hypocotyls of buckwheat seedlings contain cyanidin (Karstens, 1939), cyanidin derivatives (Troyer, 1958) and rutin (Margana and Margana, 1969). They also found that the cotyledons of buckwheat in addition to containing anthocyanins also contain rutin, orientin, homo-orientin, vitexin and saponaretin. The flavonoid, quercetin, quercitin, hyperin and rutin have been isolated and identified from immature grains of *F. esculentum* (Sato and Sakamura, 1975). The flavonoids, quercitin, kaemferol, kaempferol-3-rutinoside, rutin and quercitin-3-rutinoside-7-gala-cytoside have been isolated from the grains of *F. tataricum*. Tahir and Farooq (1987) identified quercetin from the grains of *F. kashmirianum*.

2.4.5 **Comparisons between types of buckwheat**

A comparative study has been made of the chemical composition of grains in four buckwheat cultivars (*F. esculentum, F. sagittatum, F. kashmirianum* and *F. tataricum*). It showed that grains of *F. esculentum* had the lowest content of phenolics, relatively low fat, free sugar and protein content but a higher starch content compared to the other three species. *F. esculentum* had a lower albumin-globulin content and also a lower glutelin content but a higher content of residual soluble proteins. The content of prolamins was generally low in each of the four species. Low content of phenolics in the groat fraction of *F. esculentum* accounts for its better palatability compared to the other three species which possess astringent taste.

2.5 AGRONOMY

Buckwheat can be grown in the areas which have rather poor soil conditions in various regions in the world. In general, most cultivars of buckwheat in each region can be grouped into early-, mid-, or late-maturing forms of which there are well-adapted cultivars in each area. Buckwheat in Asia is easily cultivated twice a year as summer cropping buckwheat and autumn cropping buckwheat. The autumn cropping buckwheat, especially cultivated in mountain areas or high land, is known generally to have superior taste and fragrance due to the comparatively higher temperatures in the day time and cooler temperature at night. Most summer-type cultivars for summer cropping are not so sensitive to day length (long-day photo period) and can bloom easily in conditions of 14.5 h day length, whereas the typical autumn-type cultivars for autumn cropping are very sensitive (Nagotomo, 1959). In Japan, the typical autumn-type cultivars are only in southern areas (Matano and Vjihara, 1979), and summer-type cultivars grown in northern Japan are believed to have differentiated rapidly from southern autumn-type cultivars through natural and artificial selection in course of their introduction into northern Japan (Minami and Namai, 1986).

A wide range of agro-ecotypes is available for testing from a number of countries. Much more work is needed on agronomy. For instance there is evidence that irrigation can increase yields in Mongolia by 100% due to the poor soils but irrigation is not normally practised elsewhere.

2.5.1 **Sowing and seed rate**

The seed-bed has to be prepared for effective control of weeds, conservation of moisture and provision of firm soil near the surface. Sowing dates vary from year to year depending on the preceding spring conditions which may contribute to difference in yield.

Marshall (1969) reported that sowing before mid-June was of no advantage in Pennsylvania (USA). In a 2-year study in Canada, Ali-Khan (1972) found that yields were highest from mid-June plantings. Gubbels (1977) showed that an early sowing date was of considerable advantage for higher yields in buckwheat.

Table 2.7 Grain yields kg/ha of summer and autumn buckwheats

Seeding and cultivation	*Spring planting grain yield*	*Summer planting grain yield*
Broadcast 120 cm	198	86
Drill seeded 60 cm row width	253	97
Drill seeded 20 cm row width	262	148
Cultivars: Spring planting: Shinnong 1 Summer planting: Suwan Jaere		

Source: Byung Han Choi *et al.* (1992).

Seed is sown after the field is ploughed and planked. Seed depth should be 4–6 cm, but 6–7 cm on very sandy soils. The seed rate varies from 35–40 kg/ha (Campbell and Gubbels, 1978) for row cultivation and 40–50 kg/ha when broadcast as in traditional cultivation. Ali-Khan (1973) observed that varying row spacing had no appreciable effect on grain yield, but this may depend on time of sowing. Table 2.7 shows the grains yields in relation to seeding and cultivation for two selected varieties in Korea.

Even closer rows are sometimes used in Asian countries and this increases sowing rates to as high as 70–90 kg/ha.

However, depending on heaviness of soil and need for weeding and fertilization, drilling and suitable row spacing can easily be adopted where commercialized and mechanical cultivation is desirable.

2.5.2 **Time to maturity**

Due to the wide range of agro-ecotypes which typify current cultivation in Asia there is a range of growth duration. As latitude and altitude increase, the growth cycle tends to shorten. It is convenient to consider cultivars as early (about 70 days), mid (70–90 days) and late (about 90 days). However successive multiplication of one of these cultivars outside its ecological area can readily change the flowering time.

Flowering in buckwheat begins 5 or 6 weeks after sowing. However the influences are indeterminate so flowering continues for up to a month. Days to flowering is strongly correlated with grain yield, plant height and grain-filling days (Sherchand, 1992). There is a very wide range in these characters. For instance, in Nepal, sweet buckwheat collected from lower elevation than bitter was earlier to flower and later to mature: hence these sweet forms have longer grain-filling periods. The means were 31.7 days to flower for sweet forms, 39.6 days for bitter for the autumn crop and 56.5 and 60.5 respectively for the earlier crop (Sherchand, 1992).

Table 2.8 Time to maturity in a range of (a) geographical regions, and (b) upland areas in India

(a) *Region*	*Latitude*	*Days from sowing to harvesting*
India	20–30°N	130
China and Japan	30–50°N	100–130
Russian Far East	43–50°N	90–100
Siberia and Central Russia	50–60°N	70–90
Belorussia	52–57°N	60–70

(b) *Regions*	*Sowing*	*Harvesting*
North–West Hills	June–July	September–October
North-East Hills	August–September	December–January
Nilgiri Hills	April–May	August–September
Palni Hills	January	April

This variation in local cultivars also provides a wide range of days to harvest around the mean times and dates shown in Table 2.8.

2.5.3 **Yields**

In many Asian countries good yields are considered to be 1.0–1.5 t/ha with important effects of fertilizers. However, better agronomy and better varieties alone could easily raise the yield to 1.5 2.0 t/ha.

2.5.4 **Seed weight**

The two cultigens tend to differ in 1000 seed weight. For common buckwheat 1000 seed weight is 25–35 g; for bitter buckwheat 1000 seed weight is 15–25 g. Each can be classified into small-, medium- and large-seeded forms. Some local cultivars tend to have even smaller 1000 seed weights, as low as 12 g.

2.5.5 **Fertilizers**

Apart from mulching being beneficial for seeding establishment, application of fertilizer of 20 N kg/ha and 20 N and 20 PO_2O_5 kg/ha application is beneficial. Although buckwheat has excellent tolerance to acid soils, a slight application of lime is extremely beneficial. In India, application of 200 kg/ha superphosphate and 50 kg N is beneficial for high yield. In the hills of India farmyard manure at the rate of 15–20 q/ha is recommended to raise a good crop. When grown on soils of poor fertility, buckwheat responds well to fertilizer application. It has been estimated that a buckwheat crop removes 47 kg nitrogen, 22 kg phosphorus and 40 kg potassium from the soil for each hectare planted and gives a yield of 1600 kg/ha (Campbell and Gubbels, 1978).

Application of nitrogenous fertilizers to buckwheat promotes vegetative growth but encourages lodging. Application of either ammonium nitrate or urea increases the content of dry matter, chlorophyll, proteins and soluble sugars. While ammonium nitrate increased the grain yield, urea is not always effective (Ganyushina, 1972). Sokolov and Semihov (1983) found that when labelled N was supplied as an ammonium salt during maturity of grains, it accumulated principally in the embryo. Sokolov and Semihov (1983) found that the local application of nitrogenous fertilizers in the form of a band at a depth of 30 cm prior to sowing gave maximum grain yield with minimal losses of nitrogen. Potzyinski (1984) observed that ammonium nitrate applied decreased the arginine content of grains. Marciszewski *et al.* (1984) observed some increases in the flavonoid content when nitrogen was given at 250 kg/ha but further increase of nitrogen fertilizer decreased the flavonoid content in buckwheat straw.

Boganovic (1987) reported that the average yield without fertilizer application was 1.1 t/ha; with 40N + 45P + 45K it increased to 1.25 t/ha and with 30N + 60P + 60K at sowing and 30N + 30P as a top dressing it increased to 1.25 t/ha. The maximum yield of 1.36 t/ha was obtained when 50N + 60P + 60K was applied as a basal dose along with 227 g boric acid mixed with 440 g urea in 650 l of water per hectare given as foliar spray which appeared to be the optimum dose for obtaining maximum economic yield.

Several studies indicate that phosphorus is efficiently absorbed by buckwheat and utilized in the production of dry matter (Kalra, 1971; Strong and Soper, 1973, 1974; McLachlan 1976). Thus uptake of total phosphorus by buckwheat plants treated with phosphate fertilizer was about three times more than that in plants which received no additional phosphate (Kalra, 1971). In calcareous soils phosphorus was found to be absorbed better from pellet treatment than from spread treatment; the reverse was true for non-calcareous soils. Studies carried out by Strong and Soper (1973) with excised roots of buckwheat indicate that these could efficiently absorb phosphorus even when the latter was present in high concentration. Besides, a large portion of phosphorus was taken up from pellet application and the roots proliferated extensively within the zone of applied phosphorus (Strong and Soper, 1974) presumably so because of the ability of the roots to acidify the rooting medium (McLachlan, 1976).

Gonchavik and Kozlova (1972) found that buckwheat plants responded more favourably to potassium fertilizers applied as sulphate than as chloride. Yarosh and Artemeva (1977) have recommended NPK combination at 40, 60 and 40 kg/ha, whereas Singh and Atal (1982) recommended NPK application at 50, 40 and 40 kg/ha for high herbage yield and good grain quality.

Application of micronutrient elements has also been found favourably to affect the yield and quality of buckwheat (Elagin and Kargaltsev, 1972, 1974). Several reports indicate that boron application improves grain yield (Shustova, 1962; Elagin and Kargaltsev, 1972; Chertko *et al.*, 1974). Gubbels (1980) concluded that foliar application of boron and calcium to buckwheat grown on soils containing adequate supply of these elements has no effect on yield or weight per grain. In chloroplasts isolated from magnesium-deficient buckwheat plants, photosystem II electron flow was not significantly affected but these chloroplasts had a markedly higher photosystem I activity compared to chloroplasts from normal plants (Baszynski *et al.*, 1980).

2.5.6 Harvesting and threshing

Ripening of buckwheat seeds is not uniform and delay in harvesting can result in seed-shattering. Due to the indeterminate growth, flowers, green grains and mature grains are present on the plants at the same time. Harvest is usually swathed first and then threshed. Generally swathing is done when 75% of the grains are mature. Careful handling of the crop is very important because grain-shattering can result in losses up to 22% (Campbell and Gubbels, 1978).

A moisture content of about 16% or less is necessary for safe storage of buckwheat grains. Temperature should not exceed 43 °C if the grains require drying.

2.5.7 Seed production

Seed production has to guard against genetic drift which is strong when multiplying landraces and there is the need for isolation. Additionally plant density has major effects on behaviour of plant pollinators and the floral morphology also has to be taken into account concerning the pm or thrum plants. Petelina (1950) considered that large-grained varieties should be grown in plots at an isolation distance of 200 m spacing hill (or small row) cultures without any pollinator cages, and sib-crossing within each hill (or small row) can occur by natural insect

Table 2.9 Insect visitors and seed-set *F. esculentum*

Number of visitors	*Number of flowers observed*	*Number of seeds harvested*	*Mean % seed-set*
0	8	2	25
1–2	5	2	40
3–4	8	7	87
5–6	6	6	100
7–8	5	5	100

pollinators (Namai, 1990). This sib-crossing method was established through field experiments on the pollination biology of spacing hill cultures of buckwheat, arranged at various intervals from each hill with about 10 pin or thrum plants planted alternately.

The startling relationship between the number of insect pollinators visiting a flower and percentage seed set in common buckwheat is given in Table 2.9.

2.5.8 **Pests and diseases**

Pests

Important pests of buckwheat are bruchids (*Acanthecelids obtectus*), cutworm (*Cirphis* sp.), aphid (*Myzus persica*), grain moth (*Cephitinea* sp.) and storage beetle (*Mycetophagus* sp.) which damage the crop and can be controlled by periodic treatment with pesticides during the crop season.

Fungi, mycoplasma and viruses

Several fungi have been reported to attack buckwheat in India and elsewhere. In India the following have been reported: smut (*Sphacelotheca fagopyri*) at Keti, and leaf spot (*Septoria polygonicola*) at Ootacmund in Niligiris. Other fungi include leaf blight (*Cercospora fagopyri*), *Phytopthora fagopyri* on root and basal stems, powdery mildew (*Erysiphe polygoni*) in China, brown spot leaf (*Ascochyta italica*) in Japan, rust (*Puccinia fagopyri*) in Korea and root and collar rot (*Sclerotinia libertiana*) in Japan. Pathogens reported are: aster yellows caused by mycoplasma; stem rot due to *Botrytis cinerea*; root rot due to *Fusarium* sp., *Botrytis* sp., and *Rhizoctonia*; chlorotic leaf spot due to *Alternaria alternata*; stipple spot disease caused by *Bipolaris sorokiniana*; downy mildew caused by *Peronosphora* sp. (Morrall and McKenzie, 1975; Zimmer, 1974, 1978; Singh and Atal, 1982). A few virus diseases are also reported. These include Brazilian curly top of tobacco and tomato in Brazil, aster yellow virus in Canada and mosaic virus in Russia.

Alekseeva *et al.* (1988) reported that on the basis of a 6-year study of 2900 varieties and wild forms under artificial infection with fungal pathogens such as *Peronospora fagopyri* and with viruses, resistant material was identified together with traits associated with resistance. Important to resistance of fungal disease was the number of stomata on the upper epidermis of leaves, while for resistance to viruses the hairiness of plant organs was important. Varieties with fungal resistance had three to five times less stomata/mm^2 as compared to susceptible ones. The stomata were more compact in resistant forms. Sap of fungus-resistant species *Fagopyrum cymosum* showed phytocidal activity against *F. fagopyri* when applied to leaves and stems of *F. esculentum*. Resistant forms had a relatively high rutin content in all plant parts.

Alekseeva (1988) further used electron microscopy and serological techniques in the study of a mosaic virus infection. Symptoms included a reduction in plant height by 36–52% and a reduction in yield by 75–85%, the production of low-quality grains, deformation of leaves, retarded development of reproductive organs and the presence of necrotic lesions on older leaves and chlorosis or yellow leaves. The virus was found to be a rod-shaped rhabdovirus. The most common vectors were *Aphis euonymi, Psammotethix striatus* and *Aphalava exilis*. Infection was most severe under late sowing conditions when over 3000 breeding forms of buckwheat (including wild relatives) were screened for resistance. Immunity was observed in *F. cymosum* and *F. tataricum* var. *himalianum*. Among cultivars of *F. esculentum* high resistance was observed in Aclita, Lada, Gallega, Kaliminskaya, Orbita, Prikamskaya, Radekhovskaya and Uluchshennaya. Breeding material obtained by mutagenesis was also resistant.

2.5.7 **Productivity constraints in buckwheat**

The major productivity constraints in developing areas of Asia identified in the farmers' fields (Joshi, 1988) are (i) non availability of improved varieties for cultivation; (ii) lack of information on agro-techniques for high seed production; (iii) heavy lodging and seed-shattering; (iv) little or no use of inorganic fertilizers; (v) lack of public awareness about the nutritive value of the grain (and greens); (vi) uneven fields in the hills; (vii) under-exploitation of honey production; (viii) inadequate plant stands; and (ix) a high percentage of flower abortion (see section 2.3.4).

2.5.8 **Farming systems**

In smallholder agriculture buckwheat is grown as a rain-fed crop and enters into cropping systems often as a second crop following maize or

Table 2.10 Diversified multiple cropping systems for buckwheat in RO Korea

Region	*Cropping system*
Central northern	S potato – A buckwheat, S tobacco – A buckwheat, barley – A buckwheat, S maize – A buckwheat, Sm buckwheat – A vegetable, Sm buckwheat – A buckwheat, S vegetable – A buckwheat – buckwheat vegetable, Buckwheat vegetable – Sm buckwheat – buckwheat vegetable – A buckwheat – buckwheat vegetable
Central northern	Barley – A buckwheat, Rape – A buckwheat, Sesame – A buckwheat, S vegetable – A buckwheat, Sm buckwheat – A vegetable, Sm buckwheat – A buckwheat, S vegetable – A buckwheat – buckwheat vegetable, Sm buckwheat – buckwheat vegetable – A buckwheat – buckwheat vegetable

S – Spring, Sm – Summer, A – Autumn

Source: Byung Han Choi (1992).

barley, but in many areas of India, China, Nepal and Korea it is grown as a main crop. It is also grown as an intercrop, often with vegetables, as in Korea.

Diversified multiple cropping is becoming much more researched in Asia and use of buckwheat for grains or a tender shoot vegetable is shown in Table 2.10.

2.6 POST-HARVEST TECHNOLOGY

In Japan the standardization test to certify the quality of buckwheat grain has been fixed. The buying and selling of domestically produced buckwheat follows the regulations set up by agricultural production impection law, which used second grade buckwheat as its standard. The standard fixed for this class of grains is that it should have not more than 15% moisture, it should have 85% sound grains and it must not have more than 20% damaged, immature seeds with 0% foreign material.

2.6.1 **Processing**

Buckwheat is made into flour using either the traditional methods employing stone grinders or the more modern type of industrial mills.

Noodles

Buckwheat flour is in demand by noodle-making factories, especially in Japan, to produce noodles which are sold in their pre-cooked, boiled, dried and packaged or instant stages, as well as being made directly for sale to the buckwheat noodle restaurants, small shops or stands. There are 6211 noodle-making factories and 41 556 restaurants serving buckwheat noodles in Japan.

In addition to buckwheat flour, buckwheat is used in making buckwheat tea, confectionaries and liquor as well as eating buckwheat in its grain form similar to 'kasha'.

The buckwheat kernel is made up of four basic component parts: the outer protective hull; the seed-coat or aleurone layers; the endosperm; and the embryo. Buckwheat flour is obtained from these component parts, but the flours desired from each of the individual parts exhibit their own unique characters. The outer protective hull is generally discarded or used by peasants for stuffing pillows, etc. The seed-coat part contains the most protein in the buckwheat kernel. As compared to wheat, buckwheat has no gluten, so that it is quite difficult to make buckwheat into noodle form. Within buckwheat are water-soluble proteins such as globulin and albumin. During the initial mixing of water with buckwheat flour, these proteins combine with water to produce a stickiness during the noodle-making process. When using only cold water to make noodles, this stickiness becomes one of the most important factors in giving cohesiveness to the noodles.

Concerning this point, in the milling process, water dampening and tempering cannot be carried out with buckwheat as with wheat because the above-mentioned water-soluble protein elements are inclined to dissolve in the water and choke up the sieves. If it were possible to carry out this water dampening process with buckwheat, transportation and storage would be much easier as a result of containing less water.

The endosperm part consists mainly of starch having high viscosity to give an elastic texture for making noodles. Various blends and mixtures of flours taken from these parts are combined at the mill as well as during the noodle-making process. According to the blend or mixture used, the quality of the noodle will differ considerably. For example, as the percentage of flour from the endosperm is increased, the amount of water used will increase accordingly. As the percentage of flour from the seed-coat is increased, the amount of water necessary will decrease. Generally speaking, buckwheat noodles served in restaurants contain from 20–50% wheat flour and the buckwheat noodles made in factories contain even higher percentages of wheat flour. Wheat flour helps to make the noodle-making process easier because it makes the dough stickier, and it also helps to provide a smooth feeling when eaten.

Noodle-making at the household level is simple. Blend the flour mixture, add water, about 35–45% of the weight of the flour in a wooden bowl, to make firm dough. Roll out with a rolling pin on a wooden board, pressing the rolling pin down with both hands. When the dough becomes about 3 mm thick, fold it up. Cut it about 1.5 mm wide. The noodles should have a uniform length of about 30 cm.

Buckwheat noodles are boiled in big pots. They are allowed to cook for about one min, taken out of the pot, washed thoroughly in cold water, then strained before serving.

Other cooking uses

In Europe and North America the flour is mixed with flour of other cereals and used for pancakes, biscuits and breakfast cereals as well as noodles.

2.6.2 **Processing of green plants for rutin**

Rutin is obtained from fresh or dried leaves and flowers. Green plants should be processed as quickly as possible after harvesting, as there is a rapid loss of rutin in slow field drying. A process for the rapid dehydration of leaves, using rotary alfalfa driers, has been developed for obtaining leaf meal rich in rutin and this is stable under ordinary storage conditions for at least a year. By choosing drying conditions specifically designed to minimize loss, the overall loss of rutin can be reduced to about 25%. The solvents employed for the extraction of rutin are hot water, dilute isopropyl alcohol or ethanol, and hot isopropyl alcohol. By distilling off the alcohol from the extract and straining the fat which settles out, a watery solution is obtained from which crude rutin crystallizes on standing and cooling. It is dissolved in boiling water, treated with silica gel to remove the red pigment, and rutin crystallized out from the filtrate. About 28% of rutin in the meal is recoverable by this process.

During the extraction of rutin from buckwheat the fatty matter is a by-product. It is dark in colour, almost solid at room temperature and contains an unusually high percentage (14.9%) of unsaponifiable matter. It contains lecithin (2%). The principal fatty acids present are linolenic, oleic, linoleic and palmitic. The unsaponifiable fraction contains carotene, xanthophylls, phytol, B-sitosterol and one of the cicosanols.

2.7 GENETICS AND BREEDING

Although breeding and release of varieties has been ongoing, especially in Canada, USA, Russia and Japan, crop improvement in many other countries has relied on simple selections from landraces, the status of buckwheat being as an underutilized crop.

Some of the important improved varieties of buckwheat currently under cultivation in different parts of the world are in India (Himpriya and VL Uggal-7), Nepal (IV-78 and ARU-50), Japan (Ibaraki Pref. and Hitachi Akisoba), Canada (Manor and Tokyo), USA (Mancan and Pennline 10), Russia (Siva, Victoria and Malysh) and Poland and Yugoslavia (Siva).

Out of the three cultivated species, *F. esculentum* is dimorphic and self-sterile and, hence, cross-pollination is the rule. *F. tataricum* is self-fertile and produces flowers of only one kind. The breeding system of *F. cymosum* is unknown but it may reasonably be assumed to share with *F. esculentum* a heteromorphic incompatibility system. *F. tataricum* is self-compatible and homomorphic. It appears, therefore, as though one cultigen, *F. tataricum*, has moved towards inbreeding under domestication. In recent years, *F. tataricum* has become more popular due to the presence of rutin in its leaves.

Common buckwheat is a self-incompatible, sexually propagated crop and as such does not lend itself readily to improvement by breeding. Sharma and Boyes (1961) have studied the heteromorphic sporophytic type of incompatibility involved and suggested that the inheritance of style length is controlled by a single gene. The allele for short style (S) is dominant to the allele for long style (s). However, short-styled flowers also differ from long-styled ones in having longer stamens, larger pollen grains, the thrum style incompatibility reaction and the thrum pollen incompatibility reaction. They postulated a supergene for this complex of characters, similar to that proposed for *Primula*. They further suggested that the supergene could be broken down into subunits by crossing over. Although the mechanism is not known, there is evidence of genetic change at the S locus. Thus homomorphic, highly self-compatible diploid lines have been isolated but their value in breeding has yet to be determined as severe inbreeding depression occurs.

Tartary buckwheat is self-compatible. Attempts to transfer the self-compatibility of *F. tataricum* to *F. esculentum* have proved unsuccessful. A study of pollen tube growth in the interspecific cross indicated that pollen tubes of *F. esculentum* reached the base of *F. tataricum* styles whereas those of *F. tataricum* were inhibited in *F. esculentum* and no hybrid seed was obtained (Morris, 1951). Garber and Quisenberry (1927) studied inheritance of style length in *F. esculentum* and reported the Ss systems referred to above. Sugawara (1956) reported that the optimum temperature for pollination was observed to be 20 °C. With increase in temperature, the initial rate of germination rises, but pollen tube growth tends to become arrested before fertilization is effected. Garber and Quisenberry (1927) found that large glassine bags and muslin cages were more satisfactory for selfing. Short-styled plants proved to be somewhat more fertile than long-styled plants. It would thus appear that, for breeding purposes, it is possible to obtain selfed seed in the common buckwheat.

2.7.1 **Breeding aims**

While describing methods for producing buckwheat varieties for high and regular yields, Petilina (1950) recommended selection for definite stem and leaf characters. Non-branching single stem forms with the stem divided into two long branches had a better fruit-setting capacity than forms with short branches or with more than two long branches. Selection was also recommended for forms in which the leaves dry at harvest time. Such characters are important especially for mechanical harvesting.

2.7.2 **Breeding methods**

Pisarey and Vivogranova (1952) showed that mass selections for different types of seeds in a heterogeneous population could be effective. They obtained a promising population by continued open pollination of 20 local and cultivated varieties and the subsequent selection of the most productive plants. Fesenko and Antonov (1981) outlined various methods used in breeding buckwheat in Russia. The authors described a method of evolving synthetic varieties. The method consists of selecting 300–500 best plants of the breeding material for yield and grain quality in the first year, inbred in the second, subjected to test cross by the top cross method in the third year and tested for yield in years 4 to 6. In the seventh year, 4–10 best families were then combined into a synthetic. This synthetic variety is multiplied and tested for yield in years 8 to 10. State variety trials were conducted. Likewise plants may be selected from the first or second generation for a second cycle of recurrent selection. This breeding programme is said to increase yield by 15–20% and was used to select cv. Lovchanka from the initial variety Boyatyr.

Campbell (1983) mentioned that cv. Manor is the second large-seeded sweet buckwheat licensed in Canada. It is adapted to a larger growing area than cv. Mancan and has a higher yield potential. Mancan is a mid-season cultivar developed at the Agriculture Canada Research Station in Manitoba. It has a seed size 26% larger than cv. Tokyo and is the first large-seeded buckwheat licensed in Canada. It is especially suited to south Manitoba. CM221 was developed from a cross between a semi-dwarf spontaneous mutant of breeding line CD7464 and the larger seeded line CM108. CM221 is about 80 cm tall compared with 110 cm for CM108. The semi-dwarf character proved to be controlled by a single recessive gene and to be caused by a shortening of the first six internodes in CM221. Plants of CM221 showed an early initiation of branching, especially from cotyledonary axils, resulting in a highly branched upright growth habit. CM221 is resistant to lodging. It has reduced internodes with solid stems resulting in increased resistance to wind and hail damage as compared to normal hollow stemmed plants.

Marynenko (1987) reported that the selection was carried out for more determinate populations with an increased number of inflorescences, limited branching and an increased number of vegetative nodes on the main stem. Selection for the last trait gave the greatest increase in yield.

Koblev (1987), while breeding buckwheat for resistance to lodging in the USSR, reported that the higher yielding Orlovskii Karlik KM3 is 20% shorter than its parent cv. Orel Dwarf KM3 and has high resistance to lodging. As a result of breeding work over a period of 4 years, involving crossing Orlovskii Karlik KM3 with other cultivars, mass selection and controlled pollination, the varieties Malysh 7 and Malysh 10 were developed. These combine lodging resistance with high grain quality and high yields, exceeding the standard by 0.31 to 0.45 t/ha.

De Jong (1972) cites several workers who have reported modification of incompatibility through X-radiation of grains. Esser (1953) found some homomorphic self-compatible plants in a population of autotetraploids. Attempts to transfer self-compatibility of *F. tataricum* through inter-specific crosses have proved unsuccessful (Morris 1947, 1951; Ruszkowski, 1980). Adachi *et al.* (1981) have postulated that the breakdown of incompatibility of autotetraploid buckwheat would be in the pistil and not in the pollen.

Fesenko (1968) discovered a mutant with a determinate growth habit. This character is apparently controlled by a single recessive gene. Thus the genotypes of normal plants are DD and Dd, while that of the determinate type plant is dd. Experiments have been conducted to register the useful agronomic characters associated with buckwheat having determinate growth habit (Luthar *et al.*, 1986).

An unstable gene system affecting chlorophyll production has been described in buckwheat (Ali-Khan, 1971) and is controlled by a major gene with multiple alleles. Tatebe (1972) studied leaf variegation in buckwheat and found three kinds of variegated plants: yellow, yellowish-white and pale green.

Not much is known about the inheritance of most characters in buckwheat. Morris (1947) reported that the winged condition of the grain is dominant over the non-winged condition. Her results are contradicted by those of Marshall (1969), who obtained true breeding winged lines as a result of inbreeding non-winged forms.

Several Russian breeders have investigated the use of heterosis in buckwheat (Taranenko, 1986). Since self-fertilization usually fails, Zheleznov (1967) has produced inbred lines. In various inbred lines, Marshall (1979) observed favourable effects of heterosis for plant height, dry matter yield, grain yield and grain weight. Gorina (1969) reported that heterosis was lost by the F2 generation in most hybrid populations.

Owing to indeterminate growth, the grains borne on older racemes are prone to shattering. Efforts have been made to develop varieties that are resistant to shattering (Fesenko, 1963, 1966). Fesenko and Koblev (1983) have reported the application of a short-stemmed buckwheat in breeding for resistance to lodging. Potentially useful forms of buckwheat, including varieties with high adaptability to environment, have resulted from experimental mutagenesis, (Alekseeva, 1984; Srejovic and Neskovic, 1981, 1983). Bohanec (1985) observed that buckwheat can be successfully propagated in tissue culture. Recently experiments have been conducted to obtain protoplasts (Adachi, 1986).

In the USSR, Orlovskii Karlik KM3, a donor of shortness and lodging resistance, was crossed with high-yielding but lodging-susceptible varieties. In comparison of length of stems and internode number, only the length of the third internode was a reliable marker under field and glasshouse conditions. Selection for short third internode increased the number of lodging-resistant, short forms in the hybrid population to 90% following three cycles of selection in the field and glasshouse (Fesenko *et al.*, 1986). Anokhiva (1987) mentioned that the determinate forms have unstable yield and that selection for determinate habit should be accompanied by selection for yield. Marshall (1970) reported that *F. sagittatum* Pennline 10 is an inbred line derived from a highly self-fertile S_5 plant isolated from the open-pollinated buckwheat collection Pa 53. It is homomorphic and has been released for breeding purposes in the USA.

Alekseeva (1988) described the work involving treatment of seeds of the *Fagopyrum esculentum* cv. Viktoraya with radiation and chemical mutagens in 1974 in the USSR. He observed that the selection for larger grains (the best mutant lines having 1000 seed weights of 41–48 g, 12–19 g heavier than the control) and high yield from the M2 onwards led to release of the mutant variety Podolyanka in 1984. It is high yielding (2950 kg/ha), matures 17–18 days earlier than Viktoriya and has short stems (height 60–80 cm) with seven nodes and only two or three primary branches (no secondary branches). There are 10–12 inflorescences, the first branch occurring at the second node and the first inflorescence at the fourth node.

2.7.3 **Polyploidy**

In the USSR and USA some work on induction of polyploidy has been carried out. Sando (1935) reported a colchicine-induced tetraploid buckwheat and he further reported that the tetraploid tartary buckwheat had higher rutin content than the common diploid tartary (Sando, 1956). Saharov *et al.* (1944) produced tetraploids by colchicine treatment. There was a pronounced decrease in fertility, but selection for high fertility was possible on account of the diverse materials treated with colchicine. In the progeny of the best tetraploid of variety Bolshevik, they obtained two to four times more seeds than diploids. The tetraploids were vigorous with higher percentage germination. The increase in seed size ranged from 42–85%. The tetraploids and diploids did not cross. Saharov *et al.* (1944) also obtained tetraploids in variety Bolshevik. A valuable economic character of the tetraploid was lack of seed-shedding. The tetraploids were capable of self-pollination in contrast to the diploid, in which self-pollination was rare or absent.

Studies on the growth of tetraploid *F. esculentum* showed that in the early growth stages, the tetraploids were larger but after flowering the process was reversed (Murakami and Usaburo, 1950). Tetraploid seeds contained less total sugars and tetraploid seedlings had higher total sugars.

Dorshenko (1988), while studying the possible ways of applying meiosis selection in tetraploid buckwheat improvement, reported on selection for both normal meiosis (as measured by the frequency of abnormal microspore tetrad) and high yield over 3 years in tetraploid cultivars. In Bolshevik 4 and Iskra the mean frequency of abnormal tetrads declined from 24.0 to 10.9% and from 19.4 to 10.3% respectively. When selection was for reduced frequency of abnormal tetrads and high yield in plants of Bolshevik 4, which has less than 10% abnormal tetrads in primary inflorescences, the mean frequency of abnormal tetrads fell from 17% to about 9% after only two cycles of selection. When selection was for high frequency of abnormal tetrads, this frequency stabilized at about 30–50% after 13 cycles. Among the population selected for low abnormality was a highly self-fertile plant which was used to produce monomorphic long-styled populations. Analysis of low-, moderate- and high-yielding groups from this population showed that the highest yielding group (forming 159 grains/plants) had the lowest frequency of abnormal tetrads (9%).

Although tetraploid plants produced higher dry matter and showed a higher nitrogen uptake throughout their growth period, there was a lower grain yield compared with diploid plants (Tsuzuki *et al.*, 1983). A tetraploid variety Pennquad was released in Pennsylvania (USA) in 1967 (Marshall, 1967, 1968). Pennquad has a grain weight approximately 40% greater than Tokyo but its yield under Canadian conditions averaged 30% less than Tokyo (Ali-Khan, 1970). Adachi *et al.* (1981) found that the yield of induced autotetraploid buckwheat Miyadi no 1 was superior by at least 20–30% at the places studied by them. Some varieties that have been developed recently in Japan include Miyazakioot-subu, Miyazakizairai, Sando-Soba, Hashikamiwase and Shinnano-ichigou (Nishiyama *et al.*, 1983).

2.7.4 **Germplasm collections**

Due to the crop improvement interests of the past 25 years there has been substantial activity on screening germplasm collections. For instance, the national programme in China collected over 2000 local forms between 1950 and 1958. Unfortunately these were lost but in the 1980s materials were re-collected. In India the national programme started systematic collecting of buckwheat genetic resources in 1974 and material was collected in Yugoslavia in 1975. Collecting work was accelerated about a decade later by scientists in Poland, Nepal, democratic Korea and Japan. Other explorations reported cover Bhutan, the former USSR and Myanmar.

Current holdings of buckwheat germplasm are recorded for genebanks in the following countries, with total number of accessions in parenthesis: India (539), Nepal (300), Russia (2100), China (1511), Japan (510), Germany (562), USA (132) and Yugoslavia (361). There are also small collections held by other genebanks in Canada, South Africa and Pakistan.

In 1991 IBPGR and the National Institute of Agrobiological Resources, Japan, held a regional workshop to discuss genetic resources activities on buckwheat. Participants from Asia recommended further collecting of material especially in Pakistan, some pockets in the Himalayas and north-east India, Bhutan and Myanmas.

Characterization and documentation of the germplasm collections is well underway, and IBPGR has agreed to promote a database to merge individual ones. At the present time, national programmes will continue to conserve materials and to make them available without the need for any global collection.

Standard descriptors for documentation have been developed, largely from those developed by the Indian researchers at the National Bureau of Plant Genetic Resources, Regional Station at Shimla. Screening and recording data has resulted in a great deal of useful information being available to breeders. This includes ranges, means and coefficients of variation on 11 characters for 408 populations in India (plant height, number of branches, internodes and leaves, leaf length and width, days to flower and to maturity, number of seeds/cyme, 1000 seed weight and yield/plant). Similar data are available from Nepal and China and the latter has also screened for vitamins E and P contents.

2.8 TRADE

China is the second largest country in area and production of buckwheat after the former USSR. The other major buckwheat-producing countries are the USA, Canada, Brazil, Japan, India, Nepal and Bhutan.

Japan is the major country importing buckwheat because it only meets 20% of its total required domestic demand. China, Canada, USA, Brazil and South Africa are the major suppliers to Japan. In the other countries buckwheat produce is for domestic consumption. Buckwheat production in different countries is given in Table 2.11 and imports to Japan in Table 2.12.

In India, buckwheat production is declining in view of introduction of high-yielding varieties of elite crops. However, in Bhutan, the cultivation and conservation is increasing and competing with potato cultivation. The reasons for its favour are its ability to grow on poor soils at low temperature on high mountains, early maturity and excellent top soil binding habit. In alpine regions of India and Nepal the buckwheat production is not affected where the crop-growing season is very short and buckwheat cultivation fits well in view of harsh climatic conditions.

In India the local produce does not find any central market for trading. Most of the produce is kept for home use and a limited quantity

Table 2.11 Production of buckwheat in selected countries

Country	*Area (ha)*	*Production in tonnes*
Nepal	43 000	23 100
Japan	27 800	21 000
India	60 000	35 000
Canada		30 000
Brazil		23 000
USA		26 000
South Africa		2400
China		90 000

Table 2.12 Imports of buckwheat to Japan

Supplier Country	*Tonnes*	*Price per 45 kg bag in Tokyo (¥)*
China	39 000	5800
Canada	27 000	5000
Brazil	23 000	4200
USA	16 000	
South Africa	2400	

enters the local markets which in turn is sent to big cities on religious days of fast, being one of the few permitted foods for fast days. The current price of buckwheat grains is around 15 rupees/kg and the flour is being sold at 20 rupees/kg in retail shops.

In Korea the total production of buckwheat is around 880 t/year. Its marketing is mainly for food processing companies for manufacturing noodles and to restaurants which want a traditional food. No figures for trade imports/exports are available.

2.9 HIMALAYAN FARMERS' PERCEPTION OF THE CROP

In India, Bhutan and Nepal the farmers' perceptions of the crop greatly differ due to diversity in agro-ecological regions and local preferences. The farmers of the alpine Himalayas have no option but to cultivate buckwheat. In the high and mid hills farmers find it less remunerative compared to elite crops. However, some of the farming community has a special preference for its seed and vegetable use as well as for honey production. In the lower hills the farmers have the least preference for buckwheat except for its vegetable use in a few scattered pockets. Whatever produce is provided to big cities during religious days of fast comes from higher and alpine regions of the north-west Himalayas.

The farmers' preference and perception of buckwheat cultivation and adoptions are listed below.

1. It is a traditional food of the people of the high Himalayas where the grains are used in preparation of different culinary items and alcoholic drinks.
2. It has medicinal value to check radiation hazard in the high Himalayas.
3. Farmers in some areas have a great preference for a dual-purpose crop (grains and green vegetable).
4. Its great soil erosion checking ability on the steep mountain slopes is recognized.
5. In certain pockets of the Himalayas where apple cultivation is remunerative buckwheat cultivation has been taken up as a source of honey and greater bee activity due to which apple production has increased. They require early flowering varieties for three or four crops in a year of honey.
6. In some areas farmers have a special preference for buckwheat cultivation in view of its improvement of soil texture valuable in crop rotation.
7. The ideal perception and preference of farmers in the Himalayas is to have a high-yielding variety with large seeds to earn higher economic returns.

2.10 FUTURE TASKS

Apart from crop improvement for higher yields and targeting trials for enhanced sustainable production in marginal environments, the outlines of quality research on buckwheat have been provided by Polish and Yugoslav researchers in the 1988 issue of *Fagopyrum*, the newsletter of the International Buckwheat Research Association. This publication is too long to reproduce here but it does draw attention to the need to enhance processing and products, to manipulate biochemical contents to reduce anti-nutritional factors, to improve screening methods in terms of quality evaluation and to improve agronomy.

Since this paper was written two useful articles on seed composition have appeared. These are: Obendorf *et al.* (1993) and Muzza (1993).

REFERENCES

Adachi, T. (1986) Is it possible to overcome the low yield of buckwheat by means of biotechnology? Some trials in plant genetic manipulation, in *Proceedings of 3rd International Symposium on Buckwheat*, Poland, pp. 106–16.

Adachi, T., Arie, K., Yabuya, T. and Nagatoma, T. (1981) On the incompatibility reaction and the seed set between dimorphic heterostyly and different ploidy in common buckwheat. *Fagopyrum*, **1**, 9–10.

Adachi, T., Kawabata, K., Matauzaki, N., Yabuya, T. and Nagatoma, T. (1983) Observation of pollen tube elongation, fertilization and ovule development in autogamous autotetraploid buckwheat, in *Proceedings of 2nd International Symposium on Buckwheat*, Miyazaki, pp. 103–13.

Alekseeva, E.S. (1984) Experimental mutagenesis as a method of breeding work in buckwheat. *Fagopyrum* **4**, 23–9.

Alekseeva, E.S. (1988) Application of chemical mutagens and radiation in breeding buckwheat for larger seeds. *Mutagen Breeding Newsletter*, **32**, 17–18.

Ali-Khan, S.T. (1970) Cooperative Tests of Buckwheat. *Report of the Canada Committee on Grain Breeding*, Ottawa.

Ali-Khan, S.T. (1971) Inheritance of leaf variegation in buckwheat. *Can. J. Genet. Cytol*, **13**, 736–41.

Ali-Khan, S.T. (1972) *Growing Buckwheat*. Publ. 1949, Agriculture Canada, Ottawa.

Ali-Khan, S.T. (1973) Effect of row spacing and seeding rate on yield of buckwheat. *Agron. J.*, **65**, 914–15.

Anokhiva, T.A. (1987) Breeding buckwheat for restricted growth process. *Selektsija i. Sememovods USSR*, **1**, 18–20.

Bailey, L.H. (1958) *Manual of Cultivated Plants*, Macmillan, New York.

Baszynski, T., Warcholowa, M., Krupa, Z., Tukendorf, A., Krol, M. and Wolinska, D. (1980) The effect of magnesium deficiency on photochemical activity of rape and buckwheat chloroplants. *Z. Pflanzenphysiol.*, **99**, 295–303.

Belova, L.A., Nechaev, A.P., and Severimenko, S.M. (1970) Fatty acid composition of buckwheat lipids. *Izv. Vysshucheb. Zaved. Pishch. Tekhnol.*, **1**, 32–3.

Bogataj, L.A. (1987) Study on the development and contribution of different plant organs to buckwheat dry matter production. *Fagopyrum*, **7**, 21–33.

Bogataj, L.K. (1988) A simple model for allocation of assimilates at different stages of buckwheat plant development. *Fagopyrum*, **8**, 23–6.

Bogataj, L.A. and Gaberscik, A. (1986) Analysis of net photosynthesis response curves for buckwheat. *Fagopyrum*, **6**, 6–8.

Bogataj, L.A. and Knava, M. (1985) Studies on the production of dry matter in the community of buckwheat with particular reference to leaf area. *Fagopyrum*, **5**, 7–12.

Bogdanovic, M. (1987) Effect of varietal geneotype fertilizer and sowing rate on grain yield of buckwheat. *Badovi Poljoprivednog Fakutela Univerzitata*, **4** (Sarajeru 35), 5–12.

Bohanec, B. (1985) A modified method for induction of shoot formation for micropropagation of buckwheat *in vitro*. *Fagopyrum*, **5**, 13–14.

Byung Han Choi, Keun Yong Park and Rae Kyeong Park (1992) Buckwheat genetic resources in Korea, in *Papers presented at IBPGR Workshop on Buckwheat Genetic Resources in East Asia*, BPER, Rome, pp. 45–52.

Campbell, C.G. (1983) Manor buckwheat. *Canadian J. Pl. Sci.*, **63**, 1053–54.

Campbell, C.G. and Gubbels, G.H. (1978) *Growing Buckwheat*. Publ. 1968, Agriculture Canada, Ottawa.

Chertko, N.K., Ivanov, N.P. and Ivakhenko, N.N. (1974) Effect of born microfertilizers on yield and quality of buckwheat. *Vestsi Akademii Navuk*, **3**, 47–9.

Coe, W.R. (1931) *Buckwheat Milling and the By-products*. Circ. of the United States Department of Agriculture, Washington D.C.

Darlington, C.D. (1956) *Chromosome Botany and the Origin of Cultivated Plants*, Allen and Unwin, London.

de Candolle, A. (1984) *Origin des Plantes Cultiveés*, Paris (English translation, 1886, Kegan Paul).

de Vries, F.T.P. and van Larr, H.M. (eds) (1992) *Simulation of Plant Growth and Crop Production*, Centre for Agricultural Publishing and Documentation, Wageningen.

Dorzhenko, L.I. (1988) Possible ways of applying meiosis selection in tetraploid buckwheat improvement. *Fagopyrum*, **8**, 64–78.

Eggum, B., Kreft, I. and Javornik, B. (1981) Chemical composition and protein quality of buckwheat. *Qual. Plant Foods Human Nutr.*, **30**, 175–9.

Elagin, I.N. (1959) On the problem of the origin of buckwheat. *Bot. Zh.*, **44**, 1177–81.

Elagin, I.N. and Kargaltsev, Y.V. (1972) The effect of trace elements on the harvest and quality of buckwheat. *Dokl. Vses. ord. Lenin Akad. S. KH Naukim V.I. Lenina*, **10**, 23–4.

Elagin, I.N. and Kargaltsev, Y.V. (1974) Effectiveness of treating buckwheat seeds with trace elements. *Dokl. Vses. ord. Lenin Akad. S. KH Naukim V.I. Lenina*, **12**, 16–17.

Esser, K. (1953) Genome doubling and pollen tube growth in heterostyled plants. *Z. indukt Abstammu Vereb. Lehr*, **85**, 28–50.

Farooq, S. and Tahir, I. (1987) Comparative study of some growth attributes in buckwheat cultivated in Kashmir. *Fagopyrum*, **7**, 9–12.

Fesenko, N.V. (1963) Breeding buckwheat for resistance to shedding. *Selektesiya i Semenovodstvo*, **2**, 34–41.

Fesenko, N.V. (1966) Breeding buckwheat for immunity to shedding. *Rec. Sci. Work. Ord. Agric. Exp. Stn.*, **3**, 154–79.

Fesenko, N.V. (1968) A genetic factor responsible for the determinate type of plants in buckwheat. *Genetika*, **4**, 165–6.

Fesenko, N.V. (1978) Features of intraspecific polymorphism in cultivated buckwheat and ways of using it in breeding. *Nauch. fr. VNII Zenobob-i Krupyan. Kuljitur*, **7**, 103–17.

Fesenko, N.V. and Antonov, V.V. (1981) Prospects of developing buckwheat breeding. *Selektesiya i Semenovodstvo*, **5**, 22–5.

Fesenko, N.V. and Koblev, S.U. (1983) Application of short-stemmed buckwheat from Ovlovsky kaslkik for resistance to lodging in breeding. *Fagopyrum*, **3**, 13–16.

Fesenko, N.V., Koblev, S.U. and Moltkov, A.A. (1986) An attempt to hasten the production of short stemmed buckwheat varieties. *Seleklesiya i Semenovodstvo USSR*, **6**, 25–6.

Ganyushina, E.V. (1972) Effect of urea on crop and biochemical processes of buckwheat. *Vestn. Mosk. Univ. Biol. Pochvoved.*, **27**, 102–5.

Garber, R.J. and Quisenberry, K.S. (1927) Self fertilization in buckwheat. *J. Agr. Res.*, **34**, 185–90.

Giacomini, V. (1955) Information and research on the genus *Fagopyrum*. *An. Inst. Bot. Univ. Pavia, Ser. 5*, **15**, 245–98.

Gohil, R.N. (1984) Buckwheat in India – past, present and future. *Fagopyrum*, **4**, 3–6.

Gohil, R.N., Rather, G.M., Tahir, I. and Farooq, S. (1983) Comparative cytology, growth and grain composition of West Himalayan buckwheat, in *Proceedings of 2nd International Symposium on Buckwheat*, Miyazaki, pp. 87–101.

Gonchavik, M.N. and Kozlova, A.P. (1972) Effect of chloride ions introduced in potassium fertilizer on the water system of buckwheat leaves. *Vyestsi Akad. Naruk SSR. Biyl. Navuk*, **1**, 113–17.

Gorina, E.D. (1969) Heterosis in local varieties of buckwheat. *Rec. Sci. Trans. White Russ. Sci. Res. Inst.*, **13**, 190–96.

Gubbels, G.H. (1977) Interaction of cultivars, sowing date and sowing rate on lodging, yield and seed weight of buckwheat. *Can. J. Plant Sci.*, **57**, 317 21.

Gubbels, G.H. (1980) Yield and seed weight of buckwheat after foliar application of boron and calcium. *Can. J. Plant Sci.*, **60**, 721–2.

Ikeda, K., Sugio, K., Arioka, K., Kusanno, T., Chkue, H. and Oku Mo. (1983) Proteinase inhibitors from buckwheat seeds, in *Proceedings of 2nd International Symposium on Buckwheat*, Miyazaki, pp. 195–8.

Jarotzky, R. (1928) Histologische und karylogische studien in Polygonaceae. *Jahrb. Wiss. Bot.*, **69**, 357–490.

Javornik, B., Eggum, B.O. and Kreft, I. (1981) Studies on protein fractions and protein quality of buckwheat. *Genetika*, **13**, 115–21.

Joshi, B.D. (1988) Underutilized food plants of hilly regions, in *National Seminar on Under-utilized and Under-exploited Plants*, New Delhi, 32.

Joshi, B.D. and Paroda, R.S. (1991) *Buckwheat in India*, NBPGR, Regional Station.

Kalra, Y.P. (1971) Different behaviour of crop species in phosphate absorption. *Plant and Soil*, **34**, 535–9.

Karstens, W.K.H. (1939) Anthocyanin and anthocyain formation in seedlings of *Fagopyrum esculentum*. *Recl. Trav. Bot. Neerl.*, **36**, 85–179.

Kephart, K.D., Murray, G.A. and Auld, D.L. (1990) Alternate crops for dryland production systems in northern Idaho, in *Advances in New Crops*, (eds J. Janick and J.E. Simon), Timber Press, Portland, Oregon, pp. 62–7.

Koblev, S.Y. (1987) Breeding buckwheat for resistance to lodging. *Referativayi Zhurnal*, **3**, 308–12.

Komeichi, M., Honda, Y. and Hayashi, H. (1992) Genetic resources of buckwheat in Japan, in *Papers presented at IBPGR Workshop on Buckwheat Genetic Resources in East Asia*, IBPER Rome, pp. 19–31.

Kovalanko, V.I. (1986) Phylogenetic relationships in species of the genus *Fago-*

pyrum and the genetic and evolutionary aspects of their systems of reproduction. *Referativayi Zhurnal*, **12**, 65–70.

Koyama, M., Tsujizaki, Y. and Sakamuro, S. (1973) New amides from buckwheat seeds. *Biol. Chem.*, **37**, 2749–53.

Kreft, I. (1986) Physiology of buckwheat yield, in *Proceedings of 3rd International Symposium on Buckwheat*, Poland, pp. 37–50.

Krol, M. (1986) History of buckwheat cultivation in Poland, in *Proceedings of 3rd International Symposium on Buckwheat*, Poland, pp. 7–17.

Krotov, A.S. (1963) *Buckwheat*, Izdatel stvo-sol Skonozjajstvennij Literatury, Moscow–Leningrad.

Kusiorska, K. and Koszykowska, T. (1981) Nutlet setting and development in buckwheat under field conditions. *Fagopyrum*, **1**, 7–8.

Ladefoged, A.V.N. (1952) Cultivation of buckwheat in up country Patne. *Trop. Agriculturist*, **108**, 269–70.

Lawrence, W.R. (1895) *The Valley of Kashmir*, London.

Luther, Z., Korejan, D. and Kreft, I. (1986) Breeding buckwheat with determinant growth habit, in *Proceedings of 3rd International Symposium on Buckwheat*, Poland, pp. 139–43.

Marciszewski, H., Suszko-Purzycka, A. and Kalembasa, S. (1984) The influence of nitrogen fertilization on flavonoid content buckwheat straw. *Fagopyrum*, **4**, 14–16.

Margana, U. and Margana, E. (1969) A suitable chromatographic method for the quantitative assay of rutin and flavone-c-glycosides in buckwheat seedling. *Esti. Nsv. Tead. Akad. Toim, Biol.*, **18**, 139–50.

Marshall, H.G. (1967) Registration of Pennquad buckwheat. *Crop Sci.*, **7**, 400.

Marshall, H.G. (1968) *Pennquad – a new buckwheat*. Pennsylvania State Univ. Rep. 281, 1–4.

Marshall, H.G. (1969) *Description and Culture of Buckwheat*, Pennsylvania State University, Agric. Exp. Station Bull. No. 754.

Marshall, H.G. (1970) Registration of Pennline 10 buckwheat. *Crop Sci.*, **10**, 726.

Marshall, H.G. (1979) Studies on inbreeding depression, breeding behavior and heterosis with inbred lines of buckwheat. *Crop. Sci.*, **19**, 110–14.

Martynenko, G.E. (1987) Variation in determinate population of buckwheat during breeding for morphological traits. *Referativayi Zhurnal*, **13**, 65–308.

Matano, T. and Ujihara, A. (1979) Agroecological classification and geographical distribution of the common buckwheat, *Fagopyrum eculentum* M., in East Asia, *Jap. Agric. Res.*, **13**, 157–62.

McGregor, S.E. (1976) McGregor, S.E. (1976) Insect pollination of cultivated crop plants, in *Agric. Hand Book, ARS—USDA*, Washington, D.C., pp. 119–22.

McLachlan, K.D. (1976) Comparative phosphorus responses in plants to a range of available phosphorus situations. *Aust. J. Agric. Res.*, **27**, 323–41.

Minami, H. and Namai, H. (1986) Potential genetic varieties of flowering time in late-summer type cultivars of buckwheat in Kyushu region. *Jap. J. Breeding*, **36**, 67–74.

Monsurova, V.V. (1948) Comparative caryology of two buckwheat species, *Fagopyrum esculentum* and *F. emarginatum*, in *Proc. Acad. Sci. USSR*, 61, 19–22.

Morrall, R.A.A. and McKenzie, D.L. (1975) Diseases of speciality crops in Saskatchewan. Notes on buckwheat and sunflower. *Can. J. Plant Dis. Surv.*, **55**, 69–72.

Morris, M.R. (1947) Genetic Studies on Buckwheat (*Fagopyrum* species), Ph.D. thesis, Cornell University, Cornell, NY.

Morris, M.R. (1951) Cytogenetic studies on buckwheat. *J. Hered.*, **42**, 85–9.

Morrison, F.B. (1945) *Feeds and Feedings*. Morrison Publ. Co., Ithaca, New York.
Morton, B.R. (1966) A Study of Pollination and Pollen Tube Growth in Buckwheat, *Fagopyrum sagittatum*. M.S. thesis, Pennsylvania State Univ.
Munshi, A.H. (1982) A new species of *Fagopyrum* and its species delineation in Himalayas, in *Proceedings of 3rd International Symposium on Buckwheat*, Poland, pp. 162–7.
Murakami, K. and Usaburo, M. (1950) Study on physiology of artificial polyploids. *Jap. J. Breeding*, **1**, 76–80.
Muzza, G. (1993) Storage, processing and quality aspects of buckwheat seed, in *New Crops* (eds J. Janick and J.E. Simon), pp. 251–4.
Nagatomo, T. (1959) Studies on breeding buckwheat. 1. Flora arrangement and flowering habit, *Jap. J. Breeding*, **8**, 238–46.
Nagatomo, T. (1965) Buckwheat from Afghanistan and Iran, in *Cultivated Plants and their Relatives*, (ed K. Yamashita), Kyoto University, Japan, pp. 192–200.
Nagatomo, T. (1983) Buckwheat in relation to Japanese culture. *Fagopyrum*, **2**, 23–7.
Namai, H. (1990) Pollination biology and reproduction ecology for improving genetics and breeding of common buckwheat. *Fagopyrum*, **10**, 23–46.
Namai, H. (1992) Strategies for sustainable conservation and efficient utilization of buckwheat genetic resources in the world, in *Papers presented at IBPGR Workshop on Buckwheat Genetic Resources in East Asia*, IBPER, Rome, pp. 93–104.
Nicholson, J.W.G., McQueen, R., Grant, E.A. and Burgess, P.L. (1976) The feeding value of tartary buckwheat for ruminants. *Can. J. Anim. Sci.*, **56**, 803–8.
Nishiyama, K., Kuwakino, K. and Miura, M. (1983) Amino acid composition of five buckwheat cultivars, in *Proceedings of 2nd International Symposium on Buckwheat*, Miyazaki, pp. 221–2.
Obendorf, R.L., Horbowicz, M. and Taylor, D.P. (1993) Structure and chemical composition of developing buckwheat seed, in *New Crops* (eds J. Janick and J.E. Simon), pp. 244–50.
Ohnishi, I. (1983) Isozyme variation in common buckwheat *Fagopyrum esculentum* and its related species, in *Proceedings of 2nd International Symposium on Buckwheat*, Miyazaki, pp. 39–50.
Ohnishi, O. (1986) Evaluation of world buckwheat varieties from the viewpoint of genetics, breeding and origin of buckwheat cultivation, in *Proceedings of 3rd International Symposium on Buckwheat*, Poland, pp. 198–213.
Ono, T., Sato, T. and Odagiri, S. (1978) Albumins in buckwheat seed. *Agric. Biol. Chem.*, **42**, 1779–80.
Paroda, R.S. and Mal, B. (1989) New plant sources for food and industry in India, in *New Crops for Food and Industry*, (eds G.E. Wickens, N. Haq and P. Day), Chapman & Hall, London, pp. 135–49.
Petilina, N.N. (1950) Methods of producing buckwheat varieties giving high and regular yields. *Breeding and Seed Growing*, **7**, 37–41.
Petilina, N.N. (1980) Selection of elite plants in breeding buckwheat. *Selektesiya i Semenovodst*, *4*, 13–14.
Pisarey, V.E. and Vivogranova, N.M. (1952) New ways of breeding buckwheat. *Breeding and Seed Growing*, **19** (2), 35–8.
Pomeranz, Y., Marshall, H.G., Robbins, G.S. and Gilbertson, J.T. (1975) Protein content and amino acid composition of maturing buckwheat (*Fagopyrum esculentum*). *Cereal Chem.*, **52**, 479–85.
Potszyinski, M. (1984) The effect of nitrogen fertilization on the quality of buckwheat seeds. *Fagopyrum*, **4**, 39–40.

Quisenberry, K.S. (1927) Chromosome number in buckwheat species. *Bot. Gaz.*, **83**, 85–8.

Ren, S. and Liu, A. (1986) A survey of cultivated buckwheat, pollen spreading distance and the relation between pollen carried by insects and yield, in *Proceedings of 3rd International Symposium on Buckwheat*, Poland, 10–17.

Ruszkowski, M. (1980) The possibility of changing the yielding ability of buckwheat by breeding the homostyle varieties, in *Proc. Inst. Intl. Symp. Buckwheat*, Lijubijana, pp. 7–15.

Ruszkowski, M. and Noworolnik, K. (1983) The productivity of various homostyle, homomorphic lines of buckwheat, in *Proceedings of 2nd International Symposium on Buckwheat*, Miyazaki, pp. 79–86.

Saharov, V.V., Frolova, S.L. and Modsurova, V.V. (1944) Production of higher fertile tetraploid buckwheat. *C.R. (Doklad) Akademia Nauk USSR*, **44**, 254–6.

Sando, W.J. (1935) A colchicine-induced tetraploid in buckwheat. *J. Hered.*, **30**, 271–2.

Sando, W.J. (1956) *Buckwheat Culture*. USDA Farmers' Bulletin, No. 2095.

Sato, H. and Sakamura, S. (1975) Isolation and identification of flavonoids in immature buckwheat seed (*Fagopyrum esculentum*). *Nippon Nogeikagaku Kaishi*, **49**, 53–5.

Sharma, K.D. and Boyes, J.W. (1961) Modified incompatibility of buckwheat following irradiation, *Can. J. Bot.*, **39**, 1241–6.

Scherchand, K. (1992) Buckwheat genetic resources in Nepal, in *Papers presented at IBPGR Workshop on Buckwheat Genetic Resources in East Asia*, IBPER, Rome, pp. 75–86.

Shustova, A.P. (1962) Boron requirements of buckwheat. *Fiziol. Rast.*, **8**, 553–7.

Singh, H.B. and Thomas, R.A. (1978) *Grain Amaranths, Buckwheat and Chenopods*. Indian Council of Agricultural Research, New Delhi.

Singh, V.P. and Atal, C.K. (1982) Cultivation of buckwheat in the plains as a raw material for rutin, in *Cultivation and Utilization of Medicinal Plants*, Publication and Information Directorate, CSIR, pp. 337–41.

Singh, V.P. and Mall, S.L. (1977) Seed germination studies in *Fagopyrum esculentum*, in *Proceedings of Indian National Science Academy*, **43**, 37–43.

Sokolov, O.A. and Semihov, V.F. (1968) Protein samples of *Polygonum* seeds dependent on ploidy and nutrition conditions. *Fiziol. Biokhim. Sorta.*, pp. 80.

Sokolov, O.A. and Semihov, V.F. (1983) Nitrogen uptake by buckwheat in nitrogen fertilizer local application. *Fagopyrum*, **3**, 17–18.

Srejovic, V. and Neskovic, M. (1981) Regeneration of plants from cotyledon fragments of buckwheat. *Z. Pflanzenphysiol.*, **104**, 37–42.

Srejovic, V. and Neskovic, M. (1983) Vegetative propagation of buckwheat *in vitro*. *Fagopyrum*, **3**, 5–6.

Strong, W.A. and Soper, R.J. (1973) Utilization of pelleted phosphorus by flax, wheat, rape and buckwheat from a calcareous soil. *Agron. J.*, **65**, 18–21.

Strong, W.A. and Soper, R.J. (1974) Phosphorus utilization by flax, wheat, rape and buckwheat from band or pellet like application. *Agron. J.*, **66**, 597–601.

Sugawara, K. (1956) On buckwheat pollen III. The relation between pollen germination and temperature, *Proceedings of Crop Science Society of Japan*, **24**, 264–5.

Sugawara, K. (1960) On the sterility of buckwheat with special reference to the accumulation of starch in the stem, *Proceedings of Crop Science Society of Japan*, **28**, 313–15.

Tahir, I. and Farooq, S. (1983) Growth and yield characteristics of some buckwheats from Kashmir. *Geobios*, **10**, 193–7.

Tahir, i. and Farooq, S. (1987) Occurrence of phenolic compounds in four cultivated buckwheat with particular reference to grain quality. *Fagopyrum*, **7**, 7–8.

Taranenko, L.K. (1986) Heterosis in the buckwheat selection, in *Proceedings of 3rd International Symposium on Buckwheat*, Poland, pp. 145–53.

Tatebe, T. (1958) Physiological studies in fertility in buckwheat VI. An investigation on the mechanism of heterostyly. *Jap. J. Breeding*, **8**, 149–54.

Taylor, R.S. (1989) Alternative crops for Europe, in *New Crops for Food and Industry*, (eds G.E. Wickens, N. Haq and P. Day), Chapman & Hall, London, pp. 185–93.

Troyer, J.R. (1958) Anthocinin pigment of buckwheat hypocotyls, *Ohio J. Sci.*, **58**, 187–8.

Tsuzuki, E., Shimiya, M. and Shida, S. (1983) Studies on the matter production of buckwheat plants I. Difference in dry matter and yield production between diploid and tetraploid plants in buckwheat, in *Proceedings of 2nd International Symposium on Buckwheat*, Miyazaki, pp. 165–71.

Tsvetoukhine, V. (1952) *Buckwheat and its improvement possibilities*, Annls. INRA, Paris, 1, 99–115.

Wilson, H.K. (1948) *Grain Crops*, McGraw Hill, New York.

Yang Keli (1992) Genetic resources of buckwheat (*Fagopyrum*) in China, in *Papers presented at IBPGR Workshop on Buckwheat Genetic Resources in East Asia*, IBPER, Rome, pp. 3–18.

Yarosh, N.P. and Artemeva, A.E. (1977) Variability in vitamin B1, B2 and rutin content in buckwheat seeds depending on cultivar and conditions of mineral nutrition. *Fiziol. Biokhim. Kult. Rast.*, **9**, 48–52.

Yurui, L. and Zhongging, W. (1984) Production of buckwheat in the autonomous province of Inner Mongolia. *Fagopyrum*, **4**, 7–13.

Zheleznov, A.V. (1967) Use of inbred line hybridization in buckwheat selection. *Genetica*, **12**, 12–19.

Zimmer, R.C. (1974) Chlorotic leaf spot and stipple spot, newly described disease of buckwheat in Manitoba. *Can J. Plant Dis. Surv.*, **54**, 55–6.

Zimmer, R.C. (1978) Downy mildew, a new disease of buckwheat in Manitoba. *Can. Plant Dis. Rep.*, **62**, 471–3.

3

Grain amaranth (*Amaranthus* species)

J.T. Williams and D. Brenner

3.1 HISTORY AND USES

In ancient times grain amaranths were basic crops in the Americas, and they have persisted to the present in isolated pockets in traditional agriculture. Amaranths are broad-leaved plants which are striking in appearance due to various coloration of all parts of the plant. They produce small seeds in large quantities.

The major grain-producing species are *Amaranthus hypochondriacus* L. and *A. cruentus* L. indigenous to Mexico and Guatemala and extending into the south-western USA, and *A. caudatus* L. native to the Andes of Peru and extending northwards to Ecuador and southwards to Bolivia and Argentina. Although all three of these species can be used as multipurpose plants, because they have edible leaves, *A. cruentus* has become an important vegetable species especially in other parts of the world.

In addition to the three major grain species, *A. hybridus* L., a weedy species indigenous to tropical America but now spread widely throughout the tropics and even more temperate areas, is very commonly used as a vegetable. Some selections for grain make it a useful species for hybridization with *A. hypochondriacus* and *A. cruentus* because of its shorter stature and earlier maturity, and it is a species which is already an incipient domesticate.

There are many other species of *Amaranthus* which have not been cultivated but which have been or are used as pot herbs or famine foods. Several ancient writings from China, India, Egypt and Greece attest to the use of amaranths as pot herbs or as medicines.

Grain amaranths, in recent centuries, were taken to Asia, and in the

Cereals and Pseudocereals. Edited by J.T. Williams. Published in 1995 by Chapman & Hall, 2–6 Boundary Row, London SE1 8HN. ISBN 0 412 46570 1.

hill regions of India, Pakistan, Nepal, China and Tibet became popular with local peoples. They are not restricted to the hill areas since it is thought that the early introductions to the Indian subcontinent were through Sri Lanka (Sauer, 1967) from whence the crops spread rapidly.

In the Himalayas, *A. hypochondriacus* and *A. caudatus* have been widely introduced and have become a major crop in the north-west of India. They have also become a minor crop in the northern plains, and in central and southern India. Grain amaranth has recently been introduced to Thailand.

In Africa, *A. cruentus* became an important vegetable and is also a garden ornamental. *A. hypochondriacus* and *A. caudatus* have been introduced to Ethiopia and eastern Africa as a minor grain. There is some interest in west Africa in increasing grain use of amaranths.

The use of the grain in speciality foods has caused major expansion of cultivation in the USA, Canada and there is interest for the same purpose in Europe and New Zealand. In the USA the major grain production areas are the semi-arid high plains area of the west especially in Nebraska (Figure 3.1) (Stallknecht and Schulz-Schaffer, 1993). Much of the expanded cultivation and supporting research has been in parallel with a report – and subsequent action thereon – by the National Academy of Sciences of the USA (NAS) (1975), since initial research was

Figure 3.1 Commercial amaranth grain production in Nebraska, USA.

already underway at the Rodale Research Center at Kutztown, Pennsylvania, when the report was issued.

As a result of this report on underutilized crops, assistance was provided through a competitive grants programme of the US Agency for International Development for research and development of grain amaranths in Mexico and Guatemala; Peru; Nepal, Thailand; and Kenya. This support started in the early 1980s. By 1980, India had coordinated research on grain amaranths through the National Bureau of Plant Genetic Resources and an All-India Coordinated Research Programme on Under-utilized and Under-exploited Plants, both under the auspices of the Indian Council of Agricultural Research.

Other countries with active research and development on this crop include Argentina and Venezuela. China started its own programme in the early 1980s but emphasis was on use of grain forms for fodder rather than human nutrition. Countries in Latin America, Africa and Asia with substantial plantings by 1990 include Mexico, Kenya, India and China. In all other countries plantings were small or remained associated with subsistence agriculture. By 1990, Mexico, Kenya and China were producing for export as well as internal use (Kauffman, 1990).

A major aspect of the research and activities has been the exchange of germplasm. The Rodale Research Center alone has provided materials to 607 scientists and organizations in 113 countries in the period 1982–89 (Kauffman, 1992). By 1980, five countries had commercial seed suppliers and 17 countries had germplasm collections available. By the time the NAS issued a promotional book on amaranths (National Research Council, 1984) they were able to list scientists involved with crop improvement: between one and five researchers were listed in 20 countries, between 10 and 20 in a further two countries (India and Nigeria) and there were more than 40 in the USA.

By the end of the 1980s modern cultivars – rather than landrace material – had been developed in India, USA and Peru and active breeding is underway in a number of other countries, especially Argentina (Covas, 1987), China, Thailand, India and Kenya.

3.1.1 Centres of origin and variation

In the centres of origin in the Americas the grain forms of the species were selected for those with light-coloured seeds, (white, pink or gold). A summary of the variation both in the ancestral areas of cultivation and those where the crop has been introduced is provided below.

A. caudatus (Figure 3.2) originated in the Andes in the same area where potato was domesticated. It was a staple grain in pre-conquest times and there is evidence that it was much more widespread in culti-

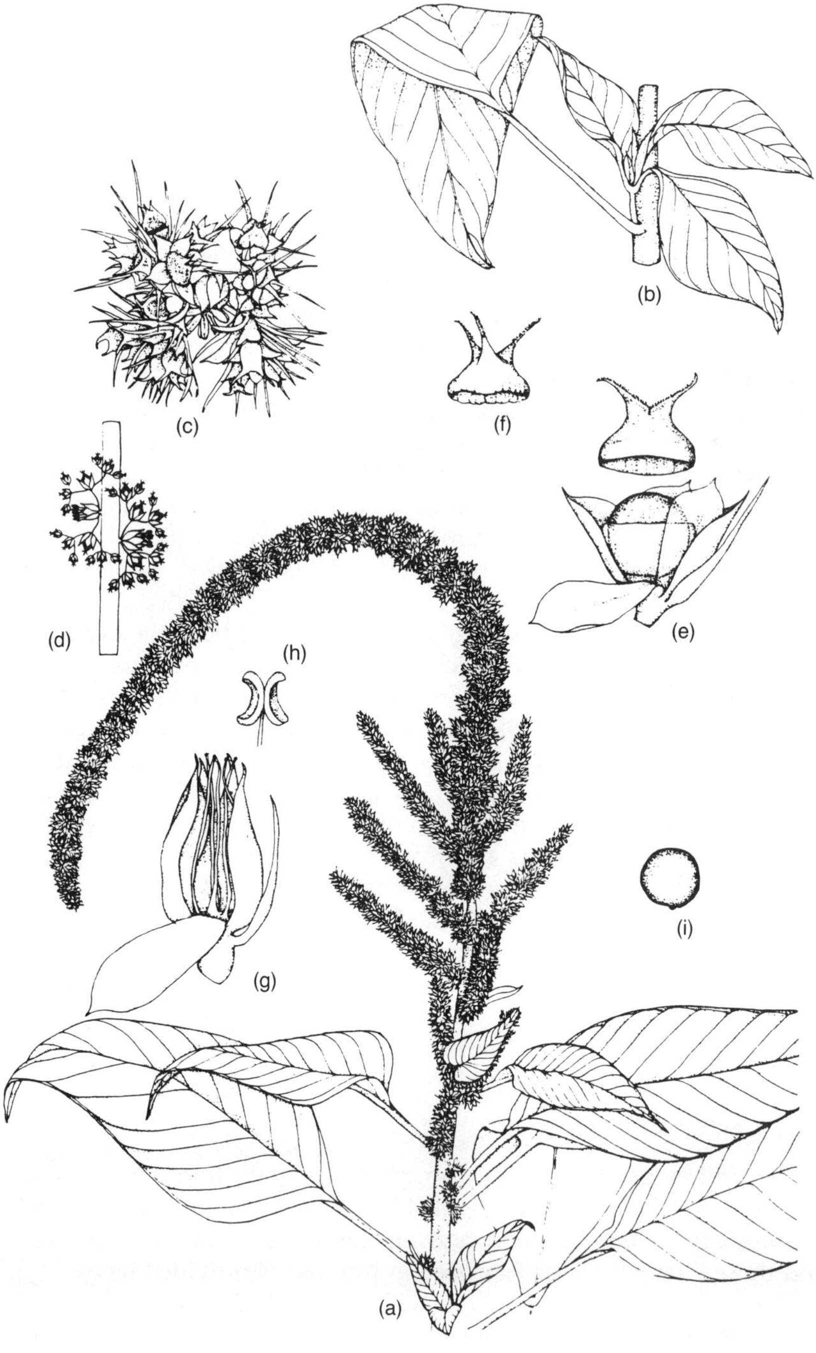
(b)
(c)
(f)
(d)
(e)
(h)
(i)
(g)
(a)

vation in the past. It is not found in the wild but is cultivated or found as an escape (Coons, 1982) especially in the highlands of Ecuador, Bolivia, Peru and north-west Argentina. In the last area, a species, *A. edulis* (syn. *A. mantegazzianus*), was the major amaranth but this is now thought to be a variety of *A. caudatus*. It has club-shaped inflorescences with the flowers in ball-like clusters whereas the species normally has long, loose inflorescences with the flowers crowded on the spikes and the inflorescence hangs down as flowering advances.

Since the inflorescences are highly coloured, horticultural varieties have been developed and are generally known as love-lies-bleeding.

Sauer (cited in Cole, 1979) records that Spaniards introduced the plant to Europe and by about 1700 it was known to be an occasional grain crop in central Europe and Russia. By the early nineteenth century it had been taken to Africa and Asia and had remained in Europe as an ornamental, and in countries such as Ethiopia as a minor and very local grain (Getahun, 1974).

The major variation is still to be found in the Andes, and this has been sampled for crop improvement. The Himalayan region has produced important secondary diversity.

Selection in the Andes led to pale-coloured seeds in grain landraces (Hauptli *et al.*, 1980). They may be white or white with a pink rim and pale seeds have been found in tombs dating from over 2000 BP (Hunziker and Planchuelo, 1971). However, there are forms with dark seeds and in the centre of diversity seven seed colours can be recognized (Hauptli *et al.*, 1980). Other morphological variation in the Andes relates to type of inflorescence and pigmentation. There are also hybrid swarms between *A. caudatus* and some weed amaranth species (Hauptli *et al.*, 1980).

Local forms are used for a variety of purposes in the Andes: grain, leaf vegetable, food colouring, medicinal purposes and, in the southern part of the species range, as animal feed. There are clear ethnobotanical data showing uses are localized and not necessarily dependent upon the type of plant as much as upon local human culture.

Massive dwindling of cultivation in the Andes has been related to three factors. First, there was opposition from the conquistadors to local crops especially if they were used in magical/religious rites. Second, changes have taken place in Amerindian culture; and third, there have

---

Figure 3.2 *Amaranth caudatus*. (a) branch with inflorescence (×23); (b) detail, axilla (×23); (c) front view of two glomerules (×4); (d) scheme of glomerules; (e) adult male flower with dehiscent cap (×8); (f) cap with three stigmata (×8); (g) male flower (×8); (h) anther (×8); (i) seed, lateral view (×6) (Courtesy of Royal Tropical Institute, The Netherlands.)

Table 3.1 Acreage of indigenous crops in the Peruvian highlands 1980s

	Total (ha)
Grain crops	
Quinoa	15000
Chenopodium pallidicaule	5000
Amaranths	500
Root crops	
Andean roots and tubers	30000
Legumes	
Lupins	5000

Source: Modified from Tapia and Mateo (1990).

been moves to modernize agriculture away from a subsistence base. Hence amaranth remains a very minor crop even in areas where it was domesticated (Table 3.1).

A. cruentus (Figure 3.3) originated in Mexico and Guatemala and by prehistoric times had reached cultivation in the south-west USA. It was domesticated for grain, as a vegetable and as a dye plant. In general, grain forms had light seeds and vegetable forms had dark seeds. From a survey of uses, *A. cruentus* appeared more widely used in Guatemala and probably originated as a domesticate in that area. It has spread out through South America as both an ornamental and vegetable and in its primary centre of diversity and and in South America it is cultivated in lowland wetter areas. In the south-west USA it was cultivated as a red dye plant, e.g. in the Hopi irrigated terraces of the Rio Grande Pueblos. Some forms appear to be pale leafy forms and there is no hard evidence for grain cultivation (Nabham, 1980).

In the nineteenth century *A. cruentus* was used as an ornamental and for leaf greens in many tropical areas of the world. In Africa it became a relatively important vegetable and Sauer recorded how its introduction outran European exploration of the continent (Cole, 1979). The species remains an ornamental in Europe. Most of the introductions to other parts of the world in the eighteenth and nineteenth centuries were dark-seeded forms.

Figure 3.3 *Amaranthus cruentus.* (a) inflorescence (×23); big leaf (×23); (c) branch with axilla (×23); (d) male flower, one tepal removed (×7); (e) female flower with dehiscing cap (×7); (f) glomerule (×7); (g) scheme of glomerules; (h) seed, lateral view (×7); (i) anther (×7) (Courtesy of Royal Tropical Institute, The Netherlands.)

(b)
(d)
(i)
(g)
(e)
(c)
(f)
(a)
(h)

More recently *A. cruentus* has been introduced to Thailand and Kenya for grain and it is of minor importance, compared with other species, for grain in India.

The archaeological record documents *A. cruentus* in the Tehuacan caves in an arid part of Mexico. Whole plants and seed were available dating from 5500 BP to the Spanish conquest. Seeds are pale in colour (Sauer, 1969).

Hauptli *et al.* (1980) found most remaining uses of *A. cruentus* in Guatemala where it is mostly a vegetable, and grain types have been lost recently; they have however persisted in Mexico.

A. hypochondriacus was most likely domesticated in central Mexico in an area farther north than the centre for *A. cruentus*. Evidence from the Tehuacan caves shows it to be a later domesticate and it appeared about 1500 BP as a light-seeded crop (Sauer, 1969). It was probably domesticated by 400 AD. From Mexico, cultivation extended to the south-west USA where it was prehistorically associated with the Salado culture and historically with the Southern Paiute (Sauer, 1967; Nabham, 1980). In Mexico it has persisted as a grain crop in the tropical highlands, e.g. the Sierra Madre Occidental.

This species was introduced to Europe before 1600 AD and has persisted as an ornamental. Grain types were taken to Asia by 1800 and within 50 years had become a staple of hill tribes in southern India and the Himalayas. By 1900 it was in cultivation from Iran and Afghanistan through the Himalayas to China and eastern Siberia. It remains the major grain amaranth species in the Himalayas, although sometimes it is also used as vegetable. Whilst cultivation was expanding in Asia, that in the primary centre was in decline.

A. hypochondriacus is the heaviest yielding of the grain amaranths (National Research Council, 1984) and wide diversity is seen in plant height, pigmentation and inflorescence morphology in the Americas, and secondary diversity has developed in the Himalayas.

3.1.2 **Cultural uses**

There is well-documented information on the use by Mexicans of ancient times making idols from amaranth seed, and later rosary beads were moulded from amaranth dough (Cole, 1979; Early, 1978). In the Codex Mendoza of 1541, two-thirds of the towns of the Aztec Empire had amaranth as a necessary part of the annual tribute to Montezuma II (Sauer, 1967). Amaranth was therefore a crop used for important ritual as well as a cash crop.

Amaranth was in the past, and has remained, a crop associated with slash and burn in tropical Aztec-speaking areas of Zongolica, Veracruz, Mexico and the Yanatile valley of south-west Peru. In temperate parts of Mexico it developed as a monocrop cash crop and in temperate parts of Peru as a smallholder crop (Early, 1990).

The commonest use of grain amaranth in the Latin American centres of diversity has been as a snack ('alegrias' in Mexico, 'turrones' in Peru) using popped seeds and molasses; and the second most important use is as flour. Other more minor uses include dye, medicines and for beer production ('chica' in Peru).

3.1.3 **Modern uses**

The major international attention which has been focused on grain amaranths for more than 15 years has been geared to commercialization of the crop and modernization of traditional cultivation as well as expansion, because of the nutritional attributes, to new areas of the developing world.

In more recent years there has been research interest in plant residues, which remain after the grain has been harvested, as animal feed. Amaranth residues are high in lysine and leucine but relatively low in sulphur-containing amino acids. A substantial literature now exists on both fodder and grain use for livestock (Mallory *et al.*, 1990). Additionally international attention has been focused on vegetable use of amaranths, especially in the tropics (National Academy of Sciences, 1975; Grubben, 1977). As breeding progresses, grain cultivars will rarely be produced for multipurpose plant use; also the vegetable species are distinct from the grain ones except for some overlap in less developed agriculture. We can expect to see much greater emphasis on grain types which are underutilized, and also on better agronomy and production of amaranth vegetables, which are, in any case, hardly underutilized in the tropics.

3.2 TAXONOMY AND ECOLOGY

3.2.1 **Taxonomy**

There are about 60 species in the genus *Amaranthus* L. (Willis, 1973), the type genus of the family Amaranthaceae. One subfamily, Amaranthoideae, contains respectively Celoseae with the cultivated vegetable genus *Celosia* and Amarantheae with the genus *Amaranthus*.

The species grown for grain and also those used for pot herbs are all herbaceous annuals. The taxonomy is still somewhat confused but mostly for the wide range of species used as vegetables. The grain species are fairly well sorted out taxonomically and the synonymy has been understood since the work of Sauer (1950; 1967), Hunziker (1952; 1987), Coons (1982) and Feine (1980). There is still some research needed to be completely clear on the progenitor species. A comprehensive worldwide taxonomic revision of the species of *Amaranthus* would aid in the use and conservation of wild germplasm. Only a checklist is presently available, based on work by James Lehmann (Brenner, 1990).

The wild progenitor species proposed have been *A. quitensis* HBK, a river bank pioneer species in upland tropical South America for *A. caudatus*; *A. hybridus* L, also a weedy river bank pioneer in the uplands of South America and wetter parts of eastern North America, for *A. cruentus*; and *A. powellii* S.Wats., a pioneer species of open situations in canyons and desert washes in the western Cordillera of the Americas, for *A. hypochondriacus*.

Relationships between these wild species are not fully known. The greatest remaining confusion in grain amaranth taxonomy is with the boundaries of *A. hybridus*. Sauer (1967) recognizes *A. quitensis* as distinct from *A. hybridus*, and Jain *et al.* (1986) have supporting evidence. Coons (1982) reduces it to synonymy with *A. hybridus*, and Kauffman and Reider (1986) include it in *A. hybridus* as the morphological type 'Sangorache'. The second confusion involves some black-seeded cultivated plants from the Himalayas that have been used in grain crop plant breeding at the Rodale Research Center and are listed (Kauffman and Reider, 1986) as *A. hybridus* the morphological type 'Prima'. The Prima types could be black-seeded forms of *A. hypochondriacus* because both have large seed-heads and are not weedy.

The grain types usually have monoecious compound inflorescences and the flowers are 5-merous with circumscissile utricles. However, Sauer (1967) has found three-anthered and also non-circumscissile *A. powellii*, and 'polymerous staminate' flowers have been found in the subspecies *mantegazzianus*. The non-circumscissile character is being used in a breeding project to reduce seed-shattering (Brenner, 1990a). The flowers are arranged in glomerules, each with an initial staminate flower and an indefinite number of pistillate flowers. The glomerules are crowded on axes which make up the spikes of the inflorescence. Each flower is subtended by a bract with a sharp point. The perianths of the flowers are made up of five free tepals; male flowers have five stamens, the female a single utricle.

Using a few of the characters listed above, the grain species can readily be distinguished. Fuller morphological descriptions are provided in section 3.3.

Bracts equal to or slightly longer than style branches
A. hypochondriacus

Bracts shorter than style branches
  Styles erect and parallel, all tepals acute
  A. cruentus

Styles spreading, inner tepals spatulate, obtuse or retuse

Inflorescence arms pendulous, erect, or spreading, usually red, pink, or green; the inflorescence arms are indeterminate and tapering, the central inflorescence branch exceeds the side branches

A. caudatus

Inflorescence compact, orange; the inflorescence branches have knobby determinate ends, the central inflorescence branch is no more exerted than the side branches

A. caudatus subsp. *mantegazzianus*

The synonymy of the species is as follows:

A. caudatus L.

subsp. *caudatus*

subsp. *mantegazzianus* (Passer.) Hanelt syn. *A. edulis* Spegazzini

A. creuntus L. syn. *A. paniculatus* L.

A. hypochondriacus L. syn. *A. leucocarpus* S. Watts; *A. frumentaceous* Buchann Hamilton; *A. flavus* L.

Many interspecies hybrids have been reported among the grain species and wild species. For instance, *A. caudatus* var. *atropurpurea* is an unstable hybrid segregate of subspecies *caudatum* and a wild species. Sauer has suggested an alternative to three independent domestications. First, a primary domestication of *A. cruentus* from *A. hybridus*. Second, secondary domestications occurred as the primary cultigen moved to new areas and crossed with other wild species, *A. powellii* and *A. quitensis*. Thus *A. hybridus* would be the prime progenitor.

3.2.2 **Ecology**

Most primitive cultivation of grain amaranths was in upland areas, such as Sierra Madre of Mexico, the Andes and secondarily in the Himalayas, but it also spread to subtropical lowland. The genepool can be found adapted to tropical, subtropical and temperate regions. Ancient cultivation was within 30 ° latitude of the equator but it is now grown at higher latitudes using strains that will initiate flowering in spite of the longer day length than that of the tropics (National Research Council, 1984).

Altitude is not a limitation. Grain amaranths grow from sea level to 3500 m. *A. hypochondriacus* grows from low to high elevations. *A. cruentus* in lower to middle elevations (0 to about 2000 m) and *A. caudatus* largely at higher elevations. *A. cruentus* is less adapted to high areas than the other two species.

Within the genus the weedy species tend to differ from the grain species in preferring bright sunlight and open places. Many are aggressive weeds, including *A. hybridus*, which is closely related to the genepool of the cultigens.

Soils

The crop grows over a wide range of soil conditions, from acid, high aluminium to alkaline soils; and from coarse to fine-textured soils. There has been an interest in drought resistance since some strains grow in semi-arid conditions. It seems to be more ideally suited to alternating wet–dry conditions so that it is sown at optimal moisture and then matures into a dry season. Soil temperature needs to be about 16–18 °C or more for germination but the later growing period can be cooler. Hence the crop finds a role as a late-season crop in a monsoon period or in areas where there is a good spring rainfall or early season irrigation.

In general grain amaranths require well-drained soils which are neutral or basic. Germplasm which has been adapted to tropical lowlands shows more tolerance of acid conditions and some specific accessions show adaptation to acidic clays of the hill areas of India (Joshi and Rana, 1991).

Some strains of amaranths can tolerate mild salinity. Initial experimentation on response to aluminium in acid soils shows that mass selection and breeding can increase the tolerance to high aluminium levels but this may be inversely related to some desirable agronomic traits in *A. hypochondriacus* and directly related to them in *A. cruentus* (Campbell and Foy, 1987).

The overall grain genepool exhibits remarkable ranges of adaptation to overall ecology: drought and saline-alkali tolerance as well as heat and cold tolerance. More research needs to be done on characterizing agro-ecological zones of amaranths, especially since substantial germplasm collections now exist which can be screened for a range of ecological tolerances and since very few data exist on ecological parameters of collection sites.

The Latin American uplands

In Latin America and the Caribbean a little over 50% of the land is classed as mountainous. If these areas are scored in relation to soils, important facts emerge in relation to crop improvement. Table 3.2 shows the soils of the upland areas of some Amerindian classification systems related to diverse cultural groups. Within the soil types in Table 3.2 these, along with the major broad ecologies related to altitude and climate, would be worthy of analysis in relation to grain amaranth cultivation. This might help to target expansion in certain areas: in

Table 3.2 Soils of upland areas of selected Latin American countries

Country	% Good soils	% Poor soils Deep	% Poor soils Shallow	% Lithosols
Peru	16	12	0	72
Bolivia	42	13	0	45
Ecuador	32	30	0	37
Colombia	14	30	7	29
Venezuela	21	62	6	11
Guatemala	35	14	51	0

Source: Novoa and Posner (1981).

Guatemala alone, 40% of the national population farm in the uplands and 25% of the populations in Peru and Ecuador.

Such further information opens interesting possibilities of addressing a number of ecological constraints to introduction of grain amaranths to other areas, because, apart from yield, crop establishment is the major constraint. This is undoubtedly related to ecology since it is due to a mixture of compounded factors: moisture, temperature, soil depth and crusting. Genotypes with specific compounded adaptations undoubtedly exist.

3.3 MORPHOLOGY AND ANATOMY

Grain amaranths are broad-leaved, erect annuals, simple or branched, up to 3 m tall with monoecious compound terminal inflorescences and bracteate flowers with five tepals.

Inflorescences are spikes with a degree of branching. The growth of the inflorescence and its branches is indeterminate and hence can reach long lengths except for *A. caudatus* subsp. *mantegazzianus* which has a determinate inflorescence. Along the spikes, flowers cluster in glomerules, which develop alternately and these are mostly on the terminal inflorescence and feebly developed in axils of branches.

There are specific morphologies of the glomerules. Each is technically a dichasial cyme where the first flower is terminal on the branch and at its base, on either side, two secondary branches develop with the second and third flowers. Both these flowers become terminal and at their base develop the next two flowers. This pattern persists to maturity in *A. caudatus* but in *A. hypochondriacus* and *A. cruentus* changes over to a monochasial branching. These arrangements are reflected in the final morphologies of the inflorescences (Pal, 1972).

The three grain species are characterized by inflorescence type and other characters associated with bracts, as well as tepal shape and form, leaf morphology and gross plant morphology (Feine, 1980).

A. caudatus

This species is up to 2(1.5–2.5) m tall at maturity, sparsely branched (depending on plant density) and the branches at right angles to the stem may start low on the plant depending on genotype. Plants tend to be overall triangular in shape and the main stem is large. Leaf shape is variable from elliptic, ovate or lanceolate to rhombic ovate and tips are acute to obtuse and bases cuneate. The inflorescence is lax and usually long and drooping (although sometimes semi-erect or erect), very long, 0.9–1.5 m, with a knobbly appearance due to the large glomerules spaced relatively far apart. Bracts are short, not as long as the style branches of female flowers and they have broad laminas and a slender midrib. Tepals of flowers are strongly recurved; outer tepals are obovate. Inner tepals are spatulate with a narrow claw. Tepal apices are mucronate or obtuse. Style branches are spreading and meet at their base in a saddle. Seeds are white, white with pink rim, red or black and are rarely more than 1 mm in diameter.

A. cruentus

This species grows up to 2(0.5–2.5) m tall, and plants are simple or branched. Leaves are elliptic, rhombate ovate to ovate lanceolate and tips are acute, obtuse or acuminate and the bases are cuneate. The inflorescence is 'soft' and branched into numerous spikes which are spreading or more or less erect. Bracts are never as long as the style branches of the female flowers and they have slender points and often exceed the tepals in length. Tepals are straight, oblong or oblong obovate with acute tips. Utricles are narrow with tapering cylinder ('tower') with style branches erect and slender. Seeds are white or yellow in grain forms and brown and black in vegetable forms.

A. hypochondriacus

This species is up to 3 m tall, variable, simple or branched. Leaves are elliptic or ovate–oblong with acute or acuminate tips and bases cuneate. Inflorescences are large, stiff and prickly to the touch and made up of terminal and lateral spikes which are thick, and erect. Bracts are large and as long as the style branches and are long-pointed, giving the prickly feel to the inflorescence. Tepals are slightly recurved with acuminate tips. Style branches are recurved meeting in a sharp cleft at a thick base. Seeds are white, gold, brown or black.

A. hybridus

This species grows up to 2(0.5–2.5) m tall and is branched. Leaves are very variable, broadly ovate, lanceolate, elliptic, ovate, oblong or rhomboid with tip acute, obtuse or acuminate. Inflorescence is rather lax with many lateral spikes. Bracts are twice as long as the tepals and exceed the style branches in length. Tepals are slightly recurved, shorter than the utricle; inner ones are slightly obovate with acute tips. Utricle narrows into a tower at the apex. Style branches are erect and meet in a shallow cleft at their base. Seeds are dark-coloured.

3.3.1 **Morphological types**

In recent development of the crop the germplasm has been categorized according to gross morphology and on how well plants adapted to conditions at the Rodale Research Center, Pennsylvania. This was developed by C.S. Kauffman and P. Wagoner and the types are denoted as 'grain types'. To a large degree these characterize groupings of landraces. The categories have proved useful in providing particular germplasm for introduction and testing. The system would need revision for detailed use in other places (Kauffman, 1992). Some of the more important grain types (GT) are listed below.

A. caudatus

1. South American GT, currently being split into several types as research progresses in Peru, largely on inflorescence type.
2. Edulis, differing from the above by determinate inflorescences.
3. Ornamental, differing from the crop forms by being slightly shorter (0.5–0.75 m) and with large drooping inflorescences which touch the ground, although there are more compact types.

A. cruentus

1. African GT, noted for bushy nature due to much branching near to the ground and many small seed-heads on side spikes. The main seed-head does not bear the majority of the grain. Shattering is also severe.
2. Mexican GT, with thin main stalk and a main central part of the inflorescence thick, erect to drooping with finger-like branches. Branching of the plant starts midway up the main stem.
3. Guatemalan GT, mostly bushy plants but branching varies widely. Often there are three or more main seed-heads at the top of the plant and no single, central spike. Shattering is severe and height varies from 0.5–2.5 m.

A. hypochondriacus

1. Nepal GT. Plants tall and relatively poorly branched or unbranched. One major seed-head and a number of side branches in the head which tend to snap as plants approach maturity.
2. Mercade GT. From Mexico. Plants tall and very bushy with many small seed-heads, but at high densities becomes single stemmed with only one seed-head.
3. Aztec GT. From Mexico FD, and States of Mexico, Morelos, Tlaxicala and Puebla. Primarily unbranched with massive plant stem.
4. Mixteco GT. From the Oaxaca area. Variable for branching and with a massive plant stem.
5. Spike types. These are segregates from the Nepal GT. Plants are small with many small stems radiating from the base.

A. hybridus

1. Prima GT. Plants short, 0.5–1.0 m tall with one massive seed-head and a number of small spikes arising in leaf axils near the ground level. The main stem is fairly thin.
2. Sangorache. Originally a dye plant. Very tall, 1.5–2.5 m, from South America. Plants have distinctive red stems and leaves.
3. Ornamental, including some miscellaneous types.
4. Weedy forms, including weedy forms from the Americas.

It is interesting to note that this practical classification includes a mix of landrace circumscription, e.g. *A. hypochondriacus* Aztec and Mixteco GTs, and also agronomic characters, e.g. *A. hybridus* Prima is short, early grain and *A. hypochondriacus* Nepal is tall, late grain and *A. hypochondriacus* Mercado is tall, bushy and late grain.

3.3.2 **Anatomy of the seed**

Basic data on seed structure (and nutrient distribution within the grain) provide information necessary for optimal fractionating during milling so that flour, grain and germ products can be produced. Seeds comprise a single-layered coat with no apparent cellular structure, adjacent to the perisperm and two cell layers adjacent to the embryo. The inner layer of this two-celled layer contains ovoid bodies in a matrix which have properties typical of lipid complexes. A small amount of endosperm remains in the mature seed, the major portion of which caps the radicle. The embryo is circular in shape with the radicle tip nearly touching the tips of the cotyledons. A typical embryo cell contains spherical bodies which are partly proteinaceous embedded in a matrix typical of lipid complexes. Perisperm is in the centre of the seed. Cells are filled with polyhedral starch granules (Irving *et al.*, 1980).

In relation to seed colour, pink seeds appear so when the red embryos are visible through a white seed-coat and translucent perisperm (Kulakow, 1987).

The starch granules of grain amaranths are 1–3 μ in diameter and the amylose content is much lower than in wheat.

Plants may be glutinous or non-glutinous. A single major dominant gene (Waxy Wx/wx) determines the non-glutinous starch (Okuno and Sakaguchi, 1982).

Physical and chemical properties of the starch grains have been studied with scanning electron microscopy (Stone and Lorenz, 1984). The attributes, swelling power, solubility, viscosity and functional characteristics, show amaranth starches alone lead to poor quality breads and cakes. However, the grain is physically such that it can be used as a protective colloid in canned foods (Yanez *et al.*, 1986).

3.4 PHYSIOLOGY OF GROWTH AND DEVELOPMENT

3.4.1 **Vegetative and reproductive growth**

Dry matter production during both phases of growth is highest in *A. hypochondriacus* followed by *A. caudatus* and *A. cruentus*. Naidu *et al.* (1982) found a positive correlation between grain protein content and leaf proteinase activity. Also nitrate reductase activity in leaves was positively correlated with grain yield.

Growth is markedly related to soil moisture. Competition for moisture in plant populations limits plant height and, when moisture is limiting, dense stands tend to result in most moisture being used for vegetative growth, leaving insufficient for grain production.

Seed yield is positively correlated with % biomass as seed and negatively with % biomass of stems. There are indications that there is some genetic control in the partitioning of biomass to seeds.

Amaranths possess C4 pathways and are exceptionally efficient photosynthesizers. It is not known how much of the photosynthate used to fill seeds comes from the leaves and how much from the flower parts. There appears to be a lack of correlation between the % of biomass as leaves and good yield and this suggests flower parts may be significant contributors (Hauptli, 1977).

Much more basic research needs to be done on response of amaranth to environmental factors such as temperature, light, water and mineral nutrition, especially since there are three metabolic types of C4 plants. The only information on this relates to the need for sodium and light: in cultures 50 μM Na is necessary for normal growth. Alanine accumulates in the dark in leaves but is reduced with light (Nable and Brownell, 1984). Sodium plays a role in nitrate uptake in C4 plants and this has been shown using *A. tricolor* (Ohta *et al.*, 1988).

3.4.2 **Day length**

Most amaranths are sensitive to day length. *A. caudatus* normally only flowers and sets seeds when day length is less than 8 h, although ornamental forms set seed at longer day lengths, and there are genotypes which flower at day lengths from 12–16 h. *A. cruentus* is much less sensitive than any other *Amaranthus* species (Weber and Kauffman, 1990). In its centre of diversity, *A. cruentus* has some strains which remain vegetative for long periods but which go to seed early if introduced to long-day latitudes. Day-neutral types are being selected.

Singh (1961) suggested that grain amaranths are sensitive to both photo- and thermo-periodism. The behaviour of a short-day grain amaranth was modelled in response to temperature and photoperiod by Angus *et al.* (1982). Temperature has a non-linear effect on rate of development prior to anthesis. There was a threshold temperature of 10 °C throughout the life cycle and before anthesis it showed a strong short-day response to light. After anthesis there is no sensitivity to photoperiod.

Short-day cultivars can have photoinductive requirements of 3, 5, 6 or 8 h photoperiod cycles for macroscopic inflorescence initiations. Water-soluble proteins appear after the induction (Kohli *et al.*, 1980).

A. caudatus is not an obligate short-day plant and can initiate inflorescence primordia under various conditions but normal development of primordia proceeds only under short photoperiods (Zabka, 1961).

3.4.3 **Germination**

All grain species germinate in the dark but germination is better with a light stimulus. Alternating temperatures are required for germination in the light. Germination seems to have one pathway at lower temperatures, 30–25 °C in the dark, and another at higher temperatures which is light-stimulated (Oladiran and Mumford, 1985).

Temperature interacts with sowing depth as shown by seedling emergence of *A. hypochondriacus*. Soil temperatures for emergence are better in the range 18.5–24 °C (Webb *et al.*, 1987).

In weedy species there may be variation in germination behaviour, as shown in *A. powellii* (Frost and Cavers, 1975), and this behaviour forms part of the ecological strategies typical of successful weeds and pioneers.

3.4.4 **Response to stress**

The plastic response of amaranth is reflected in the plant's reaction to stress. Plants respond to differing growing conditions with changes in branch numbers, branching patterns and plant size, and stress therefore markedly affects grain yield, plant biomass and stover ration (Edwards and Volak, 1980). Water stress has a major effect on dry matter partitioning (Cavagnaro, 1985).

A. hybridus has been grown in simple non-weighing lysimeters and irrigated when soil moisture content was depleted by 95, 75, 50 and 25% of field capacity. Moisture stress had adverse effects both on leaf and seed production (Chukwuma, 1988). Water stress also causes reduction in both photosynthetic capacity and stomatal conductance to water loss in *A. hypochondriacus* and it is initiated by abscisic acid (Bunce, 1987). There is patchy closure of stomata under certain environmental circumstances in *A. caudatus* subsp. *mantegazzianus* (Farquar *et al.*, 1987).

3.4.5 **Effect of age on grain**

Since popping is important in terms of utilization, age-related changes in the seeds affect the popping size. There is a tendency for size to increase with age in *A. caudatus* but less effect in the other grain species (Joshi and Rana, 1991).

3.4.6 **Dual-purpose crops**

Experimentation on *A. caudatus* grown for both seed and leaf has shown that topping at 20 cm from the ground allows sufficient dry matter in the stump for plant regrowth (Mnzava and Masam, 1985).

3.5 BIOCHEMICAL COMPOSITION

Grain composition varies according to agronomic practices but usually contents fall within the following ranges: crude protein 14–19%, fat 5–8%, starch 62–69%, total sugars 2–3%, fibre 4–5% and ash 3–4%. The ranges for % N, fibre and ash are shown in Table 3.3.

3.5.1 **Proteins and amino acids**

The amino acid content of three grain species is shown in Table 3.4. The protein is relatively rich in lysine and leucine is usually limiting compared with cereals.

Germplasm has been screened to examine the relationships between

Table 3.3 Percentage nitrogen/fibre and ash of amaranth grain (dry weight)

Species	*N*[a]	*Fibre*	*Ash*
A. caudatus	2.54	4.2	3.2
A. caudatus subsp. *mantegazzianus*	2.97	4.3	3.0
A. cruentus	3.05	4.4	3.3
	2.66	5.0	4.1
A. hypochondriacus	2.67	5.0	3.3
	2.72	5.0	3.7
A. cruentus × *hypochondriacus*	2.97	4.3	3.0

[a]N × 5.85.

Source: Saunders (1990).

Table 3.4 Reported amino acid composition in grain amaranths expressed as g/16g nitrogen

	Species and genotype			
Amino acid	*1*	*2*	*3*	*4*
Lysine[a,d]	5.3	5.5	5.1	5.5
Histidine	2.5	2.6	2.4	2.5
Threonine[c]	3.5	3.6	3.4	3.6
Cysteine	2.3	2.2	2.1	2.1
Methionine	2.4	2.3	1.9	2.6
Valine	4.1	4.6	4.2	4.5
Isoleucine	3.6	4.0	3.6	3.9
Tyrosine	2.9	3.2	2.6	3.3
Phenylalanine	3.4	4.0	3.4	4.0
Serine	5.9	5.9	5.4	6.3
Glycine	6.9	7.2	7.0	7.4
Leucine[b]	5.3	5.7	5.1	5.7

Species/genotypes: 1 = *A. caudatus*, pink, naked
2 = *A. cruentus*, not specified
3 = *A. cruentus*, not specified
4 = *A. hypochondriacus*, HH5

[a]Lysine ranges from 0.73 – 0.84% of the total protein content (Bressani, 1989).
[b]Usually considered the limiting amino acid (Singhal and Kulkrni, 1988).
[c]May be more limiting than leucine (Bressani, 1987a).
[d]6.3 g/16 g N_2 recorded for *A. hybridus* by Uzo and Okorie (1983) in Nigeria.

Sources: Betschart *et al.* (1981); Carlsson (1980).

agronomic and protein characteristics with a view to selection of cultivars with higher yields and higher nutritional value. Imeri *et al.* (1987) showed a positive correlation between yield-protein, methionine-cysteine, methionine-lysine and threonine-leucine as well as significant negative correlations between protein-cysteine, fat-methionine and cysteine-leucine. Seed weight does not appear to be a useful character since this appears weakly negatively correlated with higher yield. Seed colour may be of more interest in breeding since wild forms with black seeds may contain significantly higher protein than cultivated forms (Misra *et al.*, 1983). Some research has been carried out on the genetics of grain protein in amaranths (Pandey and Pal, 1985).

Amaranth protein is complementary when combined with other cereals. Wheat, sorghum, rice and maize have been beneficially supplemented and are especially helpful in child nutrition (Pant, 1985; Pederson *et al.*, 1987; Sanchez-Marroquin and Maya, 1985; Singhal and Kulkarni, 1988).

Glutelins are the most abundant protein as shown by a series of extraction and fractionation studies by Paredes Lopez *et al.* (1988), and these are followed by globulin and albumin and a low level of prolamine (Konishi *et al.*, 1985; Zhao and Ma, 1987).

3.5.2 **Carbohydrates**

Starch is mainly amylopectin with 5–5% amylose. The free sugar is sucrose and its content is twice that normally found in wheat and millets.

The starch makes up the bulk of the amaranth flour and has very small granules which means that it has potential industrial applications. It has high swelling power (Stone and Lorenz, 1984). The properties of the starch are shown in Table 3.5, and compared to starch of wheat, and Table 3.6 shows some comparative physical properties.

3.5.3 **Fatty acids**

Amaranth grain contains 6–10% oil which is about 76% unsaturated and rich in linoleic, oleic and palmitic acids. The oil is mostly within the germ (Betschart *et al.*, 1981; Lorenz and Hwang, 1985; Garcia *et al.*, 1987). Analysis of seed by the United States Department of Agriculture, Western Regional Research Center, showed up to 7% squalene in amaranth oil (Lyon and Becker, 1987) and up to 11% has been found in *A. caudatus* (Becker *et al.*, 1981). These levels compare with less than 1% in olive oil, wheat germ oil and rice bran oil. Squalene has applications in the cosmetic industry and is normally extracted from shark liver.

The relative proportions of fatty acids in a number of diverse grain amaranths are shown in Table 3.7.

In terms of digestibility of oils from amaranths, there is a lower digestibility than cotton seed oil and, although the oils of the grain species are fairly similar, that from *A. cruentus* gave the lowest digestibility (Garcia *et al.*, 1987).

Table 3.5 Properties of the starch of *A. hypochondriacus* compared to that of wheat

Property	*Amaranth*	*Wheat*
Amylose content (%)	7.2	22.2
Amylograph viscosities (BU)		
At peak	320	350
At 92 °C	260	270
After cooking at 35 °C	260	340
Solubility (%)		
At 60 °C	9.12	1.67
At 70 °C	14.21	2.35
At 80 °C	32.76	2.48
At 90 °C	37.43	8.21
Swelling power		
At 60 °C	1.02	4.77
At 70 °C	1.51	6.13
At 80 °C	3.44	8.20
At 90 °C	3.53	11.80
Water binding capacity (%)	71.80	127.30
Gelatinization temperature range (°C)		
Initial	62	52
Midpoint	64	55
Final	68	56

Sources: Becker *et al.* (1981); Lorenz (1981).

Table 3.6 Starch granules of grain amaranth compared with three other starches

Crop	*Granule shape*	*Granule size µm (mean)*	*Size range (µm)*
Amaranth	Round or dodecahedral	2	1–3
Rice	Polygonal	7	3–8
Wheat	Round	10	2–10
	Lenticular	20–30	
Potato	Oval	40	15–100

Source: Lehmann (1992).

Table 3.7 Relative proportions of fatty acids in grain amaranths (percentage of total fat)

Fatty acid	*Species/genotypes*					
	1	*2*	*3*	*4*	*5*	*6*
Myristic		0.06	0.05			
Palmitic	18.1	15.6	15.7	20.7	21.8	19.4
Palmitoleic		0.27	0.26			
Stearic	4.6	3.02	3.09	3.6	3.9	3.8
Oleic	26.7	21.0	21.6	35.2	28.5	25.3
Linoleic	49.4	37.1	36.2	39.9[a]	42.3[a]	50.9
Linolenic	0.9	0.87	0.90	0.7[a]	0.6[a]	0.6
Arachidanic		0.59	0.56			

Species/genotypes 1 = *A. hypochondriacus* HH4
2 = *A. hypochondriacus* A53
3 = *A. hypochondriacus* Red Top
4 = *A. cruentus* HH1
5 = *A. cruentus* HH3
6 = *A. cruentus* BP32

[a]Rapid methods of analysis have showed much higher linoleic and linolenic fractions in *A. cruentus* oil (Ayorinde *et al.*, 1988).

Sources: Carlsson (1980); Schmidt (1977).

3.5.4 **Minerals**

Seeds are excellent sources of iron, calcium and zinc as well as other micronutrients. Mineral content are shown in Table 3.8.

Iron in amaranth has high biological value in cereals, with and without fortification (Whittaker and Ologunde, 1990). Calcium levels in the grain exceed those of milk but there have been few data to indicate the bioavailability of minerals from amaranth, or to develop the best food application to release the calcium (Lehmann, 1992).

3.5.5 **Vitamins**

The ranges of vitamin contents are shown in Table 3.9. There has recently been focus on vitamin E, because tocopherols and unsaturated forms as tocotrienols (T3s) have a potent inhibitory effect on the key enzyme for chloresterol synthesis. T3s have been discovered in amaranths (Lehmann *et al.*, 1990). The ring embryo of the seed – which survives popping – harbours vitamin E forms including T3s. In the seed the vitamin E protects the oil-rich embryo from oxidation.

Table 3.8 Mineral content of amaranth grain (mg/100 mg edible portion)

Mineral	*Species/genotype*			
	1	2	*3*	*4*
Calcium	247	150	123	490
Phosphorus	500	nm	nm	nm
Iron	0.34	19.5	8.2	15.0
Mg	nm	327	263	270
Na	nm	9	6	nm
Cu	nm	0.7	0.5	0.7
Mn	nm	2.6	1.7	nm
Zn	nm	3.5	3.2	nm
B	nm	1.6	1.3	nm
Al	nm	8.4	2.4	nm
I	nm	602	498	nm
F	nm	691	565	nm

nm : not measured

Species/genotypes 1 = *A. caudatus*
2 = *A. hypochondriacus* A53
3 = *A. hypochondriacus* Red Top
4 = *A. hypochondriacus* genotype not stated

Sources: Elias (1977); Sanchez-Marroquin *et al.* (1980); Schmidt (1977).

Table 3.9 Vitamin content of amaranth grain mg per 100 g whole grain

Vitamin	*Species*		
	1	2	*3*
Thiamine	0.14	0.09	0.25
Riboflavine	0.32	0.21	0.29
Niacin	1.00	1.31	1.15
Ascorbic acid	3.00	4.90	2.80

Species: 1 = *A. caudatus*
2 = *A. cruentus*
3 = *A. hypochondriacus.*

Sources: Becker *et al.* (1981), Elias (1977), Sanchez-Marroquin (1988).

3.5.6 **Pigments**

Amaranth plants possess a range of red/green and orange/green pigmentation which can be expressed temporally or spatially with varying intensities. Amaranthine can be extracted as a dye (Lehmann, 1990) but this is unstable when exposed to light and it can change rapidly from red to brown.

In traditional agriculture amaranths are used as food colourants in

Table 3.10 Anti-nutritional factors in amaranth grain

Factor	*Whole amaranth seed*	*Other foods*
Trypsin inhibitor (TIU/mg)	0.940	Wheat 0.54
Agglutinin activity	positive	Soy bean 200 times as active
Tannins (%)	0.08–0.42	Wheat 0.17
		Millet 0.20
		Sorghum much more
Saponins, phenolics	positive	

Sources: Lorenz and Wright (1984), Saunders (1986).

particular for religious or ceremonial offerings. Ethnobotanical information is available on such uses in the south western USA, and Central and South America.

3.5.7 **Anti-nutritional factors**

Uncooked seeds contain heat labile growth inhibitors but these are inactivated by cooking or roasting. Agglutinins are present in *A. caudatus;* lectins are removed by heat processing (Singhal and Kulkarni, 1988), and lectin of *A. cruentus* has been extracted and characterized (Koeppe and Rupnow, 1988; Calderon de la Barca and Moreno, 1988).

Levels of trypsin inhibitors and tannins are so low in light-seeded amaranths that they are of no concern. Although thought to be higher in dark seeds, Suarez and Mark (1980) found no difference between different-coloured seeds.

Concentrations of anti-nutritional factors and comparisons with other foods are shown in Table 3.10.

3.5.8 **Nutritional valuc**

Due to high protein and oil contents, and balance of carbohydrates, fats and proteins, amaranth grain represents an energy food. After milling and toasting, digestion and absorption by the body is high (Morales *et al.*, 1988). Supplementation of cereal flour with amaranth can provide up to 70% of diet energy. Rice and amaranth in equal quantities approach the Food and Agriculture Organization/World Health Organization protein specifications (Singhal and Kulkarni, 1988). Furthermore, 12.7% amaranth flour added to maize meal satisfies the protein and lipid needs of young children if it is used to provide 90% of diet energy (Morales *et al.*, 1988).

More recently the United Nations organizations have begun to look at protein quality as well as quantity and amaranth protein should gain further recognition.

Table 3.11 Protein value score for amaranth in comparison with other foods (score of 100 considered ideal)

Food	*Score*
Amaranth	75
Cow milk	72
Soy bean	68
Barley	62
Wheat	60
Groundnut	52
Maize	44

Source: Lehmann (1992).

A comparison of the protein value score for amaranth and other foods is given in Table 3.11.

In smallholder agriculture, grain amaranths are often used as multipurpose plants and foliage is gathered for cooking as a vegetable. Both contents of oxalates and nitrates may become toxic if large quantities are consumed, due to binding of divalent cations (calcium and zinc) making them unavailable. Efforts are underway to breed varieties with lower leaf oxalate and nitrate (Gupta *et al.*, 1984). Saponins also occur but are not toxic in small quantities.

On the whole, oxalate and nitrate contents of leaves are lower than those in many other vegetables (Der Maderosian *et al.*, 1980) and amaranth leaves do provide significant nutrients such as iron and calcium, fibre, niacin, folic acid, vitamin C and vitamin A (Saunders and Becker, 1983).

3.6 AGRONOMY, YIELD AND PRODUCTION

Although there are considerable areas of production of grain amaranth in many parts of the world, the best agronomic practices have yet to be fully worked out. Wider use needs to include agronomic packages with germplasm sources since different varieties are extremely dependent upon local climate and soil conditions (Senthong, 1986).

3.6.1 **Sowing**

Traditionally seeds were broadcast but much better crops are achieved through seeding in rows. However, smallholder farming in many parts of the developing world will probably continue to use a degree of broadcast sowing; hence new varieties being developed for mechanized production may not be suitable in such areas.

Seeding can be achieved using farm equipment, modified to increase the seeding rate. Small farm machinery has been developed in Peru. In the USA mechanized precision seeding using grain drills or corn and soya bean planters using insecticide boxes are common. In general, less precise planters require a higher seeding rate (Weber *et al.*, 1990). Diluting amaranth seed with other ingredients such as corn grits can be used to reduce the seeding rate in drills (Robinson, 1986). It is not uncommon for seeds to be mixed with sand to produce a more uniform distribution in broadcast sowing.

In traditional agriculture, sometimes seed is sown broadcast in nurseries then seedlings are transplanted, but this is only recommended in wetter zones. The chinampa system of native Mexicans has been documented where aquaculture and intensive horticulture are combined. Seed-beds are made from canal bottom mud and cut into squares, each being seeded and seed holes covered by cow manure. About 20 days after seedling emergence mud squares are transported to highland fields where they are planted for grain. The tools and methodology date back at least to the Aztec times (Early, 1978). Vegetable amaranths are often transplanted in tropical areas such as west Africa. Certainly for grain there are no advantages in transplanted seedlings versus direct sowing (Table 3.12).

Seed-beds should be prepared so that they drain freely, are fairly level to prevent rain washing away seeds or seedlings and are relatively firm to help maintain soil moisture for germination.

Seed depth is about 1 cm and less than 2 cm; germination occurs about 5 days after sowing. Deeper sowing is advised on friable soils but increased depth reduces emergence (Weber *et al.*, 1990) and seeding rate needs to be adjusted to compensate for reduced emergence.

Regarding mechanized sowing, Weber *et al.* (1990) stress the role of adequate soil moisture at the time of planting. Since sowing depth is

Table 3.12 Comparison of yield from direct sowing versus transplanting: average of two years' data with variety Annapurna

			Yield (kg/ha)	
Plot size	*Row spacing (cm)*	*Plant spacing (cm)*	*Direct sowing*	*Transplanted*
$8m^2$	100	30	1590	1220
		20	1640	1180
		10	1870	1310
$4m^2$	50	30	2700	2250
		20	2960	2390
		10	2860	2220

Source: Joshi and Rana (1991).

shallow, the moisture has to be available near the soil surface. Pre-planting irrigation is required in some areas where soil is dry at planting time. One approach to moisture conservation in dryland areas is to open the row to remove the so-called dust mulch and the seed is then planted in the moist soil underneath. This strategy gives a head start to the germination before that of rain-sprouted weed seeds. However, planting in the furrows may cause dangers of washing-out of seedlings by surface water.

Press wheel options on planters can aid seed–soil contacts to promote rapid germination and emergence. This is discussed by Weber *et al.* (1990); options vary according to soil conditions – on heavy soils a problem with crusting may result.

Seed rates

Commercial rates are aimed at achieving 360 000 plants/ha in rows. Rates vary according to spacing and density (see below).

3.6.2 **Spacing and density**

There is a marked effect of density on yield. This can be illustrated by the results of a trial by Misra *et al.* (1985) on alkali soils (Table 3.13). At higher plant density, yield per plant decreases but unit yield per hectare increases. In this trial optimum yields were obtained from 45–60 cm

Table 3.13 The effect of plant population per unit area on yield

Spacing (cm)		*Population*	*Grain yield*
Between rows	*Within rows*	*(000/ha)*	*(g/plant)*
30	7.5	4.44	20.970
	15.0	2.22	24.480
	22.5	1.48	24.800
45	7.5	2.96	21.785
	15.0	1.48	23.923
	22.5	0.99	25.573
60	7.5	2.22	28.233
	15.0	1.11	28.270
	22.5	0.74	32.663
75	7.5	1.78	29.283
	15.0	0.89	35.310
	22.5	0.99	37.208
			LSD (0.05) = 7.313

Source: Misra *et al.* (1985).

spaced rows and 7.5 cm within rows. Under the conditions of this trial, with marginal soil conditions, about 100 000 plants/ha is suitable (Mehra and Murty, 1988).

Trials in India on non-marginal soils have shown that a row spacing of 50 cm and plant spacing within row at 20 cm is optimal (Joshi and Rana, 1991), but rows 20–30 cm apart and 10–15 cm between plants is more normal, although not recommended in drier zones.

Although commercial rates are up to 360 000 plants/ha and plant spacing in rows with 3.5 cm between plants, a range of densities between 323 230 and 360 000/ha (or 130 000–145 000 plants/acre) is recommended (Wagoner, 1983). A 3-year study at Rodale Research Center reported by Wagoner (1983) used a range of seven grain types: *A. cruentus* (Mexican, African and Mexican × African), *A. hypochondriacus* (Mercado and Nepal), *A. hybridus* (Prima) and *A. caudatus* (Ornamental). In general, varieties within a single grain type tend to exhibit similar trends in plant morphology (height, branching, stalk diameter) as density increases.

Since density affects branching, the right density response is critical if mechanical harvesting is to be used. Higher density is needed to ensure most plants produce a single stem.

3.6.3 **Irrigation**

Irrigation increases yield but is normally uneconomic, although in some areas it is needed, e.g. in the coastal desert of Peru, and it is normal in Nebraska, USA. In most cases grain amaranth is a rain-fed crop. In irrigated areas, amaranth can be introduced into the farming systems to reduce irrigation costs and to reduce the potential for soil salinization (Weber *et al.*, 1988). Irrigation right after seeding has to be avoided because of the dangers of damping-off seedlings. Also excessive irrigation can result in rank growth and lead more readily to lodging.

3.6.4 **Time to maturity**

Grain amaranths reach maturity 3–5 months after sowing. In traditional agriculture maturity can be as long as 6–7 months, e.g. small plots in Mexico for both *A. cruentus* and *A. hypochondriacus*.

There is environmental influence on time to maturity associated with the environmental effects on plant height and plant architecture. As selection and breeding continue, both early and late maturing forms have been identified within the morphological types and can be used depending on the local cropping system.

The variation in maturity in response to diverse environments is shown in Table 3.14.

Table 3.14 Days to maturity as influenced by environment

Genotype	*Days to maturity* 1	2	3	4	5	6
A. hypochondriacus						
Mercado		90	113	50	125–149	88–128
Nepal	145–190	60–70	113–127	45–50		
A. cruentus African	80–100		30	55	112	145
A. hybridus Prima		70	73	45	97	170

Sites
1 = RRC, Pennsylvania, USA: temperate
2 = Kenya, sea level: tropical
3 = Kenya, high altitude and cold
4 = Kenya, medium altitude, semi-dry
5 = Thailand, upland at Nang Hoi
6 = Thailand, upland at Chang Kian

Sources: Gupta (1986); Senthong (1986); Wagoner Haas (1980).

3.6.5 **Yield**

As with all traditional crops, yields vary widely: 500–2000 kg/ha, but they may be as low as 200–250 kg/ha in Zambia (FAO, 1988), or even as low as 100 kg/ha.

Up to 3000 kg/ha can be obtained with good agronomic practices and good seed sources, but 1500 kg/ha is more of a norm. 4000 kg/ha have been obtained in Montana, USA (Cramer, 1988), and 6000 kg/ha using improved varieties and traditional agriculture in Peru (Sumar *et al.*, 1986). In the north-west hills of India, yields average 2250 kg/ha but most cultivars are late maturing, and also tall with a tendency to lodge and a degree of seed-shattering. Selection is underway for higher yielding, earlier and more dwarf types since suitable germplasm has been identified (Joshi, 1985).

As with all crops, elite lines tested under experimental conditions may produce good yields over several years but vary in the field according to local adaptation. For instance, cv. Annapurna at the experiment station at Shimla, India, yielded on average 3500 kg/ha over 5 years (range 2900–3960) but planted in various hill areas yielded on average 2338 kg/ha (range 772–3982).

3.6.6 **Fertilizers**

Growth is encouraged by fertilizers with high potassium. On soils of moderate or poor fertility, NPK 10:10:20 at 400 kg/ha is used, and split application yields good results. In India NPK 20:30:20 meets most requirements (Joshi and Rana, 1991). Use of urban refuse of up to 25 t/ha is practised in parts of West Africa.

Nonetheless there is only a limited information base on the fertility requirements of amaranths (Kauffman and Weber, 1990), and additional research is needed. Experimental data from the University of Minnesota, USA, showed increasing yields of *A. cruentus* in response to nitrogen, but only up to 90 kg/ha, and lodging became prevalent at 55 kg and over. There were no responses to phosphorus and potassium but this was probably due to high nutrient levels in the soil.

The yield response to nitrogen application is not clear. In another study in the USA, using *A. cruentus* and *A. hypochondriacus*, three rates of nitrogen, 0, 100 and 200 kg/ha, showed a two-fold increase in yield at 100 kg/ha but little increase at 200 compared with 100. Also, in a second year, no response to nitrogen was found (Endres, 1986).

Bressani *et al.* (1987a) grew *A. cruentus* (USA and Guatemala GT), *A. hypochondriacus* (USA) and *A. caudatus* (Peru GT) with NPK 12:24:12 at 0, 30, 50 and 90 kg/ha./ Fertilizer had little effect on yield but increased the seed protein contents in the lowest yielding accessions. Carlsson (1980) had also pointed out that nitrogen fertilization at 200 kg/ha was beneficial on the amino acid composition of amaranth grain.

Joshi and Rana (1991) cite the experiences of R. Schmidt with amaranth in Ethiopia, Nepal and Indonesia which suggest that amaranth is not well adapted to soils markedly deficient in available phosphorus since this detrimentally affects seedling establishment. Hence soluble phosphorus fertilizer might have to be used in seeding in such situations.

Despite the results of fertilizer trials being inconclusive, it is fair to state that grain amaranths grow well in soils containing widely differing levels of nutrients. Almost certainly, vegetable amaranths require higher soil fertility.

One major constraint in current breeding is to develop shorter plants so that excessive lodging does not become critical when plants grow taller as a response to fertilization. Plants of height 1.0–1.5 m are required rather than 1.5–2.5 m. Also shorter stature would permit modern grain combine-harvesting.

3.6.7 **Seed weight**

One thousand seed weight falls within the range 0.4–1.2 g. There is a tendency for American landraces to bulk in the range 0.5–0.7 g and for Indian landraces to bulk in the range 0.6–0.8 g. Improvement includes selection for larger seed size in the range 0.7–0.9 g to facilitate mechanical sowing.

Pandey (1984) examined yield-related traits in F1 and F2 generations of a six plant dialled cross of *A. hypochondriacus* and apparently epistasis exists for 1000 seed weight in the F1. In the F2 it appeared to be controlled by over-dominance effects. Additive genetic variation was more important in the F1 and non-additive in the F2.

One thousand seed weight (as well as other characters such as protein %, seed yield, plant height and inflorescence length) show high genotypic coefficient of variation, high heritability and high genetic advance (Joshi, 1986).

3.6.8 **Seed treatments**

A number of storage fungi affect seeds. Some are internal, while others are external. *Aspergillus flavus, A. niger* and *Rhizopus stolonifera* are the major storage fungi of seeds in India (Sharma *et al.*, 1981; Naseema *et al.*, 1983). Thiram was superior on vegetable seeds as treatments, and Agroson and Ceresan were satisfactory on grain forms.

3.6.9 **Intercropping**

A number of observations has been made on grain amaranths in traditional intercropping subsistence systems, e.g. in Mexico, where amaranth can be the major plant and bean and squash are secondary, or maize can be the major crop with amaranth spaced sporadically through the field in the ratio 8:2 or 9:1 (Feine, 1980).

In the Himachal Pradesh hills and also the hills of Uttar Pradesh of India, companion crops include French bean, maize, *Eleusine* millet, and occasionally black gram, soya bean and taro.

3.6.10 **Pests and diseases**

Pests

Worldwide, the most important pests are lepidopterous larvae and *Lygus* bug, both cosmopolitan and feeding on a wide variety of plants, and also weevils. The following pests have been reported:

(a) *Lepidopterous larvae*

R. *Spodoptera frugipeda* Fall army worm
 * *S. exigua*
 * *S. eridamia* Southern army worm
R *Trichoplusia ni* Cabbage looper
 Heliothis zea Corn earworm
 Ostrinia nubialis European corn borer/corn earworm

R = known resistant germplasm
* = largely on *A. hybridus*

(b) Lygus *bugs*

A number of species of *Lygus* are sucking insects which feed on flowers of amaranths, killing parts of them and also feeding on immature grain which dries up and has to be removed during cleaning. There are, however, some germplasm lines with resistance (Wilson, 1989). *L. linearis*, the tarnished plant bug, is the major threat to seed production in the USA, Mexico and India, and has also been recorded from East Africa.

(c) *Weevils*

Larvae of amaranth weevil, *Conotrachelus seniculus*, bore into roots and sometimes stems, thus making plants susceptible to lodging but also allowing fungal and bacterial rotting organisms to enter. Other weevils causing similar effects are *Hypolixus trunculatus* (especially in India, Pakistan and Thailand) and *H. nubilosus* in Egypt. In some cases weevils normally associated with other crops can cause severe yield reduction in amaranth, e.g. spotted sunflower stem weevil, *Cylindrocopturus adspersus*, in North Dakota, USA.

Pests of less importance, but requiring vigilance, are grasshoppers, flea beetles, aphids (especially cowpea aphis, *Aphis craccavora*), blister beetles and leaf bugs. Many of these are general feeders.

Nematodes

Meloidogyne root nematode species have been recorded in Africa and also in South and South-east Asia. Some resistance to *M. incognita* is known in India, but this nematode and also *M. javanica*, is generally controlled through mixed cropping. *A. caudatus* is especially susceptible, *Nacobbus aberans* has been recorded on *A. hypochondriacus* in Mexico, Argentina and Nigeria (Siddiqua and Alan, 1988; Santacruz Ulibarri and Marban Mendoza, 1986).

Nematodes have been widely reported affecting weedy and vegetable amaranths. Although not serious in grain crops there is a potential for build-up and more serious effects. Some screening has been carried out, especially for vegetable forms (Babatola and Awoderu, 1986).

Fungi

On the whole there are no major fungal diseases of grain amaranths except for damping-off by common soil fungi, *Pythium*, *Rhizoctonia*, and also stem canker associated with *Phoma* and *Rhizoctonia*. Wet rot due to *Choanephora cucurbitarum* can be important in Mexico and Africa.

Control of damping-off is by good agronomy: adequate soil drainage, avoidance of high nitrogen, and planting at warm temperatures conducive to germination. Sealy *et al.* (1988) screened a range of amaranth accessions for resistance to damping-off by *Pythium myriotylum* and suggested that those showing weighted per cent mortality of under 20% can be used in resistance breeding.

No major leaf fungal diseases are reported for the Americas but species of *Colleotrichum* have been recorded causing leaf spot in USA (*C. domatium* and *C. capsici*). Light leaf spot caused by *Alternaria alternantherae* has been recorded on ornamental *A. caudatus*. White rust, *Albugo bliti*, can be serious in South-east Asia and in local areas in Mexico and has affected *A. hybridus* in Florida, USA. Leaf spots related to *Cercospora* and *Alternaria* species do not cause major problems.

Viruses and bacteria

A. hybridus has been affected by tobacco ring spot virus in the USA and this is probably seed-borne. In addition both *A. caudatus* and *A. hypochondriacus* have been affected by a mosaic virus with affinity to cucumber mosaic virus (Ram, 1980; Sharma and Chowfla, 1987). Other viruses reported have been alfalfa mosaic, beet western yellows and ivy vein clearing virus.

Most *A. caudatus* germplasm, especially the South American morphological types, has severe virus-like symptoms when grown in the field in the USA, but are disease-free when grown in glasshouses (Kauffman and Reider, 1986). A virus might be transmitted from local plants by insects or there might be another cause unrelated to viruses.

The bacterium, *Xanthomonas amaranthicola*, has been reported to cause a leaf spot disease in India (Reddy *et al.*, 1980).

3.7 POST-HARVEST TECHNOLOGY

3.7.1 **Drying and storage** (Modified from Weber *et al.*, 1990)

For storage, grain should be dried to 12–13% moisture. Depending upon weather conditions at harvest, amaranth may need little or no drying. However, to meet quality standards for marketing, correct storage is necessary and some degree of drying is usually needed.

Since amaranth seeds are small, grain driers have to be adapted using fine mesh nylon. A scale for amaranth moisture has been developed in the USA by the Dickey-John Corporation.

When drying using forced air the depth of the grain should not be too deep because more pressure is needed to move air due to the seed size of amaranth.

Grain-cleaning is usually done in two steps; removal of the major part of the trash by the grower and final cleaning by specialized equipment, often at local centres. Dried, cleaned grain has to be kept in rodent-proof, airy situations to maintain quality. Bagging requires very fine mesh or heavy duty paper sacks.

The Amaranth Institute, USA, has recently developed a set of standards to ensure post-harvest handling meets the food quality requirements of the food industry. The following contamination has to be avoided.

1. Foreign materials: dockage should be less than 1%, and zero tolerance for rodent frass.
2. Companies check for coagulas positive *Staphylococci*, *Salmonella*, *Escherichia coli*, total coliforms, yeasts and moulds. Presence of these organisms can be directly related to cleaning and storage at the grower level. Moulds may be indicated by odour and this is also checked by industry.

3.7.2 **Colour of grain**

Black seeds associated with weedy amaranths or results of pollen contamination, or planting a vegetable type for grain, are undesirable for the food industry. Seeds are not uncommonly mixed white/gold, or white/translucent brown. Industry wishes as much homogeneity as possible.

Weed seeds are impurities which are also undesirable. Hence agronomic practices should attempt to eliminate weedy amaranths as far as possible and also related species such as chenopods, although small quantities are acceptable (5–10 seeds/g).

3.7.3 **Cleaning for the food industry**

Seed is processed by a machine passing material over sets of sieves to separate fractions by density. Air aspirators and fine and rough screening fractions are used to separate these from the seed lot. Particles of similar sizes and densities can be cleaned out using a gravity table. Arrowhead Mills Inc., Texas, USA has pioneered amaranth seed lots of increased purity, and information is available (Hendershot, 1986).

3.7.4 **Harvesting**

Grain amaranth produced in Latin America and Asia is largely harvested by hand. It is sun-dried and threshed and winnowed by hand.

This practice is not affected by non-uniform maturity: the central inflorescence usually matures and dries before the numerous smaller ones of branches lower on the side.

In the USA density experiments and selection of germplasm permit much greater uniformity and machine harvesting. Machine harvesting is also a goal in Mexico. However, a killing frost is usually needed for mechanical harvesting since most varieties do not dry down when the seeds mature; plants retain high moisture contents in stems and leaves. After a killing frost, plants dry down in under 10 days.

The key to harvesting is sufficient dryness of inflorescences to permit grain combine use and row harvesting units to prevent loss due to any shattering. Shattering loss may be severe if frost only kills part of the inflorescence. Synchronous dry-down, present in maize and small grain cereals, is a research goal for amaranth. So too is non-shattering, although this latter character appears difficult to isolate (Kauffman, 1986).

3.7.5 **Processing**

Processing for food use involves three major methods: milling, puffing and extrusion. At the village level in developing countries, grains are also soaked and ground as a modified milling method.

Table 3.15 Protein quality of raw and cooked amaranth grain measured as protein efficiency rating

Species	*Raw*	*Cooked*
A. caudatus	2.06	3.63
Amaranth 'species'	1.73	3.07

Source: Bressani *et al.* (1987a).

Table 3.16 Effect of processing on protein quality of amaranth grain

Processing	*Net protein ratio*	*Protein digestibility (%)*	*Available lysine (%)*
Raw amaranth	2.36	76.4	0.91
Cooked amaranth	2.80	72.9	0.93
Toasted amaranth	1.54	59.5	0.58
c.f. casein	3.57	92.5	–

Source: Bressani *et al.* (1987a).

All cereal grains are processed for consumption: water soaking and boiling for rice, milling and baking for wheat or alkaline cooking with maize. The processing results in some loss of nutritive value, but for amaranth, wet cooking (or drum drying) increases the protein quality. However, excessive thermal processes reduce the quality of amaranth grain through protein loss or quality (mainly lysine) and reduction in protein digestibility (Tables 3.15 and 3.16). Hence processing should attempt to maximize amaranth nutritional value.

Milled product processing

Becker *et al.* (1986) tested milling in a selection of mills. The stone mill was the only one to give the desired periplasm/bran fractions and further experiments showed that mill gap settings of 0.027 in. only remove the embryo but that decrease in settings left kernels. Amaranth seed is effectively milled at 13.8% moisture producing flour fractions of predictable composition, e.g. whole kernel flour.

Various milling procedures have been recommended by Betschart *et al.* (1981) and Sanchez-Marroquin (1988) in order to obtain a high protein fraction (32–40% protein) and a starchy fraction with 7–10% protein. Combination of a mill and an air separator, or use of a pearler, both with various mesh sizes, were selected for the high protein fractions.

Abrasive milling can produce a germ-seed coat fraction containing about 75% of the total seed oil and 40% protein: this may have an unique market niche (Betschart *et al.*, 1981).

Whole flour can be processed as shown in Figure 3.4.

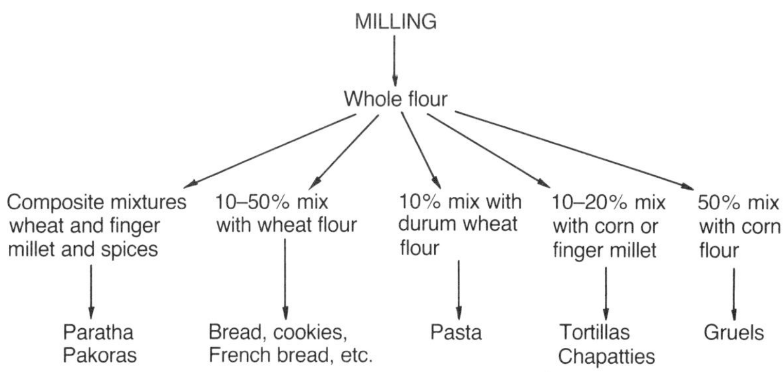

Figure 3.4 Processing of flour of grain amaranths.

Popping

When popped, amaranth seeds taste like nutty popcorn. They can be used as a snack food or as a cereal with milk and sugar. The major use of grain amaranth in Mexico is to make 'alegria' candies and the first step is toasting the seeds to pop them and making them twice their size. A molasses syrup is prepared (cakes of brown sugar boiled in water, or honey, with pinches of anise, until thickened) and poured over the popped seed, kneaded and rolled into shapes. In northern India, popped grains are mixed with sugar syrup to make 'laddoo' sweets, or mixed with honey and flattened. 'Pinole' in Mexico is made by popping seeds then grinding them into flour to which sugar is added and the mixture eaten as a sweet.

Popping amaranth reduces reactive lysine value. Home popping tends to have the highest loss due to heat-damaged grains having to be discarded. Studies on flours of amaranth, and mixture of wheat flour with toasted, popped and raw amaranth flour showed, through amylograph analysis, some changes due to the popping which are not fully understood (Sanchez-Marroquin, 1988).

Extrusion

Whole flour can be blended with flours of other cereals and processed through extrusion cooking; or high protein fractions can be separated then blended with other flours and then processed through extrusion cooking. Pilot studies have shown that small industrial plants using this processing can produce baby foods, cereals and snacks, or protein drinks especially as hypoallergenic foods. A general processing for baby food products is shown in Figure 3.5 based upon air separation fractioning and extrusion.

Work has been carried out in Guatemala on extrusion-cooked amaranth flour, derived from *A. caudatus* and *A. cruentus* and using a Brady extruder. Functional properties based on water absorption index, water retention, amylographic viscosity, damaged starch and available lysine showed that extrusion cooking increased water absorption, water retention and damaged starch, decreased viscosity and did not affect the lysine availability (Mendoza Montano and Bressani, 1987).

Other processing

Flaking is of minor consequence but is used to produce 'hojuelos' in Peru. Flakes are prepared by a vapour boiling process. Slightly soaked grain is crushed between two iron mills previously heated to 200 °C. As with all processing, toasting and flaking improve the digestibility of the protein and carbohydrate. A study in Peru compared the composition

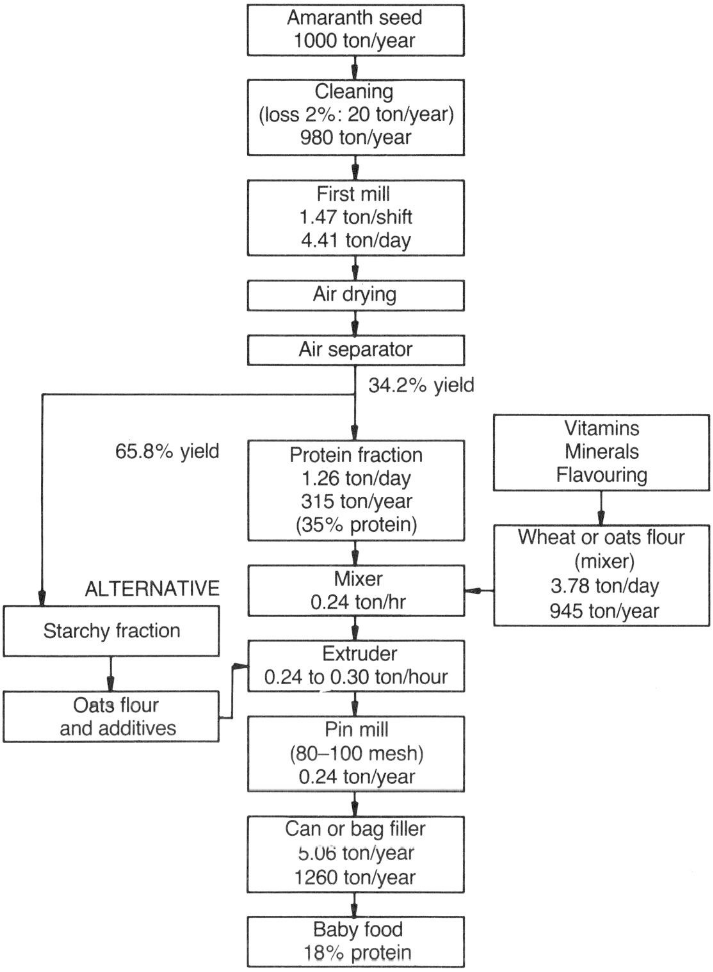

Figure 3.5 Processing of flour of grain amaranths (after Sanchez Marroquin, 1988).

of the products of *A. caudatus* when processed into flakes, popped or made into toasted flour (Table 3.17).

3.7.6 **Speciality uses**

Amaranth has a small role in special products for people with allergies to wheat-based foods. It is possible to combine amaranth with starches such as arrowroot or tapioca for production of pancakes, crackers, cakes and pie crusts (Jones, 1990).

Development of sourdough breads, which are lactic acid bacteria-

Table 3.17 Analysis of processed *A. caudatus* (per 100 g)

	Flaked	*Popped*	*Toasted flour*
Calories	385	399	389
Protein N × 6.25 g	13.24	13.43	14.59
Carbohydrate (g)	64.4	67.8	64.5
Fat (g)	8.2	8.2	8.1
Fibre (g)	3.43	3.55	6.48
Ash (g)	1.95	2.30	2.35

Source: Morales (1986).

fermented as compared to yeast breads, also increases the lysine content of the protein. Amaranth is being experimented upon for this purpose by a commercial company in the USA (Wallenborn, 1980).

Research is in progress on perfecting technologies in the development of new low-cost baby foods. Studies at the Centre for Economic and Social Studies for the Third World, at San Jeronimo Lidice, Mexico, have examined the possibility of incorporating seed germinates and sprouts into baby foods along with sprouts of alfalfa and chickpea, because sprouting increases the lysine content. Other items under development include, brittle, lollipops and marzipans incorporating amaranth. Important research and development in this area is ongoing at the Institute of Nutrition of Central America and Panama (INCAP) in Guatemala City.

Use of amaranth starch is superior to corn starch in salad dressing formulations and in thickening applications where viscosity is important (Singhal and Kulkarni, 1988). Stability of amaranth starch under heating is usually high but acidic conditions can limit its application (Lehmann, 1992).

The fine starch may find application in low-fat or fat-free products for desserts; also in coating food products, an emerging area of application (Lehmann, 1992).

3.7.7 **Use in animal nutrition**

Amaranth can be used as a forage crop and is grown in part for that purpose in China. Cultivation is similar to that for leafy vegetables with cuts made periodically through the growing season. Vegetable forms, rather than grain types, are used.

In Peru the stover is grazed or milled for use as a feed supplement after seed-heads have been harvested. There is also a substantial body of information on grain used as animal feed (Mallory *et al.*, 1990). Recent interest has focused on raw and processed grain (and vegetative parts of grain forms) for poultry (chickens, laying hens and broilers; Japanese quail; Spanish sparrow), rabbits, lambs and sheep, goats, cattle and pigs and experimental nutrition research has been carried out using rats.

Apart from nitrate and oxalate contents of leafy vegetable forms, fibrous agricultural residues and processed grain have been evaluated. In the case of broilers, extruded grain was assessed by Tillman and Waldroup (1988, 1988a, 1988b). In one experiment maize-soya diets were supplemented with 0, 10, 20, 30, 40 and 50% grain amaranth and formulated to provide nutrient requirements for starter, grower and finisher periods. Extruded grain amaranth was satisfactory up to 40% in broiler diets without any adverse effects on body weight, feed utilization, nor carcass or giblet yield. Values for apparent metabolizable energy (AME) were 3382 kcal/kg for male broilers 15–18 days old and if the AME is nitrogen-corrected the value was 3267 kcal/kg. Apparent amino acid availabilities were: methionine 85%, lysine 79%, arginine 92.9%.

For laying hens extruded amaranth can be effectively incorporated into maize-soya meal layer rations (Tillman and Waldroup, 1987). Raw and milled grain were satisfactory for growing chickens, but not popped amaranth (Acar *et al.*, 1988). They were satisfactory for quail (Saunders and Parrish, 1987). Processing is advantageous because raw grain contains heat-labile growth depressing anti-nutrients which affect chickens but not quail (Vohra *et al.*, 1989).

Use of grain amaranth crop residues for cattle was shown not to affect dry matter intake nor milk yield and also increased milk fat (Chairatanayuth, 1986). Grain amaranth 'straw' can be used for up to a level of 65% of the rations provided to lambs (Cervantes *et al.*, 1983).

3.8 GENETICS AND PLANT BREEDING

In recent years genetic research on amaranths has been carried out in India and in the USA; the former was largely based on cytology, crossibilities and hybridization and the latter was largely in relation to breeding and genetic structure of populations.

3.8.1 **Cytology and crossibilities**

All grain amaranths are diploid with 2n = 32, 34. Aneuploidy has been a factor leading to variation; aneuploids occur naturally and there have been hybridizations, especially where crop species and wild species grow together.

For *A. caudatus*, x = 16, for *A. cruentus*, x = 17 and for *A. hypochondriacus*, x = 16. For the putative progenitors, *A. hybridus*, *A. quitensis* and *A. powellii*, x = respectively 16, 16 and 17. One chromosome count for an accession of *A. caudatus* showed 2n = 30 (Xu, 1987).

Hybridization studies show few barriers between *A. cruentus*, A. hypochondriacus and *A. hybridus* but crosses between these and *A. caudatus* often result in inviable progeny (Pal and Khoshoo, 1972). However, crossing has been performed and some viable progeny obtained (Kauffman and Weber, 1990). Apparently success or otherwise in crossing depends on the specific populations used (Pal and Khoshoo, 1974), hence by the use of appropriate populations, it is possible to transfer genes among all cultivated grain species.

In relation to wider crossing, grain species can be crossed with a number of non-grain, non-progenitor weedy species such as *A. retroflexus* and amphiploidy has been induced in the progeny.

Tetraploids have been induced in all grain species (Misra *et al.*, 1971; Pal and Khoshoo, 1977). Tetraploids tend to be shorter plants and non-lodging. Tetraploids of *A. caudatus* were also found to contain more protein and increased lysine than diploids. In subspecies *mantegazzianus* induced tetraploids also showed increased threonine. Polyploidy has not yet been used in breeding.

3.8.2 **Breeding systems and pollination and methodology**

Breeding systems vary from high rates of selfing (over 90%) to up to 30% outcrossing. The degree of outcrossing varies with environmental factors (Figure 3.6) but there is a degree of genetic control enabling it to

Figure 3.6 Insects also assist outcrossing.

be modified to suit breeding procedures for either inbreeding or outbreeding species.

Stigmatic surfaces are receptive to pollen before the anthers mature and pollen maturation proceeds from the bottom to the top of the spikes. Since the major transfer of pollen is through wind some stigmas can obviously receive extraneous pollen when no pollen is available in the specific glomerule. Precautions must be taken to control pollination when seed supplies are increased. Pollination control is done with white paper bags over the inflorescence of field-grown plants, with plastic barriers within greenhouses, and with physical isolation of large field plantings (Weber and Kauffman, 1990).

For controlled pollinations (Figure 3.7), Rodale Research Center has used potted plants in glasshouses and emasculation before anthesis. Male and female flowers can be distinguished before anthesis because staminate flowers swell before the anthers are exserted. Pistillate flowers remain small until after pollination. Using headset magnifiers it is not difficult to remove staminate flowers.

For pollination, lower flowers which may have already selfed are removed and apical meristems are also removed. Flowers in about 1–2 cm are emasculated and pollinated. A gelatin capsule in which the desired pollen supply is placed is delivered and also used as protection.

Figure 3.7 Controlled pollination and seed production of amaranth germplasm at NCRPIS in Ames, Iowa.

Seed-set is better when pollinations are done on up to 3 successive days after emasculation.

More recently, since the availability of male sterile material, cross-pollination projects can be carried out more easily (Brenner, 1993).

3.8.3 **Breeding methodology**

Some methodologies used in crop management of amaranths have been outlined as follows.

Selection from germplasm holdings

Promising strains identified in germplasm collections, based on yield characters and earliness, are subjected to multilocation trials and compared to local checks. This method is widely used in India. Heritability estimates and expected genetic advance were found to be high for 1000 seed weight, inflorescence length and plant height, followed by seed yield and percentage protein (Joshi, 1986).

Mass and recurrent selection

Using two populations representing *A. cruentus* and *A. hypochondriacus*, S1 family analysis was carried out. Mass and recurrent selections for yield:height ratio and harvest index, plant weight and days to flowering were applied and selections advanced to the second generation by mass selection. The populations showed high direct and indirect selection gains indicating additive effects confirmed by variant component analysis for harvest index. In both populations mass selection was more efficient than S1 selection. Statistically proven useful parameters were plant height, head weight, threshing % and yield:height ratio (Ayiecho, 1986).

Simple mass selection of landraces over 3 cycles showed that selection gain was largest in the first cycle when 5% of the tallest high yielders were selected. Realized heritability estimates for plant height were 0.22 and 0.49 for two landraces investigated and 0.09 for yield in both (Vaidya and Jain, 1987).

Crossing and selection of recombinants

F2 populations have significant inbreeding depression compared to F1 hybrids. When F2 was subjected to bidirectional mass selection for time of first anthesis (2 cycles) and leaf length (1 cycle) responses to selection were asymmetrical. However, selection gains were significant and heritability estimates were in the range 0.08–0.19 for leaf length and 0.35–0.66 for anthesis time. Correlations between these parameters and yield showed that selection for optimal flowering time is likely to result in rapid yield improvement (Kulakow and Jain, 1987).

Grain protein contents in F1 and F2 of a diallel cross of six genotypes of *A. hypochondriacus* indicated that hybrids exceed the mean parental values and some exceeded the best parent. Heterosis was evident (Pandey and Pal, 1985).

Following the identification of gene markers, being bract length, plant and seed pigment patterns, leaf characteristics, time of flowering, determinate versus indeterminate growth and perisperm and starch of the grain, the potential for development of hybrids was advanced during the 1980s. Markers can be used to identify self-pollination in the F1 generation, and also several sources of cytoplasmic male sterility pave the way for additional hybridization technologies (Kauffman and Weber, 1990).

Interspecific hybridization

Since the grain species can be crossed this technique is simply an extension of the crossing between landraces of the same species. Rodale Research Center has produced a series of advanced breeding lines using diverse parents, e.g. *A. hypochondriacus* from Mexico and *A. hybridus* from Pakistan, and this became the major type grown in commercial plantings in the USA in 1988. It has an unbranched habit at high densities and reasonably early maturity and it performs better than selected Mexican landraces.

Segregating progeny of F2 and F3 generations are grown in widely diverse countries in Africa, Asia and the Americas and selection pressure at the particular locations results in evaluation and selection of progeny.

3.8.4 **Male sterility**

Genetic male sterility has been detected in *A. hypochondriacus* but not yet in *A. cruentus* and *A. caudatus* nor in other species (Brenner, 1993). Male sterile plants are identifiable by staminate flowers with shorter stamens and collapsed anthers due to degeneration of pollen during microsporogenesis. Peters and Jain (1987) developed a gene-cytoplasmic male sterile system that includes fertility restoration in the final cross so that growers could use F1 hybrids that have full male fertility.

The US National Plant Germplasm System has assembled male sterile accessions: five of these have been provided from Ingrid Peters of Brazil, six are selections from cv. K343 and one is a selection from cv. Plainsman. The male sterile accessions from cv. K343 have white or black seeds and were grown from rare black seeds in a predominantly white seed lot. The accession from Plainsman was first grown from a rare translucent seed in an otherwise normal seed lot (Brenner, 1993). The materials described by Peters and Jain were detected in Indian landraces. Gudu and Gupta (1988a) also found similar lines in Nepalese landraces.

There is great interest in the USA for hybrids. Making pollinations to produce hybrid seed is difficult except when male sterile mother plants are used.

3.8.5 **Breeding aims**

The following characteristics have been documented for the USA (1) where commercial cultivars amenable to mechanized harvesting are required, for India (2) where improvements in yield and quality are the most important, and for Mexico where both commercial production (3) and enhanced traditional cultivation but with yield increases are needed (4).

	1	2	3	4
Synchronous dry-down	+			
Reduced plant height	+	+	+	
Reduced branching	+			
Non-shattering seeds	+	+		
Food quality enhanced	+	+	+	+
High yields	+	+		+
Resistance to lodging			+	
Growing cycle of 120 days			+	
Late harvest				+

(Joshi and Rana, 1991; Kauffman, 1986; Maldonado and Estrada, 1986).

3.8.6 **Tissue culture for amaranth improvement**

Tissue culture was developed for seedling explants to provide leaf discs which callus and permit cell suspension cultures to be made. Embroyoids were also produced. Additionally, hypocotyl segments can be used as explants and also protoplasts have been isolated (Flores and Galston, 1980; Flores *et al.*, 1982). Despite these developments there were difficulties in regenerating plants from cultures but Bagga *et al.* (1987) reported on morphogenesis from *in vitro* cultured hypocotyl explants and, more recently, regeneration from inflorescence explants has been reported (Chakrovarty, 1993). This recent work opens the door for applying genetic engineering to amaranths.

In vitro flowering was reported by Tisserat and Galletta (1988) and further work at the Jawaharlal Nehru University, New Delhi by

Chakrovarty and co-workers, I.D. Arya and S.K. Sopary, suggests possibilities of transgenic plants. Other scientists in the same institution are identifying and cloning genes (Raina and Datta, 1992).

3.8.7 **Amaranth genetics**

A number of genetic loci have been identified especially for seed-coat colour, seedling colour, mature plant colour, leaf markings and pigment distribution. Other Mendelian characters that have been studied by S. Jain and co-workers include seed-shattering, sex expression, floral initiation, plant architecture and hybridization barriers.

3.8.8 **Germplasm collections**

Germplasm collecting took place in India and Central and South America in the 1970s. In 1981, IBPGR issued a report on the genetic resources of amaranths which laid out priorities for both vegetable and grain forms. After that time field collecting of primitive germplasm resulted in materials from Argentina, Bolivia, Ecuador, Peru, Guatemala and Mexico; Nepal, Bhutan, Thailand and Indonesia; Ethiopia, Zambia, Kenya and Nigeria.

IBPGR also, in the 1980s, designated the US National Plant Germplasm System and the National Bureau of Plant Genetic Resources in India to hold collections of the materials in long-term storage on behalf of the international community of users.

A meeting on germplasm of amaranths with special reference to vegetable forms had been held at Rodale Research Center in September 1979 and some researchers were asked to co-ordinate germplasm activities at the continental level. Soon afterwards the set of descriptors and their states were developed for use in documenting collections.

There now exist important collections of both grain and vegetable forms and major support for the field collecting to amplify these collections has been provided by Rodale Press, Rockefeller Foundation, IBPGR and NBPGR (India). AVRDC, Taiwan, China and the Mayaquez Institute of Tropical Agriculture, Puerto Rico have also donated significant germplasm to major collections.

Studies on germplasm collections

By 1984, Rodale Research Center had published a catalogue of 1133 accessions (Kauffman and Reider, 1984) and the grouping of accessions in grain types was advanced. The Indian materials collected by Joshi numbered 2250 accessions and these have been scored for 40 characters (Joshi and Rana, 1991) and the data incorporated into a catalogue.

Figure 3.8 *Amaranthus australis* (giant anaranth), in the germplasm collection at the NCRPIS in Ames, Iowa.

The North Central Regional Plant Introduction Station (NCRPIS) in Ames, Iowa maintains a collection of 3137 germplasm accessions from many parts of the world (Figure 3.8). Of these, 1357 are available for researchers. The remainder will be made available when fresh high-viability seeds are regenerated from small planting stocks.

The highest priority for germplasm acquisition are the wild species. Only 26 of the estimated 60 species in the genus are in the collection of the NCRPIS. The remainder should be acquired and made available for germplasm screening and other uses.

Further research on South America materials has been carried out by Peruvian Universities, especially Cuzco (and the materials there cover Peru, Bolivia and Ecuador) and by the project of the Peruvian Ministry of Agriculture (INIPA) on Andean crops in Lima. In the 1980s national genetic resources programmes in Bolivia, Ecuador and Colombia also collaborated and assistance was provided by IICA (a regional organization with a programme on Andean cultigens), IDRC (Canada) and IBPGR.

Mexican materials have also been looked at in more detail by a group of researchers at the University of Chapingo (Espitia, 1987) and Guatemalan materials have been the subject of research in the University of San Carlos (Juarez Gonzalez, 1984).

All these collections represent landrace materials and more detailed studies of the genetic structure of some populations have been carried out at the University of California, Davis by S.K. Jain and co-workers to compare variability within and between populations. Data show that the New World landraces are highly differentiated between populations (Hauptli and Jain, 1984) and each population behaves as a collection of nearly homozygous lines. Indian populations have also shown evidence of high levels of heterozygosity for quantitative characters (Vaidya and Jain, 1987) but less differences than between New World populations as measured by isozymes (Jain *et al.*, 1980). Such data confirm the needed approach in breeding through inter-population hybridization and selection to recombine and exploit population differences.

Allozyme studies have proved useful as assays of genetic variation (Jain *et al.*, 1980a) using the following: alcohol dehydrogenase, glutamine oxalacetic transaminase, leucine amino peptidase, acid phosphatase, malate dehydrogenase, shikimic acid dehydrogenase and phosphoglucoseisomerase (Hauptli and Jain, 1984a), and Gudu and Gupta (1988) examined peroxidase, esterase, phosphatase and leucine amino peptidase.

Morphological characters analysed by multivariate analysis have also proved useful in clustering accessions and identifying useful ones for breeding (Fatokun, 1985).

Electrophoresis of seed proteins contained two variants of 52 and 56 kilodaltons. Accessions with only 52 predominate and only a few have 56 or both. This pattern does not correlate with taxonomy but the 38 accessions examined was too small a number to derive data on variation in the genepool (Fairbanks *et al.*, 1990).

3.9 FUTURE PROSPECTS

In areas of traditional cultivation, yields have been low and farmers have often chosen other crops which are more productive and which fit better in terms of timing into cropping cycles. This situation has been exacerbated by the fact that seeds of improved varieties were not available and the negative attributes of the crop – seed-shattering and lodging in particular – led to grain amaranths not being more widely grown. However, this situation is changing rapidly.

Another problem has been that amaranths were often grown on marginal lands with low-level subsistence management. Whereas the first modern varieties were selected for better agricultural lands, this situation is also changing as landrace materials are being tested more widely and segregation of elite lines tested in a number of marginal habitats.

Also, areas of the developing world need to apply appropriate technology to develop suitable farm machinery for sowing, threshing and winnowing grain amaranth. This is beginning but has a long way to go.

Expansion of cultivation will be justified when there is new demand for the produce. A good example of this is seen in Guatemala where an Amaranth Food Company has developed products for feeding small children and pregnant women. In developed countries amaranth continues to supply a speciality market; however, research on processing and product diversification has expanded in both Mexico and Guatemala.

In all probability, grain amaranths will never compete with cereals as a staple crop but its improvement in hill areas is fully justified and indeed ongoing in many countries.

The relatively large amount of research over the past 20 years in many parts of the world has transformed the crop from one that is primitively cultivated to one that is commercial, and cultivation is expected to expand to satisfy new market niches for high-quality protein foods. As a result, in the USA, the Amaranth Institute has been founded for growers, researchers, manufacturers and any other people involved with amaranths. As cultivation expands and yields are increased, the grain price should become lower and that trend is already being seen in the USA. Obviously price has to become competitive with that of cereals, both in terms of price per hectare and available protein per hectare.

REFERENCES

Acar, N., Vohra, P., Becker, R., Hanners, G.D. and Saunders, R.M. (1988) Nutritional evaluation of grain amaranth for young chickens. *Poultry Science*, **67**, 1166–73.

Angus, J.F., Mackenzie, D.E., Myers, R.J.K. and Foale, M. (1982) Phasic development in field crops 3. The pseudocereals, buckwheat and grain amaranth. *Field Crops Research*, **5**, 305–18.

Ayorinde, F.O., Clifton, J. Jr, Afolabi, O.A. and Shepard, R.L. (1988) Rapid transesterification and mass spectrometric approach to seed oil analysis. *J. Amer. Oil Chem. Soc.*, **65**, 942–7.

Ayiecho, P.O. (1986) Quantitative studies in two grain amaranth populations using two selection methods. Ph.D. thesis, Univ. of California, Davis.

Babatola, J.O. and Awoderu, J.B. (1986) Screening of *Amaranthus* germplasm against root-knot nematode, *Meloidogyne incognita*. *Ann. Appl. Biol.*, **108**, 140–1.

Bagga, S., Venkateshwarulu, K. and Sopory, S.K. (1987) *In vitro* regeneration of plants from hypocotyl segments of *Amaranthus paniculatus*. *Plant Cell Reports*, **6**, 183–4.

Becker, R., Irving, D.W. and Saunders, R.M. (1986) Production of debranned amaranth flour by stone milling. *Lebensm. Wiss. u.-Technol.*, **19**, 372–5.

Becker, R., Wheeler, E.L., Lorenz, K., Stafford, A.E., Grosjean, O.K., Betschart, A.A. and Saunders, R.M. (1981) A compositional study of amaranth grain. *J. Sci. Food and Agric.*, **38**, 347–56.

Betschart, A.A., Irving, D.M., Shephard, A.D. and Saunders, R.M. (1981) *Amaranth cruentus*: milling characteristics, distribution of nutrients within seed components, and the effect of temperature on nutritional quality. *J. Food Sci.*, **46**, 1181–7.

Brenner, D. (1990) *Amaranth Catalog*. USDA, Regional Plant Introduction Station, Ames, Iowa.

Brenner, D. (1990a) Seed shattering control with indehiscent utricles in grain amaranths. *Legacy*, **3**(1), 2–3.

Brenner, D. (1993) Hybrid seed for increased amaranth grain yield. *Legacy*, **6**(1), 9–11.

Bressani, R., Gonzales, J.M., Elias, L.G. and Melgar, M. (1987) Effect of fertilizer application on the yield, protein and fat content, and protein quality of raw and cooked grain of three amaranth species. *Qualitas Plantarum*, **37**, 59–67.

Bressani, R., Gonzales, J.M., Zuniga, J., Breuner, M. and Elias, L.G. (1987a) Yield, selected chemical composition and nutritive value of 14 selections of amaranth grain representing four species. *J. Sci. Food and Agric.*, **38**, 347–56.

Bunce, J.A. (1987) Species-specific responses to water stress of gas exchange parameters mimicked by applied abscissic acid. *Can. J. Bot.*, **63**, 103–6.

Calderon de la Barca, A.M. and Moreno, L.V. (1988) *Amaranthus cruentus* lectin: purification, stability and some biochemical properties. *J. Food Biochem.*, **12**, 117–26.

Campbell, M. and Foy, C.D. (1987) Selection of grain amaranth species for tolerance to excess aluminium in an acid soil. *J. Plant Nutrition*, **10**, 249–60.

Carlsson, R. (1980) Quantity and quality of *Amaranthus* grain from plants in temperate, cold and hot, subtropical climates – A review, in *Proceedings of the Second Amaranth Conference*, Rodale Press, Emmaus, PA, pp. 48–58.

Cavagnaro, J.S. (1985) Effect of water stress and dry matter partitioning in four species of *Amaranthus*. M.S. thesis, Univ. of California, Davis.

Cervantes, S., J.M., Riquelme, V., E. and Gonzales, M., S. (1983) Evaluacion de la paja de alegria (*Amaranthus hypochondriacus*) como forraje para ruminantes, in *Memorias de la XVI Reunion de la Asociacion Mexicana de Produccion Animal*, Chapingo.

Chairatanayuth, P. (1986) Inclusion of amaranth crop residue in diets for cattle. Ruminant feeding systems utilising fibrous agricultural residues, in *Proceedings 5th. Annual Workshop of the Australian–Asian Fibrous Agricultural Residues Research Network* (ed. R.M. Dixon), International Development Programme of Australian Universities, Canberra, pp. 131–5.

Chakrovarty, T. (1993) Tissue culture for amaranth crop improvement. *Legacy*, **6**, 5–7.

Chukwuma, G.O. (1988) Water deficit levels and yield of African spinach. Dept. Agric. Eng., University of Nigeria (mimeographed).

Cole, J.N. (1979) *Amaranth from the Past For the Future*, Rodale Press, Emmaus, PA.

Coons, M.P. (1982) Relationships of *Amaranthus caudatus* (morphology, mixed species stands, including *A. hybridus*, crosses, hybrids, genetics). *Econ. Bot.*, **36**, 129–46.

Covas, G. (1987) Fitomejoramiento de los aramantos, in *Actas de la Primeras Jornadas Nacionales sobre Amarantos*, (eds A.D. Goldberg and G. Covas), Univ. Nac. de la Pampa, La Pampa, Argentina, pp. 50–5.

Cramer, C. (ed.) (1988) Montana releases new amaranth line. *Amaranth Today*, **4**(2-3), 6.

Der Marderosian, A.D., Beutler, J., Pfendner, W., Senft, J., Chambers, J., Yoder,

R., Weinsteiger, E. (1980) Nitrate and oxalate content of vegetable amaranth. Rodale Research Report No. 4, Rodale Institute Research Center, Kutztown, PA.

Early, D. (1978) *Hautli*: The revival of Aztec amaranth, an appropriate technology food. Paper presented at 77th Annual Meeting American Anthropological Association, Los Angeles.

Early, D.K. (1990) Amaranth production in Mexico and Peru, in *Advances in New Crops* (eds J. Janick and J.E. Simon), Timber Press, Portland, OR, pp. 140–2.

Edwards, A.D. and Volak, B. (1980) Grain amaranth: Optimisation of field population density, in *Proceedings of the Second Amaranth Conference*, Rodale Press, Emmaus, PA., pp. 91–5.

Elias, J. (1977) Food value of amaranth: greens and grain, in *Proceedings of the First Annual Seminar*, Rodale Press, Emmaus, PA., pp. 17–35.

Endres, C. (1986) Influence of production practices on yield and morphology of *Amaranthus cruentus* and *Amaranthus hypochondriacus*. M.S. thesis, University of Arkansas.

Espitia, R.E. (1987) Caracterizacion y evaluacion preliminar de germoplasma de *Amaranthus*, in *Memorias del Coloquio Nacional del Amaranto* (eds R. Trujillo, T., G. Suarez Ramos and J.M. Cervantes Sanchez), Queretaro, Mexico, pp. 113–26.

FAO (1988) *Traditional Food Plants. A Resource Book for Promoting the Exploitation and Consumption of Food Plants in Arid, Semi-arid and Sub-humid Lands of Eastern Africa*. Food and Agricultural Organization, Rome, Italy.

Fairbanks, D.J., Robinson, L.R., Maughan, P.J. and Ballen, J.G. (1990) Electrophoretic characterization of amaranth seed proteins. *Amaranth*, Aug. 1990, 189–90.

Farquar, G.D., Hubich, K.T., Terashima, I., Condon, A.G. and Richards, R.A. (1987) Genetic variation in the relationship between photosynthetic CO_2, assimilation rate and the stomatal conductance of water loss. *Progress in Photosynthesis Res.*, 1987, pp. 209–12.

Fatokun, C.A. (1985) Multivariate studies of the variability in cultivated amaranths. *Beitrage zur Tropischen Landwirtschaft und Veterinarmedizin*, **23**, 267–75.

Feine, L.B. (1980) An ethnobotanical observation and collection of grain amaranth in Mexico, in *Proceedings of the Second Amaranth Conference*, Rodale Press, Emmaus, PA, pp. 111–18.

Flores, H.E. and Galston, A.W. (1980) Tissue culture of amaranth: Initial results and prospects, in *Proceedings of the Second Amaranth Conference*, Rodale Press, Emmaus, PA, pp. 105–8.

Flores, H.E., Thier, A. and Galston, A.W. (1982) *In vitro* culture of grain and vegetable amaranths (*Amaranthus* spp.), *Amer. J. Bot.*, **69**, 1049–54.

Frost, R.A. and Cavers, P.B. (1975) The ecology of pigweeds (*Amaranthus*) in Ontario I. Interspecific and intraspecific variation in seed germination among local populations of *A. powellii* and *A. retroflexus*. *Can. J. Bot.*, **53**, 1276–84.

Garcia, L.A., Alfano, M.A. and Bressani, R. (1987) Digestibility and nutritional value of crude-oil from three amaranth species. *J. Amer. Oil Chem. Soc.*, **64**, 371–5.

Getahun, A. (1974) The role of wild plants in the native diet in Ethiopia. *Agro-Ecosystem*, **1**, 45–56.

Grubben, G.J.H. (1977) *Tropical Vegetables and their Genetic Resources*. International Board for Plant Genetic Resources, Rome, Italy.

Gudu, S. and Gupta, V.K. (1988) Electrophoresis as an aid in the identification of various species and cultivars of grain amaranths. *Acta Hort.*, **218**, 231–8.

Gudu, S. and Gupta, V.K. (1988a) Male-sterility in the grain amaranth (*Amaranthus hypochondriacus* ex-Nepal) variety *Junila. Euphytica*, **37**, 23–6.
Gupta, V.K. (1986) Grain amaranths in Kenya, in *Proceedings of the Third Amaranth Conference*, Rodale Press, Emmaus, PA, pp. 114–19.
Gupta, V.K., Kermali, I. and Ngugi, D.N. (1984) Breeding for low antinutrient factors in vegetable amaranths. *Amaranth Newsl.*, **2**, 2–3.
Hauptli, H. (1977) Agronomic potential and breeding strategy for grain amaranths, in *Proceedings of the First Amaranth Seminar*, Rodale Press, Emmaus, PA, pp. 71–9.
Hauptli, H. and Jain, S.K. (1984) Genetic structure of landrace populations of the New World grain amaranths. *Euphytica*, **33**, 857–84.
Hauptli, H. and Jain, S.K. (1984a) Allozyme variation and evolutionary relationships of grain amaranths (*Amaranthus* spp.). *Theor. Appl. Genetics*, **69**, 153–65.
Hauptli, H., Lutz, R.L. and Jain, S.K. (1980) Germplasm exploration in Central and South America, in *Proceedings of the First Amaranth Seminar*, Rodale Press, Emmaus, PA, pp. 117–22.
Hendershot, A.T. (1986) Food quality expectations of a food company, in *Proceedings of the Third Amaranth Conference*, Rodale Press, Emmaus, PA, pp. 69–83.
Hunziker, A.T. (1952) *Los Pseudocereals de la Agricultura Indigera de America.* Univ. Buenos Aires, Cordoba and Buenos Aires.
Hunziker, A.T. (1987) Taxoonomia de las especes de amarantos cultivados y de los silvestres relacionades, in *Actas de las Primeros Jornadas Nacionales sobre Amarantos* (eds A.D. Golding and G. Covas), Univ. Nac. de La Pampa, La Pampa, Argentina, p. 10.
Hunziker, A.T. and Planchuelo, A.M. (1971) Sobre un nievo hallazgo de *Amaranthus caudatus* entumbas indigenas de Argentina. *Kurtziana*, **6**, 63–7.
Imeri, A., Gonzalez, J.M., Flores, R., Elias, L.G. and Bressani, R. (1987) Genetic variability, and correlations between yield, grain size, chemical composition and protein quality of 25 cv. of amaranth (*Amaranthus caudatus*). *Archivos Latinoamericanos de Nutricion*, **37**, 132–46.
Irving, D.W., Betschart, A.A. and Saunders, R.M. (1980) Morphological studies on *Amaranthus hypochondriacus*, in *Proceedings of the Second Amaranth Conference*, Rodale Press, Emmaus, PA, pp. 60–1.
Jain, S.K., Kulakow, P.A. and Peters, I. (1986) Genetics and breeding of grain amaranth: Some research issues and findings, in *Proceedings of the Third Amaranth Conference*, Rodale Press, Emmaus, PA, pp. 174–91.
Jain, S.K., Vaidya, K.R. and Joshi, B.D. (1980) Collection and evaluation of Indian grain amaranths, in *Proceedings of the Second Amaranth Conference*, Rodale Press, Emmaus, PA, pp. 123–8.
Jain, S.K., Wu, L. and Vaidya, K.R. (1980a) Levels of morphological and allozyme variation in Indian amaranths: a striking contrast. *J. Heredity*, **71**, 283–5.
Jones, M.H. (1990) Amaranth in a wheat free diet, in *Proceedings of the Third Amaranth Conference*, Rodale Press, Emmaus, PA, pp. 89–90.
Joshi, B.D. (1985) Collection, evaluation and conservation of grain amaranth, buckwheat and chenopod. Paper presented at Third Workshop of AICRP on Under-utilised and Under-exploited Crop Plants, New Delhi.
Joshi, B.D. (1986) Genetic variability in grain amaranth. *Indian J. Agric. Sci.*, **56**, 574–6.
Joshi, B.D. and Rana, R.S. (1991) *Grain Amaranths: The Future Food Crop.* National Bureau of Plant Genetic Resources, Shimla, India.
Juarez Gonzalez, J.R. (1984) Caracterizacion prelimar de 16 muestas de bledo

(*Amaranthus* spp.) de las regiones del occidente, centro y oriente de Guatemala converacion del germoplasma. Thesis, University of San Carlos.

Kauffman, C.S. (1986) Thoughts on the development of improved varieties of grain amaranth, in *Proceedings of the First Amaranth Seminar*, Rodale Press, Emmaus, PA, pp. 168–73.

Kauffman, C.S. (1992) Realising the potential of grain amaranth. *Food Reviews*, **8**, 5–21.

Kauffman, C.S. and Reider, C. (1984) *Rodale Amaranth Germplasm Collection*, Rodale Press, Emmaus, PA.

Kauffman, C.S. and Weber, L.E. (1990) Grain amaranths, in *Advances in New Crops* (eds J. Janick and J.E. Simon), Timber Press, Portland, pp. 127–39.

Koeppe, S.J. and Rupnow, J.H. (1988) Purification and characterisation of a lectin from the seeds of amaranth (*Amaranthus cruentus*). *J. Food Sci.*, **53**, 1412–17.

Kohli, R.K., Sawhney, N. and Sawhney, S. (1980) Photo-induced changes in proteins associated with floral induction in *Amaranthus*. *Plant and Cell Physiol.*, **21**, 1483–90.

Konishi, Y., Fumata, Y., Ikeda, K., Okuno, K. and Fuwa, H. (1985) Isolation and characterisation of globulin from seeds of *Amaranthus hypochondriacus* L. *Agric. Biol. Chem.*, **49**, 1453–9.

Kulakow, P.A. (1987) Genetics of grain amaranths II. The inheritance of determinance, panicle orientation, dwarfism and embryo color in *Amaranthus caudatus*. *J. Heredity*, **78**, 293–7.

Kulakow, P. and Jain, S. (1987) Genetics of grain amaranths. IV. Variation and early generation response to selection in *Amaranthus cruentus* L. *Theoret. and Appl. Genetics*, **74**, 113–20.

Lehmann, J.W. (1992) Grain amaranth. *International Food Ingredients*, **3**, 26–37.

Lehmann, J.W., Qureshi, A.A. and Putnam, D.H. (1990) Genetic variability of vitamin E isomers in grain amaranths. *Agronomy Abstracts*: Annual Meeting, Crop Science Society of America.

Lorenz, K. (1981) *Amaranthus hypochondriacus* – Characteristics of the starch and baking potential of the flour. *Starke*, **33**, 149–53.

Lorenz, K. and Hwang, Y.S. (1985) Lipids in amaranths. *Nutrition Reports International*, **31**, 83–9.

Lorenz, K. and Wright, B. (1984) Phytate and tannin content of amaranth. *Food Chemistry*, **14**, 27–34.

Lyon, C.S. and Becker, B. (1987) Extraction and refining of oil from amaranth seed. *J. Amer. Oil Chem. Soc.*, **64**, 233–6.

Maldonado, A.U. and Estrada, L.A. (1986) Amaranth genetic variation and utilization in Mexico, in *Proceedings of the Third Amaranth Conference*, Rodale Press, Emmaus, PA, pp. 120–4.

Mallory, E., Blair, M. and Weber, L. (1990) *Forage and Grain for Livestock, Bibliography of Cultivated Amaranths*, Rodale Press, Rodale Research Centre, Kutztown, PA.

Mehra, P.N. and Murty, A.S. (1988) Some promising life support species for marginal (alkali) soils, in *Life Support Plant Species* (eds P.S. Paroda, P. Kapoor, R.K. Arora and B. Mal), National Bureau of Plant Genetic Resources, New Delhi, pp. 86–91.

Mendoza Montano, C.E. and Bressani, R. (1987) Nutritional and functional characteristics of extrusion-cooked amaranth flour. *Cereal Chem.*, **641**, 218–22.

Miller, T.E., Wing, J.S. and Huerte, A.R. (1984) The agricultural potential of selected C4 plants in arid environments. *J. Arid Environments*, **7**, 275–86.

Misra, P.S., Pal, M. and Pandey, R.M. (1985) Studies on the effect of varying plant density on the growth and yield of grain amaranth (*Amaranthus hypochondriacus* L.). *Amaranth Newsl.*, **3**.

Misra, P.S., Pal, M., Mitra, C.R. and Khoshoo, T.N. (1971) Chemergic studies on some diploid and tetraploid grain amaranths. *Proc. Ind. Acad. Sci.*, **B 74**, 155–60.

Misra, P.S., Pandey, R.M. and Pal, M. (1983) Amino acid composition in *Amaranthus*. *Fitoterapia*, **54**, 135–9.

Mnzava, N.A. and Masam, A.M. (1985) Regeneration potential, leaf seed yield of vegetable amaranth (*Amaranthus cruentus* L.) as a function of initial topping heights. *Acta Hort.*, **153**, 151–160.

Morales, E. (1986) Digestibility and utilization of grain amaranth protein and energy by small children, in *Proceedings of the Third Amaranth Conference*, Rodale Press, Emmaus, PA, pp. 157–66.

Morales, E.J., Lembcke, J. and Graham, G.G. (1988) Nutritional value for young children of grain amaranth and maize-amaranth mixtures; effect of processing. *J. Nutrition*, **118**, 78–85.

Nabham, G.P. (1980) Amaranth cultivation in the U.S., Southwest and Northwest Mexico, in *Proceedings of the Second Amaranth Conference*, Rodale Press, Emmaus, PA, pp. 129–33.

Naidu, K.R., Seethambaram, Y. and Das, V.S.R. (1982) Leaf proteinase and nitrate reductase activity in relation to grain protein levels and grain yield in four species of grain amaranth (*Amaranthus hypochondriacus*, *Amaranthus caudatus*, *Amaranthus cruentus*, *Amaranthus edulis*). *Proc. Ind. Acad. Sci.*, **91**, 433–41.

Nable, R.O. and Brownell, P.F. (1984) Effect of sodium nutrition and light upon the concentrations of alanine in leaves of C4 plants. *Australian J. Plant Physiol.*, **11**, 319–24.

Naseema, A., Balakrishnan, S. and Nair, M.C. (1983) Pathology and control of seed mycoflora of some vegetables in Kerala. *Agric. Res. J. Kerala*, **21**, 32–7.

National Academy of Sciences (NAS) (1975) *Underexploited Tropical Plants with Promising Economic Value*. National Academy Press, Washington D.C.

National Research Council (NRC) (1984) *Amaranth. Modern Prospects for an Ancient Crop*. National Academy Press, Washington D.C.

Novoa, A. and Posner, J.L. (eds) (1981) *Agricultura de Ladera en America Tropical* CATIE/Rockefeller Foundation, Turrialba, Costa Rica.

Ohta, D., Matoh, T. and Takahashi, E. (1988) Sodium-stimulated NO_3 uptake in *Amaranthus tricolor* L. *Plant Physiology*, **57**, 2223–5.

Okun, K. and Sakaguchi, S. (1982) Inheritance of starch characteristics in perisperm of *Amaranthus hypochondriacus*. *J. Heredity*, **73**, 46–7.

Oladiran, J.A. and Mumford, P.M. (1985) The stimulation of seed germination by temperature and light in agronomic *Amaranthus* species. *Biochemie und Physiologie der Pflanzen*, **180**, 45–54.

Pal, M. (1972) Evolution and improvement of cultivated amaranths. Breeding system and inflorescence structure. *Proc. Ind. Acad. Sci.*, **38**, 28–37.

Pal, M. and Khoshoo, T.N. (1972) Evolution and improvement of cultivated amaranths. 6 Cytogenetic relationships in grain types. *Theor. Appl. Gen.*, **43**, 242–51.

Pal, M. and Khoshoo, T.N. (1974) Grain amaranths. Evolutionary studies in world crops, in *Diversity and Change in the Indian Sub Continent*, (ed. J. Hutchinson), Cambridge University Press, Cambridge, pp. 129–37.

Pal, M. and Khoshoo, T.N. (1977) Evolution and improvement of cultivated amaranths. 8 induced autotetraploidy in grain types. *Z. Pflanzenzuchtung*, **78**, 135–48.

Pandey, R.M. and Pal, M. (1985) Genetics of grain protein in *Amaranthus*. *Crop Improvement*, **12**, 55–8.

Pant, K.C. (1985) Effect of heat processing (popping) on protein nutritional quality of grain amaranth. *Nutrition Reports International*, **32**, 1089–98.

Paredes Lopez, O., Mora Escobedo, R. and Ordorica Falomir, C. (1988) Isolation of amaranth proteins. *Lebensm.-Wiss. u.-Technol.*, **21**, 59–61.

Pederson, B.L., Kalinowski, L.S. and Eggum, B.O. (1987) The nutritive value of amaranth grain (*Amaranthus caudatus*) I. Protein and minerals of raw and processed grain. *Qualitas Plantarum*, **36**, 309–24.

Peters, I. and Jain, S.K. (1987) Genetics of grain amaranths III. Gene cytoplasmic male-sterility. *J. Heredity*, **78**, 251–56.

Raina, A. and Datta, A. (1992) Molecular cloning of a gene encoding a seed specific protein with nutritionally balanced amino acid composition from *Amaranthus*. *PNAS (USA)*, **89**, 11774–8.

Ram, R.D. (1980) A mosaic virus on *Amaranthus hypochondriacus* L. *Progressive Horticulture*, **12**, 15–20.

Reddy, B.N., Chakravarti, B.P. and Shekhawat, P.S. (1980) Nutritional requirements of *Xanthomonas amaranthicola*, causal organism of bacterial leaf spot of *Amaranthus*. *Indian Phytopath.*, **32**, 581–6.

Robinson, R. (1986) *Amaranth, Quinoa, Ragi, Tef and Niger: Tiny Seeds of Ancient History and Modern Interest*. Bulletin AD–SB–2949, Agricultural Experiment Station, University of Minnesota.

Sanchez-Marroquin, A. (1988) Amaranth as an enriching product in staple foods, in *Proceedings of the Third Amaranth Conference*, Rodale Press, Emmaus, PA, pp. 20–45.

Sanchez-Marroquin, A. and Maya, S. (1985) Industrial corn flour enrichment with whole amaranth flour and milling fractions in corn-based products. *Archivos Latinamericanos de Nutricion*, **35**, 518–35.

Sanchez-Marroquin, A., Maya, S. and Perez, L.P. (1980) Agroindustrial potential of amaranth in Mexico, in *Proceedings of the Second Amaranth Conference*, Rodale Press, Emmaus, PA, pp. 95–104.

Santacruz Ulibarri, H. and Marban Mendoza, N. (1986) Asociacion de *Nabbocus aberrans* a tres variedades de alegria, *Amaranthus hypochondriacus*, in *El Amaranto* (Amaranthus spp., *Alegria*): *Su Cultivo y Aprovechamiento*, (eds A. Trinidad Santos, F. Gomez Lorrence and G. Suarez Ramos), Postgraduate College, Chapingo, pp. 186–92.

Sauer, J.D. (1950) The grain amaranths: a survey of their history and classification. *Ann. Missouri Botanical Garden*, **37**, 561–616.

Sauer, J.D. (1967) The grain amaranths and their relatives: a revised taxonomic and geographic survey. *Ann. Missouri Botanical Garden*, **54**, 102–37.

Sauer, J.D. (1969) Identity of archaeologic grain amaranths from the valley of Tehuacan, Puebla, Mexico. *American Antiquity*, **34**, 80–1.

Saunders, R.M. (1990) Nutritional and starch composition studies with grain amaranths, in *Proceedings of the Third Amaranth Conference*, Rodale Press, Emmaus, PA, pp. 46–62.

Saunders, R.M. and Becker, R. (1983) *Amaranthus*: A potential Food and Feed resource, in *Advances in Cereal Science and Technology*. Vol. VI, (ed. Y. Pomeranz), Amer. Assoc. Cereal Chem., St. Paul, MIN, pp. 357–96.

Saunders, D.K. and Parrish, J.W. (1987) Assimilated energy of seeds consumed by scaled quail in Kansas. *J. Wildlife Management*, **51**, 787–90.

Schmidt, D. (1977) Grain amaranth: a look at some potentials, in *Proceedings of the First Amaranth Seminar*, Rodale Press, Emmaus, PA, pp. 121–33.

Sealy, R.L., Kenerley, C. and McWilliams, E.L. (1988) Evaluation of amaranth

accessions for resistance to damping-off by *Pythium myriotylum*. *Plant Disease*, **72**, 985–9.
Senthong, C. (1986) Yield improvement, agronomy and local use of amaranth, in *Proceedings of the Third Amaranth Seminar*, Rodale Press, Emmaus, PA, pp. 108–19.
Sharma, P.N. and Chowfla, S.C. (1987) Studies on a mosaic disease of amaranth (*Amaranthus caudatus*) in Himachal Pradesh. *Indian J. Mycology and Plant Path.*, **16**, 349–50.
Sharma, A.D., Thakur, V. and Munjal, R.L. (1981) Seed mycoflora of *Amaranthus caudatus*, its pathology and control. *Indian Phytopath.*, **33**, 242–4.
Siddiqui, M.A. and Alan, M.M. (1988) Screening *Amaranthus* species against the root-knot nematode, *Meloidogyne incognita*. *Ann. Appl. Biol.*, **108**, 140–1.
Singh, H. (1961) *Grain Amaranths, Buckwheat and Chenopods*. Indian Council Agricultural Research, New Delhi.
Singhal, R.S. and Kulkarni, P.R. (1988) Composition of the seeds of some *Amaranthus* species. *J. Sci. Food and Agric.*, **42**, 325–31.
Stallknecht, G.F. and Schulz-Schaffer, J.R. (1993) Amaranth rediscovered, in *New Crops*, (eds J. Janick and J.E. Simon), John Wiley & Sons, New York, pp. 211–21.
Stone, L.A. and Lorenz, K. (1984) The starch of *Amaranthus*: physico-chemical properties and functional characteristics (*Amaranthus cruentus, Amaranthus hypochondriacus*). *Starke*, **36**, 232–7.
Suarez-R., G. and Mark-E., E. (1980) Deposito de tanninas en la testa de *Amaranthus hypochondriacus* L. (alegria). *Agrociencia*, **42**, 35–50.
Sumar, K.L., Pacheco, N. and Aquirre, J.R. (1986) Chemical vs. organic fertilization of grain amaranth, in *Proceedings of the Third Amaranth Conference*, Rodale Press, Emmaus, PA, pp. 125–32.
Tapia, M. and Mateo, N. (1990) Andean phytogenetic and zoogenetic resources, in *Mountain Agriculture and Crop Genetic Resources*, (eds K.W. Riley, N. Mateo, G.C. Hawtin and R. Yadav), Oxford & IBH Pub. Co., New Delhi, pp. 235–53.
Tillman, P.B. and Waldroup, P.W. (1987) Effects of feeding extruded grain amaranth to laying hens. *Poultry Science*, **66**, 1679–1701.
Tillman, P.B. and Waldroup, P.W. (1988) Performance and yields of broilers fed extruded grain amaranth and grown to market weight. *Poultry Science*, **67**, 743–9.
Tillman, P.B. and Waldroup, P.W. (1988a) Assessment of extruded grain amaranth as a feed ingredient for broilers 1. Apparent metabolizable energy values. *Poultry Science*, **67**, 641–6.
Tillman, P.B. and Waldroup, P.W. (1988b) Assessment of extruded grain amaranth as a feed ingredient for broilers 2. Apparent amino acid values. *Poultry Science*, **67**, 647–51.
Tisserat, B. and Galleta, P.D. (1988) *In vitro* flowering in *Amaranthus*. *Hort. Sci.*, **23**, 210–12.
Uzo, J.O. and Okorie, A.U. (1983) *Amaranthus hybridus*: a potential grain crop for West Africa. *Nutrition Reports International*, **27**, 519–24.
Vaidya, K.R. and Jain, S.K. (1987) Response to mass selection for plant height and grain yield in amaranth. *Z. Pflanzensuchtung*, **98**, 61–4.
Vohra, P., Bersch, S., Acar, N. and Becker, R. (1990) Use of grain amaranth in diets of Japanese quail and chickens. *Nutrition Reports International*, **39**, 1227–35.
Wagoner Haas, P. (1980) The Rodale amaranth germplasm collection, in *Proceedings of the Second Amaranth Conference*, Rodale Press, Emmaus, PA, pp. 135–41.

Wallenborn, P. (1990) Prospectives for sourdough baking with amaranth flour, in *Proceedings of the Third Amaranth Conference*, Rodale Press, Emmaus, PA, p. 91.

Webb, D.M., Smith, C.W. and Shulz-Schaeffer, J. (1987) Amaranth seedling emergence as affected by seedling date and temperature on a thermogradient plate. *Agron. J.*, **79**, 23–6.

Weber, L.E., Hubbard, D., Putman, L.N. and Lehmann, J. (1988) *Amaranth Grain Production Guide*. Rodale Press/Amaranth Institute, Emmaus, PA.

Weber, L.E. and Kauffman, C.S. (1990) Plant breeding and seed production, in *Proceedings of the Fourth National Amaranth Symposium*, Amaranth Institute, MN, pp. 115–28.

Whittaker, P. and Ologunde, M.O. (1990) Study of iron bioavailability in a native Nigerian grain amaranth cereal for young children, using a rat model. *Cereal Chem.*, **67**, 505–8.

Willis, J.C. (1983) *A Dictionary of the Flowering Plants and Ferns*. 8th edn, (revised H.K. Airy Shaw), Cambridge University Press, Cambridge.

Wilson, R.L. (1989) Studies of insects feeding on grain amaranth in the Midwest. *J. Kansas Entomological Soc.*, **62**, 440–8.

Xu, Y.B. (1987) (Studies on the chromosome number of some of *Amaranthus*). *Grasslands of China*, **1987**, 48–50 (in Chinese).

Yanez, G.A., Messinger, J.K., Walker, C.E. and Rupnow, J.H. (1986) *Amaranthus hypochondriacus*: starch isolation and partial characterisation. *Cereal Chem.*, **63**, 273–6.

Zabka, G.G. (1961) Photoperiodism in *Amaranthus caudatus*. I. A re-examination of the photoperiodic response. *Amer. J. Bot.*, **48**, 21–8.

Zhao, J.M. and Ma, R.Y. (1987) (Comparative study of seed protein of amaranth grain). *Acta Agriculturae Universitas Pekinensis*, **13**, 39–46 (in Chinese with English summary).

4

Triticale (× *Triticosecale*)

Anne J. Hörlein and J. Valentine

4.1 INTRODUCTION

Triticale is an artificial crop, resulting from combining wheat and rye. The development of triticale further exploits the hybridization and subsequent polyploidy which occurred during the early evolution of wheat (involving the genera *Triticum* and *Aegilops*) by a subsequent hybridization with diploid rye (*Secale*). It combines the genomes A and B (from tetraploid wheats) or A, B and D (from hexaploid wheat) with R (from rye), the new allopolyploids being in a state of permanent hybridity, like wheat, and having new combinations of characters. In theory, all the variability available in the parental species of wheat and rye and their relatives can be tapped (Varughese, 1991).

A brief history of the development of triticale is presented in Table 4.1. A sterile plant was derived from a cross between wheat and rye in 1876; as long ago as this seems, it is still young in terms of natural evolution. Other crosses resulted in fertile plants. Natural hybrids were reported on occasions (Meister, 1921). The development of a method for chromosome doubling of sterile hybrids using colchicine (Blakeslee and Avery, 1937) resulting in fertile triticales was a major step forward in the 'evolution' of the modern triticale crop. The use of embryo culture has increased the number of plants produced from wheat × rye crosses and has potential applications for *in vitro* multiplication and use in genetic transformation systems (Stolarz, 1991; Stolarz and Lörz, 1986).

The general aim of the development of triticale has been to produce a cereal combining the 'hardiness' and vigour of rye in terms of disease tolerance, winter hardiness and tolerance of drought and poor soil conditions with the superior yield and grain characteristics of wheat. This

Cereals and Pseudocereals. Edited by J.T. Williams. Published in 1995 by Chapman & Hall, 2–6 Boundary Row, London SE1 8HN. ISBN 0 412 46570 1.

Table 4.1 Early developments in triticale synthesis

1876	First sterile plant deriving from a cross between wheat (female parent) and rye (male parent) reported in UK (Wilson, 1876)
1891	First fertile wheat – rye hybrid reported in Germany (Rimpau, 1891)
1918	Natural occurrence of male sterile F1 hybrids in winter wheat plots observed (Meister, 1921)
1918	Start of research on triticale in Russia by G.K. Meister (Müntzing, 1979)
1930s	Start of research in Sweden by A. Müntzing (1979) and Germany (Müntzing, 1979)
1935	Name triticale introduced in Germany
1937	Development of a method for chromosome doubling of sterile hybrids using colchicine (Blakeslee and Avery, 1937)
1940	Development of the embryo culture technique to rescue embryos from seeds with insufficient developed endosperm
from 1950 onwards	Intensive research and breeding work on triticale in Europe, North America and (CIMMYT) Mexico

would extend the areas of production of the higher value cereals. Occasionally, aims have been more specific. For instance, in Canada, where a research programme had been initiated at the University of Manitoba, interests included using triticale as a bridge to transfer disease resistances from rye into wheat. Other effort has been focused on the tolerance of triticale to highly acid soils, since triticales tolerate problem soils including those having high content of aluminium or boron, or being deficient in manganese or copper, which provide very poor conditions for wheat cultivation.

Rye, one of the parents, is a crop which was developed much later than wheat and barley. Originally a cereal weed, it flourished at higher latitudes and higher elevations than wheat and was able to produce some yield even on poor soils. Wheat and rye grain contain gluten, an elastic protein complex composed of glutenins and gliadins which allow breads to rise. Rye flour contains some gluten and does not rise as well as wheat flour. Triticale flour in unleavened bread acts similar to flour of soft wheat (*Triticum aestivum*) but in yeast-leavened breads it has not been a match for that of wheat. In addition, triticale grain having the complete rye genome tends to produce a sticky dough which causes some problems in mechanized bakeries due to requiring slower mixing.

The 1980s saw interest in triticale in many countries. The most recent figures estimate that 8 countries grew triticale on 1.09 million ha in

Table 4.2 Estimates of world triticale production

Country	*Growth habit*	*Area (000 ha)* 1986	1991–92
Algeria	S	–	10
Argentina	S	10	16
Australia	S	160	100
Austria	W	1	2
Belgium	W	5	10
Brazil	S	5	90
Bulgaria	W	10	100
Canada	S + W	6.5	2
Chile	S	5	10
China	S + W	25	1.5
Former Czechoslovakia	–	0.4	–
France	S + W	300	162
Germany	W	30	207
Hungary	W	5	5
India	–	0.5	–
Italy	S	15	30
Kenya	S	–	8
Luxembourg	–	1	4
Mexico	S	8	3
Morocco	S	–	10
Netherlands	W	11	4
New Zealand	–	–	2
Poland	W	100	659
Portugal	S	7	90
Romania	–	–	20
South Africa	S + W	15	95
Spain	S	30	80
Sweden	–	–	1
Switzerland	W	5	11
Tanzania	–	0.4	–
Tunisia	S	5	16
UK	W	16	16
USA	S + W	60	180
former USSR	W	250	500
Total		1090	2440

S = summer; W = winter.

Sources: Anon, 1990; Varughese, 1991; Pfeiffer (pers. comm.).

1986. By 1991–92, this had expanded to 2.44 million ha in 31 countries (Table 4.2). Whereas the major aim of international research was to produce a cereal for human food, much of the cultivation, to date, has been for forage or feed. The commercial rise of triticale has been spectacular

Table 4.3 Agronomic constraints of the early triticale varieties

Morphology		
1.	Plant height	too tall
2.	Tillering	insufficient
3.	Shrivelled grains	too many
Production		
4.	Grain yield	too low
5.	Seed-set	poor
6.	Diseases	susceptible to several
7.	Maturity	late
Others		
8.	Adaptation	poor
9.	Premature sprouting	often high
10.	Germination	often low

in 10 countries (Poland, USSR, Germany, USA, France, Bulgaria, Australia, South Africa, Portugal and Brazil).

The early 'modern' triticales, from the 1950s, had higher protein and lysine content than wheat. However, this resulted from malformed shrivelled grains, the varieties were still primitive and a number of agronomic constraints were apparent (Table 4.3). Several constraints were overcome following the apparently chance fertilization of a triticale plant by pollen from a dwarf bread wheat source which took place in the CIMMYT trials fields in Ciudad Obregon in Mexico. After breeders carried out a few generations of selection, a new breeding line, Armadillo, was developed. This had higher yield, insensitivity to day length, short and stiff straw, early maturity and better seed-set. These traits were stable and heritable. From 1970, Armadillo was in most pedigrees of all triticales at CIMMYT and used by breeders elsewhere.

In more developing parts of the world, triticale flour is valuable for baked products in which soft wheat flour is normally used. These products include cakes, biscuits, pancakes, noodles and local foods such as 'chapattis', 'tortillas', 'enjera' (Ethiopia) or 'concha' (bread made with molluscs in Mexico). For leavened bread-making triticale can be used alone but also in mixtures with wheat or maize. Triticale is also used in local brewing.

Triticale is a temperate crop and is cultivated as a winter crop in eastern Asia, the independent states of the former USSR and northern Europe, and to a lesser extent as a spring crop in Mexico and areas of South America, eastern Africa, India, Australia and southern Europe, with overlap between spring and winter crops in North America, China, South Africa and parts of Europe. Trials from CIMMYT are carried out in irrigated subtropical, mediterranean and dryland conditions as well as the tropical highlands and specific problem soils, even in non-traditional-wheat warm areas.

Previous major publications or reviews on various aspects of triticale include those of Tsen (1974), Gupta and Priyadarshan (1982), Skovmand *et al.* (1984), Gregory (1987), National Research Council (1989) and Villareal *et al.* (1990).

4.2 CLASSIFICATION

Wheat, rye, and therefore triticale belong to the subtribe Triticinae of the tribe Triticeae. (Barley is in the same tribe but in a different subtribe, the Hordeinae.) The common name was derived from the scientific names of the parental genera and was first used by E. Tschermak-Seysenegg of Austria in 1935. According to Baum (1971) × *Triticosecale* Wittmark has priority under taxonomic rules, and is now the generally accepted generic name.

Early crosses resulted in octoploids since the parental wheat was bread wheat, the hexaploid *Triticum aestivum*. The hexaploid derived from durum wheat (*T. durum* Desf.) and rye was obtained by O'Mara in 1948 (O'Mara, 1951). Both octoploid and hexaploid were often given specific epithets but as tetraploid triticales were developed experimentally and as a range of diverse *Triticum* species which were hexaploid began to be used in crosses and double crosses, recognition of resultant types by specific epithets no longer seemed useful.

Within triticale, the following classification is used basically following that of Krolow (1985).

4.2.1 **Primary triticales**

These derive directly from wheat–rye crosses after doubling of the chromosome number. Primary octoploid triticales (genomic constitution = AABBDDRR) have 21 pairs of chromosomes from hexaploid wheat (seven each of the A, B and D genomes) and seven pairs of chromosomes from rye (R genome). Primary hexaploid triticale (genomic constitution = AABBRR) have 14 pairs of chromosomes from cultivated tetraploid wheat (seven each of the A and B genome) and seven from rye.

Primary triticales are pure and unmodified by further hybridization. They are stable and homozygous due to chromosome doubling. Nevertheless primary triticales have problems in terms of chromosome abnormalities, partial sterility, shrivelled seed, low yields and poor agronomic types (Bao, 1985; Gupta and Reddy, 1991).

4.2.2 **Secondary triticales**

These derive from hybridization between primary triticales. Derivatives of crosses between wheat and triticale, as long as there has only been genetic recombination between the wheat genomes and no substitution of R chromosomes has taken place, can also be considered secondary triticales. Many are produced by crossing different ploidy levels of triticale, most commonly octoploid triticale with hexaploid triticale. Secondary triticales form the basis of much of the triticale grown today. Triticales resulting from hybridization at the same ploidy level can be referred to as recombined triticales.

4.2.3 **Substituted triticales**

A rye chromosome is substituted by a wheat chromosome. This is achieved by selfing or backcrossing of hybrids between hexaploid wheat and hexaploid triticale. The transition from triticale into wheat is continuous for these substitutional lines. If more than three rye chromosomes are substituted by wheat chromosomes, the term substitutional wheat should be used instead. Substituted triticales contrast with complete triticales which retain unchanged all the chromosomes of the rye parent. Both are important in triticale development because complete ones tend to be more productive in stress conditions, retaining the robustness of rye (Varughese *et al.*, 1986; Gupta and Reddy, 1991). Armadillo was the first important substituted triticale. In this line and its derivatives, the chromosome 2D substituted for 2R (Gustafson and Zillinsky, 1973). Winter triticales tend to have the complete rye genome.

Other categories exist, e.g. **autoalloploid triticales**, which possess an autotetraploid rye genome as in decaploid triticale. The rye genome can be retained intact but part of the wheat genome A or B can be replaced by D genome chromosomes.

Triticale-like wheats exist which are wheats with one or a few rye chromosomes. All these substitution forms are of potential benefit to breeders of all three crops: wheat, rye and triticale. Homoeologous wheat and rye chromosomes are recognizable from the genetic markers which they carry. High levels of repetitive sequence DNA provide a significant source of variation and large quantitative differences in randomly repeated sequences can be detected cytologically via changes in heterochromatic C-banding patterns.

4.3 MORPHOLOGY

Figure 4.1 shows ears and grains of wheat and rye contrasted to triticale. In gross morphology, triticale resembles wheat more than rye, but

Figure 4.1 Ears and grains of wheat and rye contrasted to those of triticale.

there are clear morphological differences as described by Baum and Gupta (1991). In particular, triticale is distinguished from wheat as having an asymmetrical glume and a symmetrical lemma with distal converging veins. Both rye and triticale have pubescent peduncles, the gene or genes responsible for this character being on the long arm of a single rye chromosome (O'Mara, 1951). Triticale varieties normally possess awns but awnless varieties have been produced.

4.3.1 **Plant height**

Lodging resistance in triticale is becoming more and more important as this crop competes with wheat on richer soils. Since in triticale, lodging appears to be associated with height rather than with poor straw length (Skovmand *et al.*, 1984), reduction of height is an important breeding aim. Triticale is shorter than rye, but taller than wheat, and shows a considerable variability for this character. In line with other cereals, however, reduced height often leads to reduced biomass and yield (Sandha *et al.*, 1991; Oettler *et al.*, 1991). Not surprisingly, effort has been developed to incorporate semi-dwarf genes into triticale.

In CIMMYT, limited progress was achieved by crossing Armadillo strains with stiff-strawed triticales (Varughese *et al.*, 1987). More significant progress has been made by using crosses with adapted semi-dwarf wheats, in the case of CIMMYT via the development of two octoploid triticales (Maya 1 and 2) (Varughese *et al.*, 1987).

Twenty-two semi-dwarf genes with recessive or dominant inheritance are known for wheat, with *Rht1* and *Rht2* genes, originating from *Triticum aestivum* cv. Norin 10, being the most important ones (Börner and Mettin, 1989).

Studies by Bodova (1991) on 34 varieties of triticale of different origin showed that 14 of these carried dwarfing genes. In winter triticale varieties, the semi-dwarf gene most frequently used is the semi-recessive gene *Rht1* from Norin 10. The *Rht1* gene is now almost universally present in selections from European breeding programmes (Gregory, 1987). Another gene, *Rht3*, derived from *T. aestivum* cv. Tom Thumb, is very potent. Its dwarfing effect in wheat is a reduction of height by up to 40% (Stoddart, 1987). It is generally less used because of negative pleiotropic effects for total biomass and grain yield (Börner *et al.*, 1987). The gene is present in the Hungarian variety Bokolo. This variety has, for the reasons given above, not been widely grown. Both *Rht1* and *Rht3* are *GA3*-insensitive and are located on chromosome 4A (Gale and Marshall, 1976). The *Rht2* gene of wheat is located on chromosome 4D which is not present in hexaploid triticales combining the complete rye genome. Other *Rht* genes are being incorporated from wheat.

Dwarfing genes from rye have also been used in attempts to reduce plant height in triticale (Zillinsky and Borlaug, 1971). Using rye cv. Snoopy with recessive dwarfing genes, they achieved little success, though Snoopy and other dwarf ryes figure in the pedigrees of CIMMYT dwarf triticales. A short, high-yielding triticale from Poland (CHD888, now named Prego) contains a semi-dwarf rye (Wolski, 1991b). A rye mutant, *EM1*, containing a dominant gene, has been used in a number of breeding programmes.

4.4 PHYSIOLOGICAL ADAPTATION

Triticale has the potential for adaptation to a wide range of environmental conditions. Because triticale has a higher nutritional value than rye, the replacement of rye with triticale on typical rye soils is of great economic importance worldwide (Aniol and Gustafson, 1984). However, triticale can also compete with established cereals in less extreme environments (Gregory, 1987).

4.4.1 **Adaptation to semi-arid regions**

Selection for tolerance to drought has been notable in semi-arid areas such as the region of the Southern Cape of South America where annual rainfall is 400–700 mm (Lopez and Garbini, 1985). Although selection procedures for water use efficiency (Bushuk, 1976) have not been specifically targeted, it is known that rye needs only 70–80% of the water required by wheat to produce the same amount of dry matter.

Some triticales have been considered tolerant to salinity (Francois *et al.*, 1988). While superior tolerance to salinity can be observed in varieties of triticale in the initial growth stages when compared to wheat and barley, this is not always translated into higher yields, possibly because of water and heat stresses at maturity (Baier, 1991).

Substantial research has been done on sandy, copper-deficient soils. Triticale extracts this copper more efficiently than wheat (Graham, 1978). Graham (1988) considered that, in relation to the ability to absorb copper, zinc, manganese and phosphorus, rye has the best ability, with triticale being intermediate between wheat and rye.

4.4.2 **Adaptation to acid soils**

Baier and Nedel (1986) stressed the potential of triticale in areas of southern Brazil with acid soils, low phosphorus availability and aluminium toxicity. Triticale varieties have been evaluated in Brazil since 1969. The occurrence of tolerance to toxic levels of aluminium, iron and manganese and efficiency of utilization of phosphorus on acid soils has been summarized by Baier (1991). In other countries, such as Poland, Kenya, Ethiopia, India, Ecuador and Mexico, triticale has also proved to be better than wheat on acid soils. Royo *et al.* (1993) suggested complete triticales were better adapted to acid soils, whereas substituted types were, in general, better suited to alkaline soils.

4.4.3 **Growth and development**

As in other cereals, the rate of development of triticale is governed by response to photoperiod and temperature, which in turn affects adaptation to environment, duration of various growth stages, earliness of flowering and maturity and the values of yield components and yield itself (Federov, 1988). Like rye, triticale has a longer grain-filling period than wheat (Gawlik and Seiffert, 1990) though the rate of grain-filling may be slower. Some winter types, for example Lasko, appear to have low or no vernalization requirement (Valentine and Middleton, 1987; Stolcova, 1990). In the first spring triticales from CIMMYT, late maturity and low yields reflected the use of northern latitude varieties which had requirements for long day lengths. Triticales have now been selected for day length neutrality and can be grown in diverse latitudes and planted at different times of the year.

4.4.4 Pre-harvest sprouting

Pre-harvest sprouting affects the maintenance of a good, firm seed after exposure to rain (Huskowska *et al.*, 1985). It is a problem in many of the areas targeted for triticale production.

Pre-harvest sprouting is commonly associated with increased levels of α-amylase in the seed. Laboratory screening for α-amylase activity helps in the selection of low sprouting lines. High premature α-amylase levels can be measured using the Hagberg falling number test (Chojnacki *et al.*, 1976; Huskowska *et al.*, 1985; Oettler, 1987, 1991). Increased α-amylase levels can be rain-induced, or even be independent of environmental conditions.

Although some recent triticale lines can maintain higher falling numbers than older cultivars (Clarke *et al.*, 1986), genetic improvement remains difficult (Wolski, 1991a). Knowledge of the genetic control of pre-harvest sprouting is still fragmentary (Oettler, 1991). Dormancy is regarded as the most important component of tolerance.

CIMMYT initiated a programme to select for sprouting resistance in 1982 by subjecting more than 200 segregating populations to conditions which induce sprouting. Materials selected for good sprouting resistance were identified in Brazil. Crosses were made between Brazilian and the CIMMYT lines to develop low and acceptable levels of sprouting.

There appears to be a relationship between triticale lines which carry the *Rht3* dwarfing gene and low α-amylase levels (Chojnacki *et al.*, 1976) but these lines are not used greatly for genetic improvement due to the negative pleiotropic effects of *Rht3*.

4.4.5 Malformed seed

A major problem which has been the subject of considerable research has been abnormal endosperm formation resulting in shrivelled grain.

The specific or test weight, the weight of grain that packs into a given volume, is a guide to the extent of shrivelling. Values for bread wheat are normally above 75 kg/hl. The first triticales ranged from 58–72 kg/hl. Through the application of strong selection pressure and the extensive use of the 2D(2R) substitution from Armadillo in breeding, specific weights of spring triticales at CIMMYT have been substantially raised to above 78 kg/hl (Varughese *et al.*, 1987). In the UK, reported levels of specific weight range from 67 for Protius to 75 for Olympus, compared with a range between 76 for Riband, currently the most widely grown winter wheat variety, and 81 for Hereward (Anon., 1993). The Polish variety Lasko has relatively small, unshrivelled grains similar in

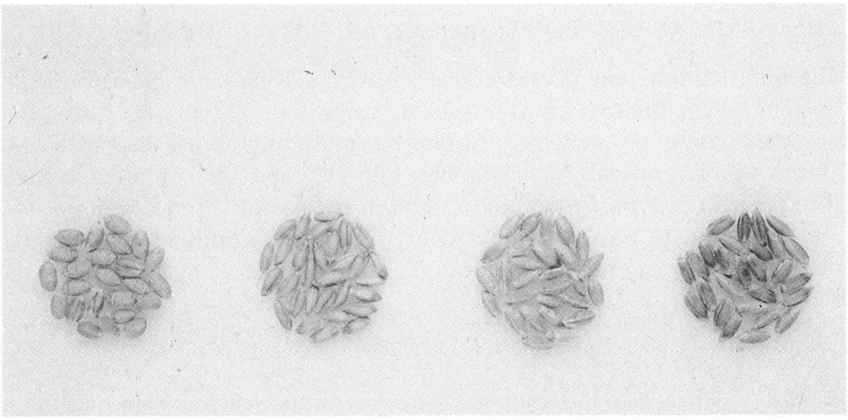

Figure 4.2 Seeds of triticale, wheat and rye.

appearance to wheat, with a specific weight of 72 in the UK. Seeds of triticale, wheat and rye are shown in Figure 4.2.

Grain shrivelling has been attributed to many causal factors, such as aneuploidy, increased α-amylase activity, late replicating terminal heterochromatin on the rye chromosomes and irregularities in the early stages of endosperm development (Varughese *et al.*, 1984). The role of heterochromatin is unclear, as in recent studies heterochromatin has been found not to be necessarily the cause for grain shrivelling (Gupta and Reddy, 1991). In abnormal endosperm development, the endosperm may not grow fast enough or it may develop internal cavities. During ripening, the endosperm collapses and gives the grain its irregular form (Thomas *et al.*, 1980). Spring triticales containing the 2D(2R) substitution have less grain shrivelling than complete triticales, though good progress has also been made in reducing shrivelling in winter triticales.

4.4.6 Seed germination and viability

Apart from pre-harvest sprouting, germination may be decreased due to mechanical damage to the projecting seed embryo (Pickett, 1988).

At room temperature, seed viability of triticale (and rye) has been found to decrease rapidly to nil after 3 years in non-sealed containers or 6 years in sealed containers (Hwang and Ha, 1991). Greater care needs to be taken than for wheat, barley and oats to store seed at low temperatures and moisture.

4.5 CHEMICAL COMPOSITION AND NUTRITIVE VALUE

The main component of the grain is starch. Levels have been found to be generally similar to those of wheat (Table 4.4).

Protein content is generally similar to, or slightly above, that of wheat (Table 4.5). In general, values lie between 11% and 14% on a dry matter (DM) basis (Leterme *et al.*, 1990), though lower and upper limits of 9% and 23% due to the effects of genotype and environment, particularly nitrogen fertilizer, have been found. Triticale has a higher content of

Table 4.4 Starch (% DM) content of triticale compared to wheat and other cereals

	Triticale			*Wheat*			
	Mean	*Range*	*n*	*Mean*	*Range*	*n*	*Location*
(1)[a]	62.2	–	–	65.1	–	–	Belgium
(2)[b]	66.5	61.9–69.5	80	70.3	–	6	Sweden
(3)	60.1	53.2–63.0	15	60.3	–	2	Czechoslovakia

[a]Rye 62.0, winter barley 57.0, oats 43.0, maize 70.3.
[b]Rye 65.6, Lasko triticale 68.1.

Key to references
(1) Leterme *et al.* (1990)
(2) Pettersson and Åman (1987)
(3) Heger and Eggum (1991)

Table 4.5 Crude protein (% DM) content of triticale compared to wheat and other cereals

	Triticale			*Wheat*			*Location*
	Mean	*Range*	*n*	*Mean*	*Range*	*n*	
(4)	15.6	10.8–19.2	3	14.0	–	1	UK
(1)[a]	13.5	–	–	13.1	–	–	Belgium
(2)[b]	11.7	9.4–16.5	80	10.7	–	6	Sweden
(3)	13.0	11.2–14.8	15	13.0	–	2	Czechoslovakia
(5)	10.8	9.5–12.2	3	12.3	–	1	Australia

[a]Rye 11.0, winter barley 10.7, oats 11.6, maize 10.5.
[b]Rye 9.5, Lasko triticale (n = 4) 12.1.

Key to references
(1) Leterme *et al.* (1990)
(2) Pettersson and Åman (1987)
(3) Heger and Eggum (1991)
(4) Charmley and Greenhalgh (1987)
(5) Batterham *et al.* (1990)

Table 4.6 Levels of essential amino acids (g/16 g N) of triticale compared to wheat and other cereals

	Triticale			*Wheat*			*Location*
	Lysine	*Threonine*	*Methionine*	*Lysine*	*Threonine*	*Methionine*	
(4)	3.1	3.2	3.0[a]	2.6	3.1	2.9[a]	UK
(3)	3.1	2.9	1.8	2.9	2.8	1.8	Czechoslovakia
(6)[b]	3.4	3.2	1.8	2.6	3.0	1.8	Sweden
(5)	3.9	3.7	1.7	3.7	4.1	2.0	Australia

[a](+ cysteine).
[b]For rye – lysine 3.7, threonine 3.3 and methionine 1.7.

Key to references
(3) Heger and Eggum (1991)
(4) Charmley and Greenhalgh (1987)
(5) Batterham *et al.* (1990)
(6) Thomke *et al.* (1987)

albumins (26–28% of protein compared to 3–5% in wheat and 34% in rye) and lower content of the prolamins (24–26% compared to 40–50% in wheat and 19% in rye) (Chung and Pomeranz, 1985). These differences in protein composition are reflected in higher levels of essential amino acids in triticale compared to wheat (Table 4.6). The amino acid composition has been found to depend on the storage problems that accumulate in the grain as protein is increased. Thus for lysine, the most limiting essential amino acid lysine/protein is higher than wheat and rye at higher levels of protein and intermediate at lower levels (Mossé *et al.*, 1988). From a nutritional viewpoint, the biological value of the triticale protein is above that of wheat (Thomke *et al.*, 1987; Heger and Eggum, 1991) and amino acid composition is predictable from a knowledge of N or protein % (Mossé *et al.*, 1988).

Triticale may have slightly higher fibre levels, particularly neutral detergent fibre, than wheat (Table 4.7). Levels of soluble β-glucans and oils (ether extract) are low. Triticale appears intermediate or similar to wheat in its levels of pentosans, alkylresorcinols and trypsin and chymotrypsin inhibitors, components which in rye have been found adversely to affect the growth of monogastric animals (Table 4.7). Metabolizable or digestible energy levels are broadly similar to those of wheat (Table 4.7).

4.6 AGRONOMY, YIELD AND PRODUCTION

In general, triticale can be managed in a similar manner to wheat. Most of this section refers to triticale in the UK unless otherwise stated.

Table 4.7 Other grain components (% DM) and energy levels (MJ/kg^{-1} DM) of triticale compared to wheat and other cereals

	Triticale Mean	Triticale Range	Triticale n	Wheat Mean	Wheat Range	Wheat n
Neutral detergent fibre (NDF)						
(4)	15.8	15.5–16.1	3	10.8	–	1
(1)	14.8	–	–	13.9	–	–
(5)	12.6	12.0–13.7	3	9.7	–	1
Acid detergent fibre (ADF)						
(4)	3.5	3.3–3.7	3	3.3	–	1
(1)[a]	4.0	–	–	3.6	–	–
Lignin						
(1)[b]	1.3	–	–	1.2	–	–
(2)[c]	2.4	–	10	1.5	–	6
Pentosans						
(7) soluble	1.5	1.3–1.6	2	1.2	0.9–1.5	2
total	7.1	6.2–7.9	2	6.6	6.2–7.1	2
(2) soluble[d]	1.1	–	10	1.1	–	6
total[d]	6.7	–	10	5.0	–	6
(5) soluble	0.9	0.6–1.5	3	0.5	–	1
total	4.7	4.5–4.9	3	4.8	–	1
Soluble β-glucans						
(7)[e]	0.52	0.40–0.66	4	0.67	0.56–0.75	4
Lipids						
(4)	2.9	2.3–3.4	3	2.1	–	1
(3)	1.5	1.1–1.9	15	1.6	1.6–1.7	2
(5)	2.4	2.3–2.7	3	2.9	–	1
Ash						
(2)[f]	1.8	1.1–3.0	80	1.6	–	6
(5)	1.6	1.5–1.8	3	1.3	–	1
Metabolizable energy						
(4) sheep	13.3	13.1–13.4	3	12.9	–	1
poultry[g]	14.3	14.0–14.5	3	12.9	–	1
(6) piglets[h]	15.1	15.0–15.1	2	15.1	–	2
laying hens	14.5	14.3–14.8	4	14.4	–	1
Digestible energy						
(4) pigs	15.3	14.6–15.7	6	16.3	16.0–16.6	2
(8)[i]	15.0	–	16	15.1	–	17
Trypsin						
(5)	0.4	0.36–0.42	3	0.13	–	1

Table 4.7 Continued

	Triticale Mean	Triticale Range	Triticale n	Wheat Mean	Wheat Range	Wheat n
Chymotrypsin (5)	0.39	0.38–0.41	3	0.26	–	1
Alkylresorcinols (9)[j]	1.0	0.7–1.3	4	0.6	–	1

[a]NDF for rye 13.2, ADF for rye 9.0.
[b]Rye 1.4.
[c]Rye 3.0.
[d]Calculated from sum of arabinose and xylose residues. Rye – soluble 2.6, total 7.5, Lasko triticale – soluble 1.0, total 5.3.
[e]Barley 2.89, oats 2.13.
[f]Rye 1.9, Lasko triticale 1.6.
[g]True ME corrected to zero N retention.
[h]Rye 14.8.
[i]Rye 14.9.
[j]Rye 2.0, Lasko triticale 0.7.

Key to references
(1) Leterme *et al.* (1990)
(2) Pettersson and Åman (1987)
(3) Heger and Eggum (1991)
(4) Charmley and Greenhalgh (1987)
(5) Batterham *et al.* (1990)
(6) Thomke *et al.* (1987)
(7) Henry (1985)
(8) Rundgren (1988)
(9) Gohil *et al.* (1988)

4.6.1 **Sowing**

In northern Europe, the optimum sowing time lies between mid-September and mid-October, although sowing can be delayed for a further month. On fertile soils, triticale should not be sown before the beginning of October as earlier sowing might cause severe lodging, while on very light drought-prone soils earlier sowing is probably advisable (Furber, 1990). Some varieties with low vernalization requirement can be sown in early spring, though yields are likely to be reduced significantly.

Recommendations for seed rate in the UK range between 130–180 kg/ha, depending on the tillering capacity of the variety being sown, sowing date and soil conditions. An optimum seed rate should be targeted at establishing 200–250 plants/m^2 at the end of the winter (McDonald, 1990).

4.6.2 **Fertilizers**

Comparisons between the typical balances between nitrogen added in fertilizer and nitrogen removed in the harvested produce have revealed triticale to absorb more nitrogen from the soil than any other UK arable crop, including winter wheat (Sylvester-Bradley, 1993).

Spring nitrogen has to be applied extremely carefully in order to prevent lodging causing harvesting problems and yield losses. Lower nitrogen rates than for wheat as well as splitting of the nitrogen application are recommended (Furber, 1990; McDonald, 1990).

4.6.3 **Growth regulators**

The use of growth regulators, such as chlormequat, mepiquat chloride and ethephon, is possible in triticale and is recommended for fertile soils. Although all varieties respond to growth regulators, the level of response depends on the variety (McDonald, 1990). However, on less fertile soils with a lower risk of lodging, the management of the crop should be possible without the use of growth regulators.

4.6.2 **Weed control**

Usually herbicides used against broad-leaved weeds in wheat are also suitable for triticale, whereas severe damage can occur after the application of herbicides against grass weeds.

4.6.5 **Pesticides**

The big advantage of triticale over wheat is its tolerance towards fungal diseases. Therefore disease control is generally not necessary and prophylactic spraying in low-risk situations is uneconomic. In 1990, only 8% of triticale crops in the UK were sprayed compared to 97% of wheat and 96% of winter barley (Davis *et al.*, 1991).

Insecticides for pest control, which are approved for wheat and barley, normally do not carry label recommendations for triticale, but their use for triticale might be possible (Furber, 1990; McDonald, 1990).

4.6.6 **Harvesting**

The harvest of triticale should not be delayed to avoid grain sprouting under wet conditions. In order to cope with the greater quantity of straw, different combine-harvester settings are required.

4.6.7 **Grain yield**

Spring lines from CIMMYT are tested in International Yield Nurseries (ITYNs) at 50–60 locations annually. The results show a high rate of progress in the crop's yield potential in the period 1969–70 to 1974–75 (Villareal *et al.*, 1990), with triticale outyielding the best bread wheats since 1972–73. Care should be taken however in comparing wheat and triticale varieties in the same trials because of inter-plot competition and differences in optimum management practices (Gregory, 1987). Mean yield in all locations in the 18th ITYN was 4.80 t/ha, while that in Ciadud Obregon (a more favourable irrigated sub-tropic location) was 7.12 t/ha, gains of 94% and 64% respectively compared to those obtained in the 1st ITYN in 1969–70.

Substituted triticales appear to do better in the irrigated subtropics, while complete triticales have a slight or definite advantage under dryland conditions, in tropical highlands and acid soils (Villareal *et al.*, 1990; Pfeiffer and Fox, 1991). According to Fox *et al.* (1990), however, there was evidence of a 'blurring' of phenotypic differences between complete and substituted triticales in recent material, where complete triticales had been found with shorter stature or a degree of earliness formerly limited to substituted germplasm. They suggested maintaining and intercrossing complete and substituted types in a breeding programme.

In Europe, winter triticale varieties produce higher yields than wheat on light soils (Ford *et al.*, 1984; Wolski, 1991a). Furber (1990) indicated triticale to yield as well as or better than wheat on sites where cereals had been grown continuously (5.4–7.5 vs 5.5–6.5 t/ha), when grown as second cereals after a one-year break (7.9–8.9 vs 7.2–8.7 t/ha) and occasionally in more fertile situations (5.6–8.9 vs 8.4–8.8 t/ha).

4.6.8 **Diseases and pests**

Fungi

Although triticale can be infected by most of the fungal diseases attacking wheat or rye, it is normally considered to have greater levels of tolerance to common wheat diseases than wheat (Skovmand *et al.*, 1984). All the diseases in triticale are caused by wheat or rye pathogens and disease reaction of the two parents is reflected in triticale, but resistance is not automatically obtained from combining the genomes of wheat and rye. For all the diseases occurring on triticale, different levels of resistance have been found. Singh and Saari (1991) presented evidence showing that highly mutable obligate parasites could cause at least as great a threat to triticale as to wheat. Concerning the rust diseases, leaf rust (*Puccinia recondita*) is the most common. Severe epidemics of stem rust (*P. graminis tritici*) have occurred in Australia and there was a high incidence of yellow rust (*P. striiformis*) in an international triticale nursery in Ecuador in 1987. On the other hand, Wolski (1991a) reported that, while most triticale strains were highly resistant to rusts and mildew, where attacks occurred they were in general late, of a slow-rusting type and were less damaging than those observed in wheat. At the very least, it seems triticale can play an important role as a diverse source of resistance to these diseases. Triticale is regarded as being immune to powdery mildew (*Erysiphe graminis*), but occasional infections occur (Singh and Saari, 1991; Wolfe and Klints, 1987).

Other frequent fungal diseases on triticale of varying importance in different areas of the world include scab (*Fusarium* spp), which can cause serious losses in yield and produce harmful toxins in the grain, blotch diseases caused by *Septoria tritici*, *S. nodorum* and *S. avenae triticea*, and leaf spot diseases (*Cochiobolus sativus* = *Helminthosporium sativum* and *Pyrenophora trichostoma* = *Helminthosporium tritici-repentis*).

Diseases of minor importance include the root and stem diseases take-all (*Gaeumannomyces graminis*), eyespot (*Pseudocercosporella herpotrichoides*), sharp eyespot (*Rhizoctonia cerealis*), together with loose smut (*Ustilago nuda*) and bunt (*Tilletia* spp.) (Skovmand *et al.*, 1984; Singh and Saari, 1991). For these diseases, triticale generally shows greater resistance than wheat.

Ergot (*Claviceps purpurea*) may occur under cool moist conditions, particularly in varieties with a very open flowering habit. It hardly affects yield but its sclerotia contain toxic alkaloids.

Occasionally triticale is infected with net blotch (*Pyrenophora teres*) and zonate eyespot (*P. giganteum*). In subtropical highlands *Phaeoseptoria vermiformis*, another leaf blotch, is frequent. Fusarium leaf blotch (*Fusarium nivale*) is occasionally a problem in the cooler areas of the subtropics, and causes snow mould in winter triticales (Singh and Saari, 1991).

Viruses and nematodes

The most common virus disease on triticale is infection with barley yellow dwarf virus transmitted by aphids. Considerable variation in tolerance exists (Collin *et al.*, 1990). Triticale varieties with resistance to cereal cyst nematode (*Heterodera avenae*) have been found. This offers the opportunity to control nematode populations using a resistant triticale variety as a break-crop on lighter land (Cook and York, 1987).

4.7 POST-HARVEST TECHNOLOGY

Although triticale offers a broad spectrum for utilization in industry and animal and human nutrition, most of its potential has not yet been explored for routine use. The main agricultural uses for triticale are first, and most important, as a grain crop and, second, for forage production (Brignall *et al.*, 1988).

4.7.1 **Milling**

Milling experiments using the triticale varieties Bokolo and Lasko showed that the grains were soft-milling and the yields of white flours comparable in colour to those from wheat were low (Flour Milling and Baking Research Association, 1985).

Lasko has milling characteristics similar to rye (Glattes and Mar, 1983). According to Kolkunova *et al.* (1982), conventional rye milling is suitable for the processing of triticale into flour.

4.7.2 **Baking**

Triticale has been used successfully in breads, noodle products, cookies, cakes, doughnuts, muffins, extruded breakfast cereals, pancake and waffle mixes (Kahn and Penfield, 1983).

White and wholemeal flours of the triticale varieties Bokolo and Lasko are suitable for making traditional sponge and plain cake and short pastry, although the products have a stronger flavour than the equivalent wheat products. These triticale flours behaved similarly to *T. durum* flours in extrusion cooking and gave bland-flavoured, crunchy, expanded product (Flour Milling and Baking Research Association, 1985).

Triticale flour is deficient in quality and quantity of gluten (Kahn and Penfield, 1983). The dough resembles rye dough rather than wheat dough (Kolkunova *et al.*, 1982). Cooper (1985) recommends using triticale flour in combination with wheat flour. Baking tests with triticale using either yeast or sour dough resulted in unsatisfactory quality products (Mar and Glattes, 1984). However, breads with acceptable quality have been produced from grain of certain varieties of triticale (Lorenz *et al.*, 1972). According to Kolkunova *et al.* (1982), triticale requires different technological approaches to wheat. Triticale partially inherits certain biological properties of rye, such as high activity of α-amylase and an unbalanced character of the components of gluten complex. The practical problems caused by high α-amylase activity would be the same for both triticale and wheat (Flour Milling and Baking Research Association, 1985), and therefore triticale can be used for the same products as wheat varieties with inferior baking quality (Lorenz *et al.*, 1972).

Improvements in the gluten strength and quality are necessary to increase the suitability of triticale for bread-making. Selection for both characters has been successful (Amaya and Skovmand, 1985). Although the rye genome is detrimental to the bread-making quality of triticale, there could be scope for introducing bread-making quality from bread wheat into secondary triticales (Peña and Ballance, 1987). A considerable range of gluten quality among different hexaploid triticales was found. The transfer of a segment of chromosome *1D* of wheat containing the *Glu-1D* encoding high molecular weight gluten subunits has been reported (Lukaszewski and Curtis, 1992).

4.7.3 **Brewing**

Triticale is an attractive grain to malt. It is superior to barley in terms of the amount of hot water extract, diastatic power and steeping period duration (Tombros and Briggs, 1984; Blanchflower and Briggs, 1988; Lersrutaiyotin *et al.*, 1991). As with wheat and rye, the absence of husk creates filtration problems of the mash. The high level of soluble nitrogen is a further disadvantage. However, the use of an unmalted adjunct with triticale malt in a grist containing barley malt would overcome these problems (Tombros and Briggs, 1984). Some malts were found to have protracted filtration times and produced hazy and viscous worts (Blanchflower and Briggs, 1988). The breeding of triticales with low content of nitrogen and non-starch polysaccharides might be worthwhile.

4.7.4 **Livestock**

Many studies have shown that triticale – as expected from its composition – can be substituted for other cereals in animal diets (e.g. Hanrahan, 1986; Hulan and Proudfoot, 1987; Burgstaller *et al.*, 1989; Smith and Pearson, 1989; see also Hill, 1991) though adverse effects of feeding high levels of some cultivars have been found by some researchers (e.g. Richter *et al.*, 1990). In economic terms, the use of triticale for livestock feeding is likely to depend on its price and the price of cereal substitutes (Théwis and Leterme, 1991) and, for feed compounders, its availability in worthwhile quantities. Their high content of essential amino acids makes them suitable for feeding to monogastric animals. Apart from farm livestock, triticale grain can be substituted for wheat middlings in the diets of fish (Hughes, 1990; Burtle, 1991).

Most triticale is used as grain for livestock feed. However, a small proportion is grown for forage production. Winter types may also be used for silage (Brignall *et al.*, 1989). Triticale was found to be a suitable forage crop on its own or in mixtures, such as triticale/legume mixtures (Carnide *et al.*, 1991). Total above-ground dry matter yields were found to be generally similar for the best triticale and wheat varieties (Ford *et al.*, 1984). Experiments on *in vitro* digestibility of small-grain cereals showed that leaf digestibility was highest in triticale and oats, whereas it was lowest in wheat and rye (Bruckner and Hanna, 1990).

4.7.5 Bio-ethanol production

Triticale is an ideal substrate for the production of ethanol as it contains high levels of amylases which convert starch into fermentable sugars (Thomas *et al.*, 1991). Renewable raw materials with a high amount of starch, like potatoes and maize, have been proposed as substrates for the triticale amylase system.

4.8 GENETICS AND BREEDING

4.8.1 **Reproductive system**

While the wheat species used as a source of germplasm for triticale are mainly self-pollinating or autogamous, *Secale cereale* is predominantly cross-pollinating or allogamous. Triticale, their intergeneric hybrid, is considered as a self-pollinated genus (Malik, 1984). In fact, considerable cross-pollination can occur depending on genotype and environment, especially if the weather is unfavourable to pollination. In 16 triticale varieties of European and North American origin, the average number of outcrossed florets per ear was found to be zero in six varieties, between one and two in nine varieties, and 18 (over a quarter of florets) in one variety (Pickett, 1988). Bagging ears to prevent cross-pollination gave good grain set, proving that the self-incompatibility system of rye does not seem to be operative in triticale.

4.8.2 **Ploidy levels**

Triticale has been synthesized at four different ploidy levels: tetraploid, hexaploid, octoploid and decaploid, with most emphasis being on hexaploid and octoploid triticale.

Tetraploid triticales, produced from crossing hexaploid triticale with diploid rye, followed by several generations of selfing (Krolow, 1973) or by backcrossing F1 hybrids of triticale × rye to the triticale parents (Baum and Lelley, 1988), have an intact rye genome (seven chromosome pairs), but a reconstituted wheat genome having seven chromosome pairs from more than one wheat genome (Gupta and Reddy, 1991).

Their main potential lies in the improvement of hexaploid triticale and diploid rye.

Decaploid triticales were achieved by producing pentaploid hybrids between octoploid triticale and diploid rye, followed by chromosome doubling (Müntzing, 1955). They suffered from cytological instability, caused by the high number of chromosomes, and are of little agronomic significance.

Compared to primary hexaploid triticale, primary octoploid triticale lines generally show an inferior performance (Oettler *et al.*, 1991).

Due to the combination of the genomes of the two different genera, wheat and rye, it is often very difficult to predict the mode of inheritance of various attributes.

4.8.3 **Synthesizing primary triticales**

For the production of primary triticale, wheat is normally used as the female parent and rye as the male parent. The crossability between *Triticum* and *Secale* is under control of two major genes, *Kr1* and *Kr2*, located on chromosomes 5B and 5A respectively. In their dominant condition, they suppress the production of seed (Riley and Chapman, 1967). Most varieties of tetraploid and hexaploid wheat carry both genes in the dominant form. The crossability problems can be overcome by the use of embryo culture. Embryos excised 2 to 3 weeks after pollination are placed on an artificial medium in order to develop roots and shoots. Afterwards the plantlets are transferred into soil. A detailed description of the embryo culture method is given by Kaltsikes (1974). The successful use of embryo culture shows that the embryo itself is capable of further development, but the endosperm or the maternal plant is unable to support it (Kaltsikes, 1974). Doubling the chromosome number is necessary to produce an amphidiploid hybrid. Colchicine is the normally used chemical, though different methods are available (Kaltsikes, 1974). Crossability and the success of embryo culture and colchicine treatment are affected strongly by the direction of the cross as well as by the genotype of the parents (Lelley and Taira, 1979; Tanner, 1981; Tanner and Falk, 1981; Oettler, 1983; Oettler, 1985).

4.8.4 **Cytology**

Identification of wheat and rye chromosomes is possible by isozymes (Jaaska, 1983) and cytological methods, such as chromosome size, arm:length ratio, and Giemsa C-banding (Merker, 1973; Seal, 1982; Hoan *et al.*, 1989). Translocations between wheat and rye chromosomes can be detected by C-banding (Lukaszewski and Gustafson, 1983) and biotin-labelling of the total genomic DNA of rye used as a probe for *in-situ* hybridization (Hoan *et al.*, 1989; Heslop-Harrison *et al.*, 1990).

Substitutions and translocations offer possibilities for increasing the variation between triticales. An important substitution present in many spring triticales is that between the chromosomes 2R and 2D alluded to earlier. Another substitution, 6D(6R), has been identified in a small number of spring and winter triticale varieties (Lukaszewski, 1988).

Despite the clear homoeology between the chromosomes of rye and wheat, few reports of pairing between wheat and rye chromosomes exist (Miller, 1984). Lukaszewski and Gustafson (1983) found on average that there was one wheat–rye translocation per four plants from triticale-wheat populations. All three wheat genomes were involved in these translocations between rye and wheat chromosomes. The frequency of translocations occurring within homoeologous groups was considerably lower than between homoeologous groups. The authors assumed that the high frequency of translocations resulted from random fusion due to misdivision of chromosomes during highly disorganized meiosis. Wheat–rye translocations can be successfully used in wheat breeding programmes, especially for transferring disease resistances, such as leaf rust, stem rust, stripe rust and powdery mildew (Lukaszewski, 1990; Ren *et al.*, 1990), but they have not yet been exploited for the improvement for triticale (Gupta and Reddy, 1991). So far, translocations involving a segment of the D genome have not been identified in advanced triticale lines, but their presence cannot be ruled out (Lukaszewski and Gustafson, 1987; Gupta and Reddy, 1991). A transfer of the *Glu-D1d* allelle from chromosome 1D to 1R using a combination of centric break-fusion translocation and 5D(5B) induced homoeologous pairing has been reported (Lukaszewski and Curtis, 1992).

Variation of chromosome number in advanced triticale lines and varieties has often been reported (Tsuchiya and Larter, 1971, Guedes Pinto *et al.*, 1984). Meiotic instability in triticale, leading to aneuploidy and decreased fertility, is a result of a combination of several factors, such as a shorter duration of the meiotic cycle compared to rye, homoeologous pairing, cytoplasmic and/or cytoplasmic-nuclear interactions, telomeric heterochromatin bands of rye and even environmental factors (Scoles and Kaltsikes, 1974; Lukaszewski and Gustafson, 1987). Off-types occurring in triticale varieties are often unstable and aneuploid (Pickett, 1988). The frequency of aneuploidy, resulting in a decrease in high-yielding triticale lines, has been estimated at between 4% and 11% (Jung and Lelley, 1985; Pickett, 1988). However, meiosis in all successful triticale varieties is much more regular than in new amphidiploids, indicating that improvements by selection are possible (Lukaszewski and Gustafson, 1987).

4.8.5 **Selection for yield and other characteristics**

As in other cereals, grain yield is the product of its individual components (number of grains per ear, number of spikelets per ear, number of grains per spikelet, number of tillers with fertile ears and thousand grain weight). From triticale, Lelley (1992) found that number of grains per spikelet was the most important yield component; breeding should aim to increase this yield component by increasing the number of fertile florets per spikelet, for which high variability exists (Salmon and Larter, 1978), while keeping a rather low number of spikelets per spike. Very good progress has clearly been made in improving the yield potential of triticale in many parts of the world.

For traits such as earliness, plant height, grain yield, thousand grain weight, test weight, spikelets per spike and kernels per spike, substantial genetic variation coupled with high heritability occur in primary triticales. Selection within primary triticale should be very effective (Oettler *et al.*, 1991). As the performance of primary triticales is generally much lower than that of secondary triticales, Lelley (1992) considered that breeding should concentrate on the improvement of existing high-yielding varieties rather than on the creation of new primary triticales. This is a controversial view in relation to the need to widen the genetic base of triticale.

Forage triticale may, like winter rye, produce high dry matter yields in early spring, or alternatively grow more slowly and give high yields later (Brignall *et al.*, 1989). The main breeding aim in winter forage triticale is to develop early high-yielding cultivars which are able to escape from the summer drought and therefore can be grown on lighter soils (Ford *et al.*, 1984). Because triticale is normally only selected for feed grain characteristics, assessment of its forage yield and quality is warranted.

4.8.6 **Breeding methods**

Triticale is normally handled in genetic improvement programmes using the pedigree selection method for self-pollinated, small-grained cereals. Special attention has to be paid, however, to the risk of cross-pollination and the occurrence of aneuploids. Yield components, such as fertility, spike length and number of tillers, vary more in triticale than in other cereals. Therefore, visual selection can be recommended as a suitable method for selecting high-yielding lines in triticale (Salmon and Larter, 1978). Yield testing in trials normally starts in the F5 generation. In order to prevent outcrossing in later generations, it is advisable to isolate plants (by bagging the ears of single plants) within the most promising advanced lines (Gustafson and Qualset, 1975; Wolski and Tymieniecka, 1980).

According to Struss and Röbbelen (1989), heterosis might cover meiotic irregularities and affect the efficiency of pedigree selection methods in early generations. Accelerated pedigree selection, which defers selection until later generations, might lead to greater progress in triticale breeding (Hörlein, 1991). In accelerated pedigree selection, initial selection is not based on individual plants but on multiple rows of lines from unselected individual plants (Valentine, 1984). These lines can be developed in the glasshouse resulting in a shortening of the breeding cycle.

Attempts have been made to develop hybrid triticale varieties. The development of those varieties depend on the availability of a suitable seed production system. Several cytoplasmic male sterility-inducing cytoplasms have been described as being suitable for triticale (Napela, 1991). However, at the moment triticale hybrid variety production remains experimental.

Triticale varieties are generally capable of achieving a high level of purity (Pickett, 1988). But, they never can be regarded as homogeneous and it is important that maintenance breeding is based on selection of individual plants or even ears (Wolski *et al.*, 1985). During seed production the undesirable risk of outcrossing requires special precautions such as isolation of seed crops from potential sources of contaminating pollen with only one generation of multiplication from basic seed (Pickett, 1988).

Regarding alternative breeding technologies, haploid production through anther culture followed by chromosome doubling allows fixation of homozygous genotypes in a relatively short time. Depending on genotype and environment, the number of regenerated green plants varied from zero to 23% (Bernard and Charmet, 1985). Anther culture of F1 hybrids leads to a larger population of aneuploidy in triticale than other cereals and it should therefore be reserved for more advanced and meiotically more stable generations (Charmet *et al.*, 1986). This somewhat devalues the use of haploid production as a method of reducing the length of the breeding process. Plant regeneration from immature embryo explants has been shown to be successful, but not as yet from protoplasts derived from embryogenic cell suspension, an important step in the genetic transformation of cereals (Stolarz and Lörz, 1986; Stolarz, 1991).

4.8.7 **Genetic resources**

In 1988, CIMMYT in Mexico accepted the responsibility for maintaining a global long-term genetic conservation collection of triticale. Genetic resources are available for researchers and breeders.

A system is also in place in collaboration with the International Board for Plant Genetic Resources to conserve and make available materials of the parents of triticale. Rye is conserved at the Polish Genebank, Radzikow. Bread wheat is conserved at CIMMYT and durum wheat at the International Center for Agricultural Research in the Dry Areas (ICARDA), Aleppo, Syria which also covers wild and primitively cultivated species related to wheat. Many other national institutions maintain cereal collections.

In low-latitude triticales, the extensive use of Armadillo from the mid 1980s and the early use by CIMMYT of wheats developed for irrigated agriculture as parents for triticale resulted in a narrow genetic base. The primary source for widening the genetic base of triticale is the development of primary triticales. Starting in 1986, CIMMYT initiated a special project to improve primary triticales and to specifically use parents possessing adaptations to specific conditions and resistances to diseases. Spring rye germplasm was also collected from around the world to broaden the impact of this parent, the contribution of rye not having received much attention from breeders.

Broadening of the genetic base was also sought through crossing adapted triticales of different pedigrees, crossing triticales with bread wheats and crossing spring and winter triticales (Varughese, 1991).

4.9 OCTOPLOID TRITICALES

Most of this chapter has referred to the triticale in its present hexaploid form. There is, however, increasing interest in octoploid triticale for direct cultivation, in particular in China and the independent states of the former USSR, and as a bridge for the transfer of genes from the A, B and D genomes of bread wheat present in octoploid triticale to hexaploid triticale.

Octoploid triticales have generally been considered much more unstable than hexaploid triticales by virtue of their higher chromosome numbers. Their likely advantage lies in the fact that they contain genes derived from bread wheat, which is more widely adapted than durum wheat. Octoploid triticales have been shown, not unexpectedly, to have better baking quality than hexaploid triticales (Stepochkina and Stepochkin, 1985; Fedorova *et al.*, 1988). Dorofeev *et al.* (1988) reported octoploid triticales to be more susceptible to leaf rust (*Puccinia recondita*) than hexaploid triticales. An extensive review of the breeding and cultivation of octoploid triticale has been published (Bao and Zhang, 1981) but is in Chinese and has not been accessed by the present authors.

Octoploid triticales have been crossed with hexaploid or tetraploid triticales to improve lodging resistance and straw stiffness (Varughese *et al.*, 1987; Wolski, 1991; Pojmaj and Wolski, 1991) and increase winter hardiness (Wolski, 1991).

C banding and *in situ* molecular hybridization has shown that D chromosomes or sequents frequently replace wheat chromosomes from the A or B genomes while mostly releasing the complete rye genome (Lukaszewski *et al.*, 1987; Gustafson *et al.*, 1989).

4.10 FUTURE PROSPECTS

The development of triticale into an economic crop within 100 years is a major achievement particularly as the main advances in international breeding have occurred after 1960. In many areas of the world, it has found a definite role as a crop for marginal conditions, particularly on light acid soils. Its largest area is in Poland where it is more productive than rye. In north-west Europe, triticale production could increase if, through the withdrawal of price support to traditional cereals, more cereals are fed to livestock in place of imported materials. Its greater disease resistance than wheat or barley is a distinct advantage in the face of stricter regulations on the application of nitrogen and pesticides brought about by increasing awareness of the environment. Lower inputs also mean reduced costs. On the other hand, a 36% reduction in the triticale area in the UK in 1993 (H-GCA, 1994) is presumably the result of conscious decisions by farmers to set aside marginal land in return for Arable Aid Payments and illustrates the susceptibility of the triticale crop to political decisions.

In order to secure the long-term future of triticale and to expand the crop, we consider that a radical re-modelling of the triticale phenotype is required. Increased yield would make it more competitive with wheat on 'normal' soils. Much shorter stiffer straw would make a tremendous difference to the acceptability of the crop in Europe where winter cereals with a long growing period are grown. The development of triticales that could be used as filler wheats by bread-makers would improve its marketability. Although triticale is mainly sold as feed grain, the distilling and malting industries are important developing markets. Growing triticale for specialist niche markets (health food shops, bakeries, etc) could be economically rewarding. Continued effort is needed to increase the genetic base of the crop, in order both to lessen the risk of genetic vulnerability and to maintain good breeding progress. Care should also be taken to maintain the good disease resistance of triticale, in which context triticales with the complete rye genome should be sought whenever possible. The development of new markets should proceed in parallel to genetic improvement. This requires different approaches in different parts of the world and a medium- to long-term view is required.

ACKNOWLEDGEMENT

The financial support of Semundo Ltd is gratefully acknowledged.

REFERENCES

Amaya, A. and Skovmand, B. (1985) Current status of hexaploid triticale quality, in *Genetics and Breeding of Triticale,* EUCARPIA meeting, Clermont-Ferrand, July 1984, pp. 603–6.

Aniol, A. and Gustafson, J.P. (1984) Chromosome location of genes controlling aluminium tolerance in wheat, rye, and triticale. *Can. J. Genet. Cytol.,* **26**, 701–5.

Anon (1990) World triticale production – some notes from Brazil. *Triticale Topics,* (International Edition), **6**, 11–12.

Anon (1993) *Recommended Varieties of Cereals 1993.* Farmers Leaflet No. 8. National Institute of Agricultural Botany, Cambridge, UK.

Baier, A.C. (1991) Triticale in the warmer areas: Is its efficiency in nutrient uptake enough to bring farmer acceptance? in *Wheat for the Nontraditional Warm Areas,* (ed D.A. Saunders), CIMMYT, Mexico.

Baier, A.C. and Nedel, J.L. (1986) Triticale in Brazil, in *Triticale International Symposium* AIAS, Sydney, pp. 270–82.

Bao, W.K. (1985) Evaluation of primary strains in breeding work of octoploid triticale, in *Genetics and Breeding of Triticale,* EUCARPIA meeting, Clermont-Ferrand, July 1984, pp. 121–4.

Bao, W. and Zhang, C. (1981) *Breeding and cultivation of octoploid triticale.* China: Guizhou People's Press.

Batterham, E.S., Saini, H.S. and Baigent, D.R. (1990) The effect of rate of feeding on the nutritional value of three cultivars for growing pigs. *Animal Feed Science and Technology,* **27**, 317–25.

Baum, B.R. (1971) The taxonomic and cytogenetic implications of the problem naming amphidiploids of *Triticum* and *Secale. Euphytica,* **20**, 302–6.

Baum, B.R. and Gupta, P.K. (1991) A taxonomic study of triticale (× *Triticosecale*), in *Proceedings of the 2nd International Triticale Symposium,* CIMMYT, Mexico, pp. 32–5.

Baum, M. and Lelley, T. (1988) A high efficient, rapid method to produce 4× triticale. *Tag. – Ber. Akad. Landwirtsch. – Wis, DDR* (Berlin), **226**, 199–207.

Bernard, S. and Charmet, G. (1985) Improvement of haploid production through *in vitro* anther culture in hexaploid triticale, in *Genetics and Breeding of Triticale,* EUCARPIA meeting, Clermont-Ferrand, July 1984, pp. 305–16.

Blakeslee, A.F. and Avery, A.G. (1937) Methods of inducing doubling of chromosomes in plants. *J. Heredity,* **28**, 392–411.

Blanchflower, A.J. and Briggs, D.E. (1988) *The malting of triticale. H-GCA Project Report No. 19,* H-GCA London.

Bodova, Z. (1991) Study of hexaploid triticale (Triticale Müntzing) on presence of Rht genes, in *Vyskumneho ustanvu rastlinneg vyrby vpiestanochr,* (ed V. Pace), pp. 5–11.

Börner, A. and Mettin, D. (1989) Genetische Grundlagen der Halmverkürzung (Dwarfinsmus) beim Weizen und Möglichkeiten der züchterischen Nutzung. *Kulturpflanze,* **37**, 29–55.

Börner, A. Lehman, C.O. and Mettin, D. (1987) Preliminary results of a screening for GA_3 response in wheats of the Gatersleben gene bank. *Kulturpflanze,* **35**, 179–86.

Brignall, D.M., Ward, M.R. and Whittington, W.J. (1988) Yield and quality of triticale cultivars at progressive stages of maturity. *Agric. Sci.*, **111**, 75–84.

Brignall, D.M., Ward, M.R. and Whittington, W.J. (1989) Relationship between growth stage and digestible organic matter in triticale. *J. of Agric. Sci.*, **113**, 1–11.

Bruckner, P.L. and Hanna, W.W. (1990) *In vitro* digestibility of fresh leaves and stems of small-grain species and genotypes. *Crop Science*, **30**, 196–202.

Burgstaller, G., Bohm, M. and Propstmeier, G. (1989) Experiments on digestibility and feeding value of triticale in pigs and ruminants. *Wirtschaftseigene futter*, **35**, 101–7.

Burtle, G.J. (1991) The utilization of small grains in diets for aquaculture, in *Proceedings of the Georgia Nutrition Conference for the Feed Industry*, Atlanta, Georgia, USA, November 1991, pp. 108–14.

Bushuk, W. (1976) *Rye: Production, Chemistry and Technology*. Amer. Assoc. Cereal Chem., St. Paul, MI.

Carnide, V., Guedes-Pinto, H., Mascarenhas-Ferreira, A. and Sequeira, C. (1991). Triticale-legume mixtures, in *Proceedings of 2nd International Triticale Symposium*, CIMMYT, Mexico, pp. 542–5.

Charmet, G., Bernard, S. and Bernard, M. (1986) Origin of aneuploid plants obtained by anther culture in triticale. *Can. J. Genet. Cytol.*, **28**, 444–52.

Charmley, E. and Greenhalgh, J.F.D. (1987) Nutritive value of three cultivars of triticale for sheep, pigs and poultry. *Animal Feed Science and Technology*, **18**, 19–35.

Chojnacki, G., Brykczynski, J. and Tymienieka, E. (1976) Preliminary information on sprouting in triticale. *Cereal Research Communications*, **4**, 111–14.

Chung, O.K. and Pomeranz, Y. (1985) Amino-acids in cereal proteins and protein fractions, in *Digestibility and Availability in Cereals and Oilseeds*, (eds Finley, J.W. and Hopkins, D.T.), American Association of Cereal Chemists Inc., St Paul, Minnesota, USA, pp. 65–77.

Clarke, J.M., McCaig, T.N. and Townley-Smith, T.F. (1986) Kernel development and changes in falling number in triticale compared with wheat. *Can. J. Plant Sci.*, **66**, 877–84.

Collin, J., Comeau, A. and St Pierre, C.A. (1990) Tolerance to barley yellow dwarf virus in triticale. *Crop Science*, **39**, 1008–14.

Cook, R. and York, P.A. (1987) Variation in triticales in reaction to cereal cyst nematode. *Tests of Agrochemicals and Cultivars*, **8**, 158–9.

Cooper, K.V. (1985) *The Australian Triticale Cookery Book*, Savvas Publishing, Adelaide, Australia.

Davis, R.P., Garthwaite, D.G. and Thomas, M.R. (1991) *Pesticide Usage Survey Report 85. Arable Farm Crops in England and Wales, 1990*. Ministry of Agriculture, Fisheries and Food.

Dorofeev, V.F., Chikida, N.N., Grigor eva, O.G. and Odintsova, I.G. (1988) Sources of resistance to brown rust in triticale. *Naucho Tekhnicheskii Byulleten Vsesoyuznogo Ordena Lenina i Ordena Druzhby Naradov Nauchno Issledovatel Skogo Instituta Rastenievodstva Imeni NI Vavilova*, **177**, 67–70.

Federov, A.K. (1988) Features of development of triticale in relation to winter hardiness and yield. *Spornik Nauchnykh Trudov po Prikladnoi Botanike, Genetike i Selektsii*, **122**, 26–31.

Fedorova, T.N., Berkutova, N.S. and Cazareva, F.N. (1988) Development of triticale breeding and seed production. Biochemical and technological features of the grain of octoploid (8 ×) triticalis. *Selektziya i Semenovodstvo, Moscow*, **6**, 12–15.

Flour Milling and Baking Research Association. (1985). *The Milling and Baking Properties of Triticale.* Results of a sponsored research contract between The Flour Milling and Baking Research Association and Guinness Crop Developments, Semundo Ltd, Farmers Seeds Federal Ltd, and Food From Britain.

Ford, M.A., Austin, R.B., Gregory, R.S. and Morgan, C.L. (1984) A comparison of the grain and biomass yields of winter wheat, rye and triticale. *J. Agric. Sci.*, **103**, 395–403.

Fox, P.N., Skovmand, D., Thompson, B.K., Braun, H.J. and Cormier, R. (1990) Yield and adaptation of hexaploid spring triticale. *Euphytica*, **47**, 57–64.

Francois, L.E., Donovan, T.J., Mass, E.V. and Rubenthaler, G.L. (1988) Effect of salinity on grain yield and quality, vegetative growth, and germination of triticale. *Agron. J.*, **80**, 642–7.

Furber, M.J. (1990) Triticale crop management, in *Triticale – a Growing Market*, (ed Semundo), Evergreen Publishing Ltd, UK, pp. 9–17.

Gale, M.D. and Marshall, G.A. (1976) The chromosomal location of a major dwarfing gene from Norin 10 in new British semidwarf wheats. *Heredity*, **35**, 417–21.

Gawlik, J. and Seiffert, M. (1990) Development of some forms of winter triticale as compared with winter wheat and winter rye cultivars. *Archiv. fur Acker- und Pflanzenbau und bodenkunde*, **34**, 337–45.

Glattes, H. and Mar, A. (1983) Triticale in praktischen Mahl- und Bakversuch. 1. Teil: Mahlversuche. *Getreide, Mehl und Brot*, **37**(12), 359–62.

Gohil, S., Pettersson, D., Solomonsson, A.C. and Åman, P. (1988) Analysis of alkyl- and alkenylresorcinols in triticale, wheat and rye. *J. Sci. Food Agric.*, **45**, 43–52.

Graham, R.D. (1978) Tolerance of triticale, wheat and rye to copper deficiency. *Nature, London*, **271**, 542–3.

Graham, R.D. (1988) Development of wheats with enhanced nutrient efficiency: progress and potential, in *Wheat Production Constraints in Tropical Environments. Proceedings of the International Conference*, Chiang Mai, Thailand, January 1987 (ed A.R. Klatt), CIMMYT, Mexico, pp. 395–408.

Gregory, R. (1987) Triticale breeding, in *Wheat breeding – its Scientific Basis*, (ed F.G.H. Lupton), Chapman & Hall, London, pp. 269–86.

Guedes-Pinto, H., Rangel-Figueiredo, T. and Carnide, O. (1984) Aneuploidy in high yielding 6 × -triticales. *Cereal Research Communications*, **12**, 229–35.

Gupta, P.K. and Priyadarshan, P.M. (1982) Triticale: present status and future prospects. *Advances in Genetics*, **21**, 255–345.

Gupta, P.K. and Reddy, V.R.K. (1991) Cytogenetics of triticale – a man-made cereal, in *Chromosome Engineering in Plants: Genetics, Breeding, Evolution*, eds P.K. Gupta and T. Tsuchiya), Elsevier, Amsterdam, pp. 335–9.

Gustafson, J.P. and Qualset, C.O. (1975) Genetics and breeding of 42-chromosome triticale II. Relations between chromosomal variability and reproductive characters. *Crop. Sci.*, **15**, 810–13.

Gustafson, J.P. and Zillinsky, F.J. (1973) Identification of D-genome chromosome from hexaploid wheat in a 42-chromosome triticale, in *Proceedings of 4th International Wheat Genetics Symposium*, Columbia, USA, pp. 225–31.

Gustafson, J.P., Dillé, J.E. and Skovmand, B. (1989) Wheat substitution in hexaploid triticale. *Plant Breeding*, **102**, 109–12.

H-GCA (1994) Final 1993 June Results Census and Revised UK Cereal Production Estimates. *H-GCA (Home-Grown Cereals Authority) Weekly Digest* 14, February 1994.

Hanrahan, T.J. (1986) Triticale as an ingredient in pig feeds. *Irish J. Agric. Res.*, **25**, 149–52.
Heger, J. and Eggum, B.O. (1991) The nutritional values of some high yielding cultivars of triticale. *Journal of Cereal Science*, **14**, 63–71.
Henry, R.J. (1985) A comparison of the non-starch carbohydrates in cereal grains. *J. Sci. Food Agric.*, **36**, 1243–53.
Heslop-Harrison, J.R., Leitch, A.R., Schwarzacher, T. and Anamthawat-Jónsson, K. (1990) Detection and characterization of 1B/1R translocations in hexaploid wheat. *Heredity*, **65**, 385–92.
Hill, G.M. (1991) Quality: Triticale in animal nutrition, in *Proceedings of 2nd International Triticale Symposium*, CIMMYT, Mexico, pp. 422–7.
Hoan, T.L., Armstrona, K.C. and Miki, B. (1989) Detection of rye DNA in wheat–rye hybrids and wheat translocation stocks using total genomic DNA as a probe. *Plant Molecular Biology Reporter*, **7**, 150–8.
Hörlein, A. (1991) Triticale breeding in UK. *Triticale Topics (International Edition)*, **6**, 5–6.
Hughes, G.G. (1990) Use of triticale as a replacement for wheat middlings in diets for Atlantic salmon. *Aquaculture*, **90**, 173–8.
Hulan, H.W. and Proudfoot, F.G. (1987) The effect of different dietary levels of triticale on the general performance of SCWL layer genotypes to 490 days of age, in *Annual Report Research Station Kentville, Nova Scotia*, pp. 118–20.
Huskowska, T., Wolski, T. and Ceglinska, A. (1985) Breeding of winter triticale for improvement of grain characters, in *Genetics and Breeding of Triticale*, EUCARPIA, pp. 641–9.
Hwang, J.J. and Ha, Y.W. (1991) Germination of rye, triticale, wheat, oats and barley after storage for various periods under different conditions. *Research Reports of the Rural Development Administration, upland and industrial crops (Korea Republic)*, **33**, 103–11.
Jaaska, V. (1983) Secale and triticale, in *Isozymes in Plant Genetics and Breeding*, Part B, (eds S.D. Tanksley and T.J. Orton), Elsevier Science Publishers B.V., Amsterdam, pp. 79–101.
Jung, C. and Lelley, T. (1985) Genetic interactions between wheat and rye genomes in triticale. 2. Morphological and yield characters. *Theor. Appl. Genet.*, **70**, 427–32.
Kahn, C.B. and Penfield, M.P. (1983) Snack crackers containing whole-grain triticale flour: crispness, taste and acceptability. *Journal of Food Science*, **48**, 266–7
Kaltsikes, P.J. (1974) Methods for triticale production. *Z. Pflanzenzüchtung*, **71**, 264–86.
Kolkunova, G.K., Maksimchuk, B.M., Moslova, N.M. and Vedernikova, E.I. (1982) Processing triticale into flour, in *Proceedings of 7th Cereal and Bread Congress*, Prague, pp. 415–18.
Krolow, K.K. (1973) 4 × triticale production and use in triticale breeding, in *Proceedings of 4th International Wheat Genetics Symposium*, MO Agric. Exp. Stn. Columbia, MO. pp. 237–43.
Krolow, K.D. (1985) Triticale (*Triticosecale* Wittmack), in *Lehrbuch der Züchtung Landwirtschaftlicher Kulturpflanzen – spezieller Teil*, (eds W. Hoffmann, A. Mudra and W. Plarre), Verlag Paul Parey, Berlin, pp. 67–77.
Lelley, T. (1992) Triticale, still a promise? *Plant Breeding*, **109**, 1–17.
Lelley, T. and Taira, T. (1979) Effect of genotype on response to colchicine treatment in wheat–rye hybrids. *Z. Pflanzennzüchtung*, **82**, 87–9.
Lersrutaiyotin, R., Shigenaga, S. and Utsunomiya, N. (1991) Malting quality of

hexaploid triticale in comparison with that of barley, wheat and rye. *Jap. J. of Crop Science*, **60**, 291–7.

Leterme, P., Tahon, F. and Théwis, A. (1990) Le triticale en alimentation animale. 1. Composition chimique. *Revue de l'Agriculture*, **43**, 617–29.

Lopez, J.R. and Garbini, S.K. (1985) *Triticale situacion actual y perspectivas de su cultivo in la Republica Argentina*. INTA, pp. 12.

Lorenz, K., Welsh, J., Normann, R. and Maga, J. (1972) Comparative mixing and baking properties of wheat and triticale flours. *Cereal Chem.*, **49**(2), 187–93.

Lukaszewski, A.J. (1988) Chromosome constitution of hexaploid triticale lines in the recent international yield trials. *Plant Breeding*, **100**, 268–72.

Lukaszewski, A.J. (1990) Frequency of 1RS.1AL and 1RS.1BL translocations in United States wheats. *Crop Science*, **30**, 1151–3.

Lukaszewski, A.J. and Curtis, C.J. (1992) Transfer of the *Glu-D1* gene from chromosome 1D of breadwheat to chromosome 1R in hexaploid triticale. *Plant Breeding*, **109**, 203–10.

Lukaszewski, A.J. and Gustafson, J.P. (1983) Translocations and modifications of chromosomes in triticale × wheat hybrids. *Theor. Appl. Genet.*, **64**, 239–48.

Lukaszewski, A.J. and Gustafson, J.P. (1987) Cytogenetics of triticale, in *Plant Breeding Reviews*, Vol 5, (ed J. Janick), Van Nostrand Reinhold Co., New York, pp. 41–93.

Lukaszewski, A.J., Apolininarska, B. and Gustafson, J.P. (1987) Introduction of the D-genome chromosomes from breadwheat into hexaploid triticale with complete rye genome. *Genome*, **29**, 425–30.

Malik, K.P.S. (1984) Seed set by spontaneous cross-pollination in triticale. *Cereal Research Communications*, **12**(1–2), 111–12.

Mar, A. and Glattes, H. (1984) Triticale im praktischen Mahl- und Backversuch. 2. Teil: Backversuche. *Getreide, Mehl und Brot*, **38**(1), 9–12.

McDonald, H.G. (1990) *Rye and Triticale in the UK*. HGCA (Home-Grown Cereals Authority) Review 21.

Meister, G.K. (1921) Natural hybridization of wheat and rye in Russia. *J. Hered.*, **12**, 467–70.

Merker, A. (1973) A Giemsa technique for rapid identification of chromosomes in triticale. *Hereditäs*, **75**, 280–2.

Miller, T.E. (1984) The homoeologous relationship between the chromosomes of rye and wheat. Current status. *Can. J. Genet. Cytol.*, **26**, 578–89.

Mossé, J., Huet, J.C. and Baudet, J. (1988) The amino acid composition of triticale grain as a function of nitrogen content: comparison with wheat and rye. *Journal Cereal Science*, **7**, 49–60.

Müntzing, A. (1955) Mode of production and properties of a triticale-strain with 70 chromosomes. *Wheat Inf. Serv.*, **2**, 1–2.

Müntzing, A. (1979) Triticale – Results and problems. Fortschritte der Pflanzenzüchtung. Suppl. to Z. *Pflanzenzüchtung*, **10**, 1–103.

Napela, S. (1991) Hybrid triticale: present and future, in *Proceedings of 2nd International Triticale Symposium*, CIMMYT, Mexico, pp. 402–7.

National Research Council. (1989) *Triticale: A Promising Addition to the World's Cereal Grains*. National Academy Press, Washington, D.C.

Oettler, G. (1983) Die Rolle des Weizens und des Roggens bei der Herstellung primärer Triticale, in *Bericht über die Arbeitstagung 1983 der Arbeitsgemeinschaft der Saatzuchtleiter, Gumpenstein*. Verlag der Bundesanstalt für alpenländische Landwirtschaft, Gumpenstein, pp. 143–54.

Oettler, G. (1985) The effect of wheat cytoplasm on the synthesis of wheat × rye hybrids. *Theor. Appl. Genet.*, **71**, 467–71.

Oettler, G. (1987) Variation of falling number in primary triticale and their wheat and rye parents, in *4th International Symposium on Pre-harvest Sprouting in Cereals*, (ed D.J. Mares), pp. 123–30.
Oettler, G. (1991) Preharvest sprouting, in *Proceedings of 2nd International Triticale Symposium*, CIMMYT, Mexico, pp. 182–4.
Oettler, G., Wehmann, F. and Ultz, H.F. (1991) Influence of wheat and rye parents on agronomic characters in primary hexaploid and octoploid triticale. *Theor. Appl. Genet.*, **81**, 401–5.
O'Mara, I.G. (1951) The cytogenetics of triticale. *The Botanical Review*, **XIX**, 587–605.
Peña, R.J. and Ballance, G. M. (1987) Comparison of gluten quality in triticale: A fractionation-reconstitution study. *Cereal Chem.*, **64**(2), 128–32.
Pettersson, D. and Åman, P. (1987) Effects of enzyme supplementation of diets based on wheat, rye or triticale on their productive value for broiler chickens. *Animal Feed Science and Technology*, **20**, 313–24.
Pfeiffer, W.H. and Fox, P.N. (1991) Adaptation of triticale, in *Proceedings of 2nd International Triticale Symposium*, CIMMYT, Mexico, pp. 54–63.
Pickett, A.A. (1988) Factors affecting seed production of triticale (× *Triticosecale* Wittm.). *Plant Varieties and Seeds*, **1**, 63–74.
Pojmaj, M.S. and Wolski, T. (1991) Breeding strategies for improving lodging resistance in winter triticale in connection with heterosis, in *Proceedings of 2nd International Triticale Symposium*, CIMMYT, Mexico, pp. 140–3.
Ren, Z.L., Lelley, T. and Röbbelen, G. (1990) The use of monosomic rye addition lines for transferring rye chromatin into bread wheat. I. The occurrence of translocations. *Plant Breeding*, **105**, 257–64.
Richter, G., Ochrimenko, C., Cyriaci, G. and Hennig, A. (1990) Studies of the use of native triticale for poultry. 1. Broiler fattening. *Archives of Animal Nutrition*, **40**, 595–606.
Riley, R. and Chapman, V. (1967) The inheritance in wheat of crossability with rye. *Genet. Res.*, **9**, 259–67.
Rimpau, W. (1891) Kreuzungsprodukte landwirtschaflicher Kulturpflanzen. *Landwirtschaftl. Jahrb.*, **20**, 335–71.
Royo, C., Rodriguez, A. and Romagosa, I. (1993) Differential adaptation of complete and substituted triticale. *Plant Breeding*, **111**, 113–19.
Rundgren, M. (1988) Evaluation of triticale given to pigs, poultry and rats. *Animal Feed Science and Technology*, **19**, 359–375.
Salmon, D.F. and Larter, E.N. (1978) Visual selection as a method for improving yield of triticale. *Crop Science*, **18**, 427–30.
Sandha, G.S., Sharma, B.D., Gill, K.S. and Dhindsa, G.S. (1991) Studies on determining the contribution of different characters to grain yield in triticale, in *Proceedings of 2nd International Triticale Symposium*, CIMMYT, Mexico, pp. 93–7.
Scoles, G.J. and Kaltsikes, K.J. (1974) The cytology and cytogenetics of triticale. *Z. Pflanzenzüchtung*, **73**, 13–43.
Seal, A.G. (1982) C-banded wheat chromosomes in wheat and triticale. *Theor. Appl. Genet.*, **63**, 39–47.
Singh, R.P. and Saari, E.E. (1991) Biotic stresses in triticale, in *Proceedings of 2nd International Triticale Symposium*, CIMMYT, Mexico, pp. 171–81.
Skovmand, B., Fox, P.N. and Villareal, R.L. (1984) Triticale in commercial agriculture: Progress and promise. *Advances in Agronomy*, **37**, 1–45.
Smith, W.C. and Pearson, G. (1989) Partial and total replacement of barley by triticale in diets for growing pigs. *N.Z. Journal of Agricultural Research*, **32**, 447–52.

Stepochkina, N.I. and Stepochkin, P.I. (1985) Grain quality in hexaploid and octoploid winter triticales and their use in breeding. *Sibirskii Vestnik Sel Skokhozyaistvennoi Nauki*, **3**, 29–33.

Stoddart, J.L. (1987) Genetic and hormonal regulation of stature, in *Developmental Mutants in Higher Plants*, (eds H. Thomas and D. Grierson), Cambridge University Press, pp. 155–80.

Stolarz, A. (1991) Cell and protoplast culture, somatic embryogenesis and transformation studies in different forms of × *Triticosecale* Wittmack, in *Proceedings of 2nd International Triticale Symposium*, CIMMYT, Mexico, pp. 286–9.

Stolarz, A. and Lörz, H. (1986) Somatic embryogenesis, *in vitro* multiplication and plant regeneration from immature embryo explants of hexaploid triticale (× *Triticosecale* Wittmack). *Z. Pflanzenzüchtung*, **96**, 353–62.

Stolcova, M. (1990) Specific features of triticale development. *Rostinna Vyroba*, **9**, 983–9.

Struss, D. and Röbbelen, G. (1989) Cytological irregularities masked by heterozygosity in triticale. *Plant Breeding*, **102**, 265–72.

Sylvester-Bradley, R. (1993) Opportunities for lower nitrogen inputs without loss of yield or quality: an agronomic and economic appraisal, in *Proceedings of the Cereals R&D Conference*, Cambridge, H-GCA (Home-Grown Cereals Authority, London, pp. 198–217.

Tanner, D.G. (1981) Parental effects of chromosome doubling for the production of primary triticale. *Can. J. Genet. Cytol.*, **23**, 191–3.

Tanner, D.G. and Falk, D.E. (1981) The interaction of genetically controlled crossability in wheat and rye. *Can. J. Genet. Cytol.*, **23**, 27–32.

Théwis, A. and Leterme, P. (1991) Le triticale en alimentation animale 3. Les ruminants. *Revue d'Agriculture*, **44**, 707–16.

Thomas, J.B., Kaltsikes, P.J., Gustafson, J.P. and Roupakias, D.G. (1980) Development of kernel shrivelling in triticale. *Z. Pflanzenzüchtung*, **85**, 1–27.

Thomas, L., Senn, Th., Pieper, H.J. (1991) Bioethanol aus Triticale. *GIT Fachz. Lab.*, **10**, 1087–92.

Thomke, S., Pettersson, D., Åman, P., Rundgren, M. (1987) Chemical characterization and nutritive evaluation of triticale in comparison with wheat and rye. *Journal of Swedish Seed Association*, **1**, 65–7.

Tombros, S. and Briggs, D.E. (1984) Micromalting triticale. *J. Inst. Brew.*, **90**, 263–5.

Tsen, C.C. (ed) (1974) *Triticale: First Man-Made Cereal.* Amer. Assoc. Cereal Chem., St. Paul, MI.

Tsuchiya, T. and Larter, E.N. (1971) Further results on chromosome stability of hexaploid triticale. *Euphytica*, **20**, 591–6.

Valentine, J. (1984) Accelerated pedigree selection: an alternative to individual plant selection in the normal pedigree breeding method in the self-pollinated cereals. *Euphytica*, **33**, 943–51.

Valentine, J. and Middleton, B.T. (1987) The effect of embryo culture on reproductive development in winter cereals. *Cereal Research Communications*, **15**, 161–5.

Varughese, J.P., Lelley, T. and Röbbelen, G. (1984) Cell division rate, cell number and nuclear aberrations in relation to seed shrivelling in triticale. *Z. Pflanzenzüchtung*, **92**, 102–16.

Varughese, G. (1991) Recognizing triticale's proper place among the world's cereals, in *Proceedings of 2nd International Triticale Symposium*, CIMMYT, Mexico, pp. 6–8.

Varughese, G., Barker, T. and Saari, E. (1987) *Triticale*. CIMMYT, Mexico.

Varughese, G., Sari, E.E. and Abdalla, O.S. (1986) Two decades of triticale breeding and research at CIMMYT, in *Proceedings International Triticale Symposium*, Australia, AIAS, Sydney, pp. 148–69.
Villareal, R.L., Varughese, G. and Abdalla, O.S. (1990) Advances in spring triticale breeding. *Plant Breeding Reviews*, **8**, 43–90.
Wilson, A.S. (1876) On wheat and rye hybrids. *Trans. Proc. Bot. Soc. Edinburgh*, **12**, 286–8.
Wolfe, M.S. and Klints, T.M.C. van (1987) Powdery mildew of wheat. *Annual Report of the Plant Breeding Institute 1986*, pp. 93–5.
Wolski, T. (1991a) Winter triticale breeding, in *Proceedings of 2nd International Triticale Symposium*, CIMMYT, Mexico, pp. 41–8.
Wolski, T. (1991b) Impact of semi dwarf rye germplasm on triticale improvement, in *Proceedings of 2nd International Triticale Symposium*, CIMMYT, Mexico, pp. 144–9.
Wolski, T. and Tymieniecka, E. (1980) The present state and main problems in winter triticale breeding in Laski and Choryn experimental stations. *Hodowla Roslin Aklimatyzacja i Nasiennictwo*, **24**, 475–85.
Wolski, T., Maczinska, L., Tymieniecka, E. (1985) Winter triticale varieties from the Choryn and Laski experimental stations, in *Genetics and Breeding of Triticale*, EUCARPIA, pp. 487–96.
Zillinsky, F.J. and Borlaug, N.E. (1971) Progress in developing triticale as an economic crop. *CIMMYT Research Bulletin 17*.

Part Two

A Neglected Cereal and Two Promising Ones

5

Fonio (*Digitaria exilis* and *D. iburua*)

N. Haq and F. Dania Ogbe

5.1 HISTORY AND ORIGINS

Fonio is an indigenous African cereal which is identified with early crop domestication in the western Sudan zone. Between 6500 and 2500 BC the Sahara was moister than at present and was occupied by pastoral nomads, and black fonio is probably associated with incipient agriculture in this region. As agriculture emerged in the savanna area north of the tropical rain forest of West Africa, three cereals became integral: fonio, pearl millet and sorghum.

Fonio (*Digitaria exilis* Stapf) (Figure 5.1) is cultivated from Cape Verde in the west to Lake Chad in the east. Black fonio (*D. iburua* Stapf) is much more limited in cultivation, being found in northern Nigeria and in an area of Togo and Benin and scattered elsewhere. Black fonio is currently almost certainly a relict of formerly wider cultivation (Figure 5.2).

Fonio is also known as Acha, Fundi, petit mil or hungry rice, and it is a cereal crop throughout the savanna zone; in parts of Guinea and Nigeria it is a staple crop. Elsewhere it is an important part of the diet. Although it is the major staple for about 4 million people, its input to the diet of millions more is hardly appreciated.

Fonio is considered to be the oldest West African cereal with cultivation dating back to 5000 BC (Purseglove, 1985). Surprisingly it has received little international attention. When the *ad hoc* panel of the Advisory Committee on Technology Invention of US National Academy of Sciences finalized a priority list of underexploited tropical plants with promising economic value in 1974 none of the minor cereals were con-

Cereals and Pseudocereals. Edited by J.T. Williams. Published in 1995 by Chapman & Hall, 2–6 Boundary Row, London SE1 8HN. ISBN 0 412 46570 1.

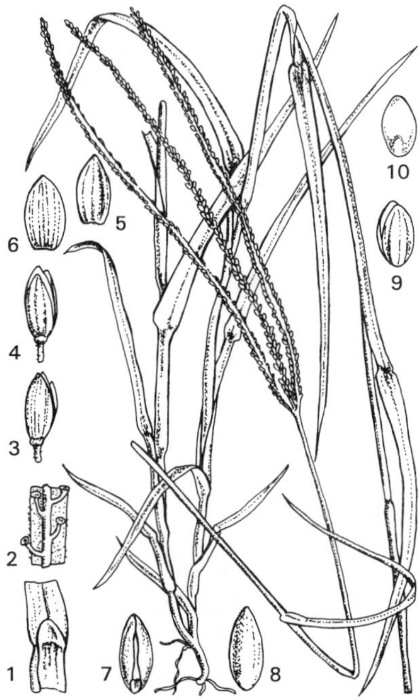

Figure 5.1 Fonio (*Digitaria exilis*).

sidered promising. As a result, further research was not spurred, as was the case for quinoa and amaranths (see Chapters 1 and 3). Nonetheless, as interest in enhancement of minor cereals has increased in recent years, especially of finger millet (*Eleusine*). The fact that in West Africa fonio occupies a similar ecological niche to finger millet in the eastern parts of the continent justifies a new look at its potential. Additionally, there had been by the mid 1980s little serious research done in Africa on finger millet improvement or agronomy with the exception of work at Serere in Uganda, and this is only more recently being rectified. The same situation which existed for finger millet exists for fonio. Another publication (National Research Council, 1992) has drawn recent attention to fonio. It is therefore a neglected crop with potential; the main drawback has probably been that information was not readily available. The NRC account, and this present one, should rectify that situation. As the crops are essentially rain-fed crops of the semi-arid tropics, screening of germplasm for better drought-resistant varieties is important, especially as low input crops grown by poor and marginal farmers.

Fonio is nutritious and its grains are rich in methionine, an amino acid vital to human health and deficient in many major cereals. The

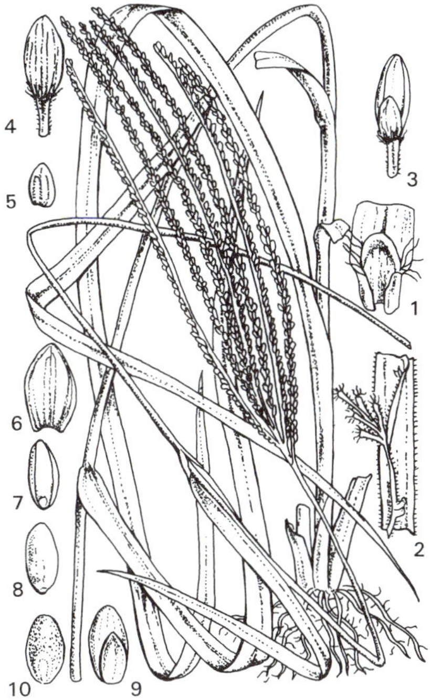

Figure 5.2 Black fonio (*Digitaria iburua*).

crop is used in a variety of ways. The small grains have an attractive flavour and are made into porridge and couscous, or are ground and mixed with other flours to make breads, or are popped, and are also used to brew beer. Portères (1955) recorded several uses which date back to historical records from the fourteenth century (Al Oman, 1921). It is believed to be important nutritionally in the area of consumption but the utilization of food is not well documented outside the area of production.

It is an important food crop of tribal people of northern Nigeria where the ethnic groups may consume it three times a day in various ways. A family of seven to ten people will consume about 2–3 kg per day in major fonio growing areas (Jideani, 1990). Fonio foods are prepared mostly from whole grain which can be thick or soft and consumed as gruel. Fonio grains are easily cohesive when cooked, due to their small size, a property needed for the consistency of the type of food called 'tuwo acha'. Tuwo acha is an unfermented porridge similar to tuwo made from other cereals. 'Kunu-acha' is fermented porridge made from the whole grain, then soaked and allowed to ferment for about 3 days. The fermented paste is boiled in water to give 'kunu acha'. Hausa porridge is made of unground fonio and mixed with honey or sugar and eaten raw after brief fermentation. Muller (1970) reported that this is the only raw cereal product eaten in the area. The whole grain is also used when cooked in water and condiments, and vegetables added and on occasion fish as well. The grain is also boiled and eaten with fish, meat or spinach or used in soup.

Irvine (1969) reported that beer produced from fonio is used as payment for harvesting crops. Local people like the taste and they consider the grain important, so much so that in many places they reserved it for chiefs (Vietmeyer, pers. comm.), other dignitaries and for special occasions. In recent times, some people have made side-by-side comparisons of dishes made with fonio and common rice and have greatly preferred the fonio (Vietmeyer, pers. comm.). It has also formed part of the traditional bride price. In some areas *D. exilis* can be regarded as the staple crop and in others it forms an important component of the farming systems and diet. Its importance is illustrated by the fact that it is of significance in human culture in some areas. The Dogon of Mali believe that the cosmos was born from grain of *D. exilis* (National Research Council, 1992). The Turuma of northern Nigeria use different varieties of *D. exilis* in specific ritual contexts. It is believed that these ritual contexts can be given relative dates (B. Sharpe, pers. comm.). A synthesis of such anthropological data with data on the botany of the crop and farming systems in which it operates may be able to illustrate the development and diversification of the crop in the area.

The grain of *Digitaria* is digested efficiently by sheep, goats, donkeys and other livestock. It is valuable feed for monogastric animals, notably pigs and poultry, because of its high methionine content. The straw and chaff are also fed to animals. Both make excellent fodder and the hay is often sold in markets. The straw is sometimes chopped and mixed with clay for building houses or walls. It is also burnt to provide heat for cooking or ash for potash.

Many species of *Digitaria* serve as fodder throughout the tropics. However, of those which are drought-resistant, and which would occupy similar ecologies to fonio, most better ones are from Southern Africa, namely *D. milanjiana*, *D. smutsii* and *D. pentzii*, or from east Africa, such as *D. macroblephara*. Only in very recent times have materials of the West African *Digitaria* species been looked at, or even screened, as for instance by the forage programme of the International Livestock Center for Africa (ILCA) located in Ethiopia, which has been expanding its farming systems research in West Africa in association with the International Institute for Tropical Agriculture (IITA) in Nigeria and the Sahalian Center of the International Crops Research Institute for the Semi-Arid Tropics in Niger.

A number of wild *Digitaria* species are gathered in West Africa for food in times of famine or food scarcity. These include *D. longiflora* and *D. fuscescens* (Pesl.) Henr., both related to the two fonio species.

Only one other species of *Digitaria, D. criciata* (Nees) Canus, is a cultivated grain species elsewhere. This is the crop raishan of the Khasi hills, Assam, in India where it is a local cultigen. The cultivars belong to variety *esculenta* Bor, which is also used for fodder. This may have been more widespread in the past since a cultivar is known from northern Vietnam (Veldkamp, 1973).

Table 5.1 Common names of *D. exilis*

Country	*Local group*	*Vernacular name*
Cote d'Ivoire	Niarafolo	Feni
	Dan	Pon, Pohin
	Wobé	Pohin
Benin	Booriba	Podji
	Sonba-Bariba	Moin, Podji
	Somba-Yaoualan	Pei, Podji
	Tamberna	Ipopera
Gambia	Manding	Findo
Ghana	Dagomba	Kabugu
Guinca	Fula	Foni, Fondc
	Sussu	Foundé
	Bassari	Frume
	Malinke	Foni, Foigné
	Kissi	Pendé, Kpando
Burk Faso	Bobo	Fin, Peni
	Dorhasi	Foni
	Dorhon-finng	Deli
Mali	Bambara	Fini
	Peul	Sérémé
	Sonorai	Foyo, Fodio, Fingi
	Dogon	Pon
Scncgal	Wolof	Findi, Ndenge
	Manding	Findi, Findo, Feindi
	Diola	Eboniaye
Sierra Leonne	Kissi	Fendergi, Fudingi
	Konno et Koranko	Fundi
	Meude	Kputi, Pote
	Sussu	Fundengi
	Timèné	Fundi, Peni, Pene
Togo	Konkomba	Epich, Ipwi
	Dagomba	Kabega
	Lamba	Afio
	Losso	Kafia

5.1.1 Origins and diversity

Almost certainly, the centre of origin of fonio is around the central delta of the Niger. Here and in neighbouring areas, Portères (1955) recognized three different types of patterns of diversity. He also thought that there was a small secondary centre of diversity in Togo.

Black fonio is closely related to the Hausa culture. There appears to be a centre of diversity in northern Nigeria based on Kano, Zaria and Katsina, but this may extend to the region of Zinder in Niger (Portères, 1976). Vernacular names also relate black fonio with the Hausa.

In the case of fonio, there are many vernacular names (Table 5.1) but they confirm the centre of origin as proposed by Portères. Murdoch (1959) uses the origin of fonio and names to confirm a nuclear Mande origin of agriculture centred on the upper Niger river. The nuclear Mande, like their dispersed kinsmen speaking dialects of the Negritic linguistic stock, assumed early cultural leadership in West Africa; and later powerful trading empires, such as that of Ghana based on the Soninke and that of Mali based on the Malinke, but all based on food production through shifting hoe cultivation and secondary animal husbandry.

5.2 TAXONOMY

The genus *Digitaria* belongs to the subfamily Panicoideae of the Gramineae and falls within the tribe Paniceae. It is widely distributed in tropical and warm temperate areas. Estimates of the number of species in the genus vary from about 230 (Clayton, 1972) to about 325 species (Henrard, 1950). Species of the genus have a small or suppressed lower glume and a chartaceous to cartilaginous upper lemma with flat hyaline margins infolding and concealing most of the palea.

Henrard monographed the genus and this still continues to serve as the basis of any taxonomic study. His treatment was an elaboration of treatments by Heckel (1901) and Stapf (1915). It subdivided the genus into four subgenera, and 32 sections, the latter often based heavily on hair types of spikelets. This resulted in the establishment of some artificial sections with glabrous spikelets, and it has since been noted that some species are in sections from which their overall description would exclude them.

Both *D. exilis* and *D. iburua* are placed in the subgenus Eu Digitaria and the section Atrofuscae. Both are annual with erect culms. Characteristic features of the species are:

D. exilis. Culms erect or ascending, nearly simple, over 45 cm high, leaf sheaths smooth, leaf blades linear, 5–15 cm long and 4–9 mm wide, and glabrous. Racemes are 2–4 digitate, sessile and slender, 5–12 cm long. Upper glume broadly oblong, subobtuse. Spikelets 1.5 mm long and pedicels of spikelets without hairs, nearly smooth. Fruits are much narrower at the base, and acute.

D. iburua. Culms erect, over 50 cm high, simple with four or five nodes. Internodes are enclosed in sheaths which are glabrous at the base of the plant. Racemes are subcomposite, 4–10 subdigitate, the lowest somewhat distant from the others which are densely crowded and suberect. Spikelets are tightly imbricate, elliptic-lanceolate, subacute at the base, acute at the tips, about 2 mm long, glabrous. Upper glume very delicate and hyaline, ovate. Pedicels of spikelets are densely short hairy, thickened upwards and longer hairs form a coronula towards the tip. Fruit is pointed and slightly compressed from the back.

In both species, the caryopsis is tightly enclosed within the brown husks.

5.2.1 Related species

D. exilis is only known in cultivation. However, it is not at all clear from which species it arose. Dalziel (1937) and others think that it is closest to *D. longiflora* pers. but Henrard considered *D. barbinodis* Henr. to be very closely related, and in fact it is often see together in cultivated fields with the cultigen in Nigeria and Togo. *D. barbinodis* is endemic to Malian Sudan. *D. longiflora* is an annual or short-lived perennial widely distributed in West Africa (Senegal, Gambia, Guinea, Liberia, Sierra Leone, Cote d'Ivoire, Ghana, Benin and Nigeria, as well as in Mali and Burkina Faso). *D. barbinodis* is also an annual widely distributed in tropical Africa. Another species closely related to both cultigens and *D. barbinodis* is *D. laeviglumis* Fernald. It is a secondary invader of many disturbed sites in tropical Africa, but, although also annual, is stoloniferous and less likely to be a progenitor.

The need for taxonomic revision and more emphasis placed on species relationships is apparent because, although some putative related species are classed in the same section of the subgenus, *D. longiflora* is placed in another section Verrucipilae.

D. iburua resembles *D. exilis* in many ways but various other species have been mentioned as possible relatives or progenitors. Stapf (1915) reported that botanically the cultigen is very close to *D. ternata* Stapf. It is widespread in the hot regions of Africa and also Asia (Portères, 1946). Others have thought that *D. tricostulata* (Hackel) Henr., distributed in South Africa and north to Kenya, is close to *D. iburua* (Zeven and de Wet, 1982), and this species falls close in relation to Henrard's classification. Other species, such as *D. atrofusca* (Hackel) Camus, although similar, can probably be ruled out in view of their distributions being in areas of the world more remote from the area of diversity of the cultigen.

The two cultigens are hexaploid, while many of the weedy species described as possibly related are diploid and some tetraploid. However, not enough is known to clarify the origin of the cultigens. More than likely, several species have been involved and there could well be three progenitors in each case, depending on the area of diversity, and some will be common to both cultigens. It would seem prudent to investigate this before further taxonomic revisions.

There appears to have been only limited isozyme studies to date (Hayward and Hecker, 1980). These studies utilized only one isozyme (esterase) from a single species, *D. milanjiana*, an important range grass from eastern Africa.

5.2.2 **Variation within fonio**

Portères (1955) was able to identify a number of botanical varieties of fonio. The most primitive types have pedicels grouped mostly by threes (occasionally by twos and fours), with spikelets in one or two rows along the racemes, there being two racemes in the inflorescence, each with 50–100 spikelets. These primitive forms have sparse, fine culms and grain matures early.

By contrast, advanced types possess inflorescences with three or four racemes. Pedicels are grouped by threes but never by fours and spikelets are in two or three rows on the racemes. Whereas the more primitive forms have 50–100 spikelets for every 10 cm, the more advanced forms have over 100, and sometimes 120–140 spikelets per 10 cm of length.

Portéres found that these botanical varieties possessed a number of somewhat diverse cultivars within each one. Additionally, the botanical varieties are differentiated by geography as well as morphology. Major ones are as follows:

var. *gracilis* Rough pedicels and leaf margins curled. Typical in the region of Kankan, Guinea.
var. *stricta* Smooth pedicels, leaf margins only slightly curled, spikelets widely spaced. Found in Guinea, Mali, the Casamance and Burkino Faso.
var. *rustica* All organs covered with coarse setulae. Robust, late maturing.
subvar. *rubra* Rich in anthocyanins and elongated spikelets with narrow tips, found in region of Kankan to Bamako and Fouta Djallan.
var. *densa* Tall robust advanced type, found in Togo.

There are up to 15 of these varieties and some are hybrids of others.

Within the varieties there may be as many as 10–15 recognizable agro-ecotypes. This enabled Portères to plot the diversity on maps, the highest diversity is seen in the upper basin of the river Niger from its source to the central delta. This approximates to the area of the old empires of Mali in the twelfth and thirteenth centuries. Interestingly, the traditional varieties represent early and late maturing forms.

5.2.3 **Variation with black fonio**

Since the distribution of black fonio is much more limited than that of fonio, botanical varieties have not been described in the same way.

5.2.4 **Ecology**

The cultural limit of fonio is on the annual isohyet of 150 mm whereas in general sorghum and pearl millet are limited by the isohyet of 200–250 mm (Portères, 1976). In relation to cultivation of fonio, early varieties are used in dry conditions, late varieties in wet conditions.

Fonio is not greatly limited by soil type, except perhaps by heavy

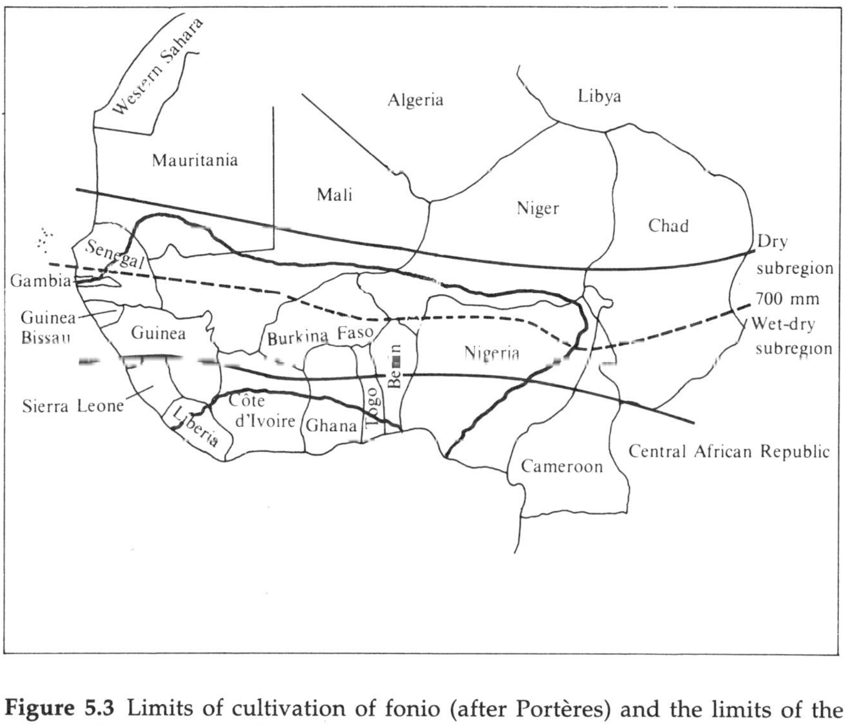

Figure 5.3 Limits of cultivation of fonio (after Portères) and the limits of the semi-arid tropics as defined by the International Crops Research Institute for the Semi-Arid Tropics (ICRISAT).

Figure 5.4 Distribution sandy acid soils in West Africa, after ICRISAT. Compare this to the distribution of fonio cultivation in Figure 5.3.

soils, nor by terrain or exposure. Cultivation occurs on sandy, gravelly soils, with or without lime, and whether fertile or relatively infertile (Figures 5.3 and 5.4). Cultivation is seen on slopes, flat terrain or river beds (Portères, 1955).

Black fonio, where cultivated, is under similar ecology to fonio. The two do not compete, and sometimes they are found in mixtures. However, black fonio is cultivated in more upland conditions, 400–1300 m. It is thought that the present areas of cultivation are not the ancestral ones, but that in Neolithic times it was known in the Air mountains when this area of the Sahara was wetter, and that following the drying it was carried further south by human migrations. As such it would have been a sub-alpine plant (Portères, 1946).

5.3 BIOCHEMISTRY

The primary product of *Digitaria* is the grain. According to the chemical composition (Table 5.2), fonio is comparable to rice in nutritive value. So too are most of the minor millets (Malleshi, 1989).

Fonio is somewhat low in calcium (finger millet having a very high content) but other minerals compare favourably with those of rice (Table 5.3).

5.3.1 **Amino acids**

The advantage of fonio in the diet is the high level of methionine (Table 5.4). The level of arginine is unlikely to affect the lysine availability. The leucine to isoleucine ratio of fonio protein is not the most favourable but is similar to many small millets.

Benito *et al.* (1986) reported that the high content of methionine-cysteine of *D. exilis* grains can make it an excellent complement to legumes. They suggested that it should be used as a genetic base to improve the protein quality of other plant foods, through breeding.

Table 5.2 Chemical composition of fonio grain compared with that of finger millet and milled rice (g/100 g)

	Fonio (range)	*Finger millet*	*Milled rice*
Carbohydrate	74.7–81.0	72.0	78.2
Protein	7.2–8.7	7.3	6.8
Fat	1.1–1.8	1.3	0.5
Fibre	1.1–1.6	3.6	0.2
Ash	1.6–2.1	n.s.	n.s.

Source: Cisse (1975); Gami (1988); Malleshi (1989).

Table 5.3 Composition of minerals in grain of *D. exilis* (g/100 g)

Ca	30
K	16
Mg	40
Na	10
P	170

Table 5.4 Amino acid content of grain of fonio compared with that of finger millet and rice (g/gN)

	Fonio	*Finger millet*	*Rice*
Arginine	0.24	0.30	0.48
Alamine	0.31	–	–
Aspartic acid	0.40–0.41	–	–
Cysteine	0.13–0.18	0.14	0.09
Glutamic acid	1.26–1.39	–	–
Glycine	0.20	–	–
Histidine	0.13	0.13	0.13
Isoleucine	0.25	0.40	0.30
Leucine	0.61	0.69	0.50
Lysine	0.16	0.22	0.23
Methionine	0.35	0.21	0.15
Phenylalanine	0.31	0.31	0.28
Proline	0.44	–	–
Serine	0.31–0.36	–	–
Threonine	0.21–0.27	0.24	0.23
Tryptophan	0.09	0.10	0.08
Tyrosine	0.023	0.22	0.29
Valine	0.34	0.48	0.38

Sources: Cisse (1975); Jideani (1990); Malleshi (1989).

Table 5.5 Calorific value of *D. exilis* and Sxx Cereals (value in K cal. pr. 100 gm)

Fonio *D. exilis*	393.94
Sorghum	394.09
Maize	409.65
Millet	413.79
Rice	397.14

Source: Oyenuga (1968).

5.3.2 **Calorific value**

Calorific values of fonio compare favourably with those of several major cereals (Table 5.5).

5.3.3 Straw value for feed

The chemical composition of the straw (Table 5.6) make it suitable for use in balanced rations of ruminants and other livestock.

Table 5.6 Chemical composition of *Digitaria* hay, straw (% dry matter)

	Fonio	*Finger millet*
Dry matter	97.0	98.0
Nitrogen free extract	55.86	50.0
Crude fibre	30.08	37.2
Ash	7.04	n.s.
Crude protein	5.35	3.40
Ether extract	1.66	1.30

Sources: Oyenuga (1968); Mudgal and Sampath (1970).
n.s. not stated.

5.4 AGRONOMY

Fonio is closely associated with traditional rain-fed cultivation, particularly in the Sudano-Sahelian climatic zone. In this zone even subsistence-orientated sorghum and pearl millet cropping systems have hardly yet been markedly changed by recent innovations in agricultural production techniques. Mixed cropping occurs which basically acts as replacement systems in time and space and provides risk avoidance for the farm families. This is important because in this zone rainfall is low, variable and undependable. Moreover, soils are often poor; many are

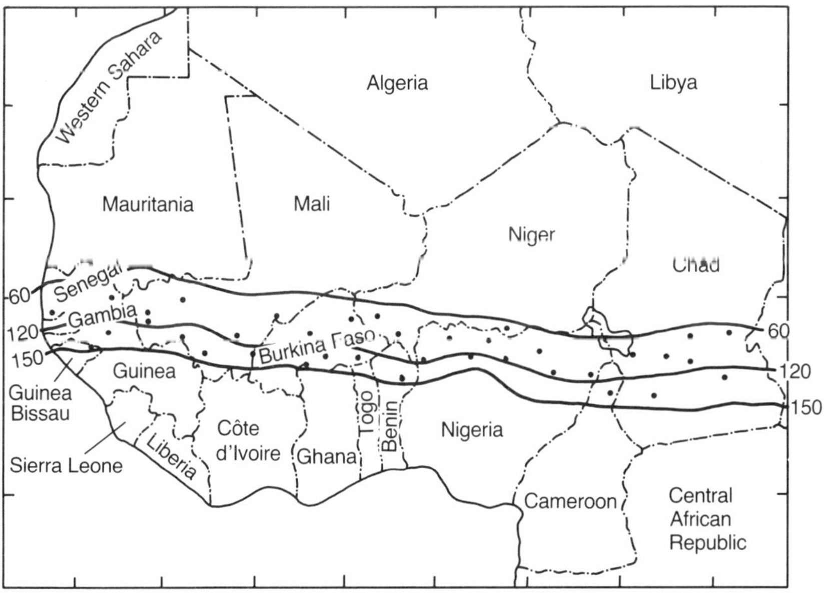

Figure 5.5 Distribution of fonio with mean length of growing season (days) as defined by ICRISAT.

sands which are acidic, low in cation exchange capacity, mineral reserves and organic matter.

One important aspect of the zone is the length of the growing season, and there are local varieties of fonio adapted to growing seasons from 60–150 days (Figure 5.5).

Since there have been no attempts to try to develop intensive culture for fonio with the use of specially enhanced varieties, it is possible to envisage a scenario for the crop's improvement through multilocational screening of germplasm and application of cultivation techniques to improve water retention and improve fertility through mulch or better use of residues. Fonio is likely to find a continuing place in development of such sustainable systems.

5.4.1 **Sowing**

Land is normally prepared at the end of dry weather and sometimes burning and cleaning of land is practised before the seed is planted. The cultivated *Digitaria* species are sown in the months of May and June, the ground being slightly opened and the soil lightly drawn over the seed with a hoe (Gani, 1988). Seeds are broadcast or sown in rows (Gandy, 1987). The seeding rate varies considerably. Portères (1955) reported the seeding rate of 30–50 kg/ha but others have recommended the rate of 15–20 kg/ha (Cisse, 1975). He reported the seeding rate of 5–10 kg/ha when planted at a spacing of 25–30 cm between rows.

About 2000 grains weigh 1 g. The germination rate is very high and seeds start to germinate at 3–4 days after planting. However, the germination rate depends on the temperature and humidity at the time of germination. Although associated with sorghum–millet areas it is grown elsewhere, although heavy soils are not suitable for the majority of cultivated varieties (Portères, 1955). *D. exilis* is common in Guinea's Fouta Djallon region and the soils are acidic clays with high aluminium content, a combination which is toxic to most food crops. Here it is grown generally with upland rice. However, selection of suitable local varieties can readily be made for all terrains, fertile to unproductive soils, plateaux, valleys and river banks.

5.4.2 **Growth and harvesting**

The crop is a poor competitor with weeds. Weeding is necessary if good yield is to be achieved. The weeding is carried out after 4–5 weeks of planting.

Flowering starts at about 45 days after planting and ripening begins around September. The maturity period varies from 90–130 days (Purseglove, 1985) but some varieties are quick maturing and a crop

Table 5.7 Harvest index of *Digitaria* spp. (in %)

Spp.	*Harvest index*
D. exilis	19.23
D. iburua	25.00
D. barbinodis	9.78

can be achieved in 60 days (Haq, 1991). The seed kernel is a tiny yellow grain weighing about 0.5 mg.

At maturity the slender stems bend towards the earth by the mere weight of the grains. They are reaped with hooked knives and tied up after cutting down in small sheaves and allowed to dry. If they are allowed to remain in the field then the grains become agglutinated because of being covered by straw of plants.

Jideani (1990) stated that harvesting contributes to the high cost of the production process as the crop is harvested by the farmer himself, his family and friends, who harvest the crop manually. The grain is threshed by hand to separate it from straw either on the farm or at the farmer's home and arranged in convenient heaps on a mat. The heaped crop is beaten with sticks until most seeds are free from the straw. The grain is encased in a protective hull which is easily removed by pounding it in a mortar using a pestle. The yellow and brown kernels are decorticated by further pounding. The bran (pericarp) is separated from the grain by winnowing (Figure 5.1).

Local machinery is being developed in Nigeria for harvesting (Jideani, 1990). Harvest indexes have been compared: black fonio has a higher index than fonio (Table 5.7).

5.4.3 **Fertilizers and biomass**

Little work has been done on the application of fertilizer. It is general belief that the plant growth is optimum with adequate supply of mineral elements either in the soil or in culture solution. Jideani (1990) reported that the application of adequate fertilizer (calcium, aluminium nitrate and superphosphate) increased the yield substantially. Gani (1988) studied mineral nutrition and was concerned with the effects of different levels of nitrogen on the growth of three species, *D. exilis*, *D. iburua*, and *D. barbinodis*. He carried out an experiment with five different levels of nitrogen (168 p.p.m., 130 p.p.m., 88 p.p.m., 46 p.p.m., and 5.6 p.p.m.) keeping the other mineral elements constant (Table 5.8). The levels 168 and 130 p.p.m. had similar effects on growth and 5.6 p.p.m. level did not support growth as plants did not flower. He then carried out a second experiment using four levels of nitrogen (130, 65, 33 and

Table 5.8 Influence of different nitrogen levels on growth and crop development of *Digitaria* species

Characters	*Nitrogen levels (p.p.m)*					*Mean*	*L.S.D.*	*F test*
	168	130	88	46	5.6		5%	
First experiment								
Number tillers/per plant	13.00	11.44	10.11	7.22	2.22	8.80	2.67	***
Root fresh weight/plant (g)	18.36	17.33	16.49	8.61	4.13	12.98	6.01	***
Root dry weight/plant (g)	1.92	1.48	1.78	0.80	0.40	1.28	0.39	***
Shoot fresh weight/plant (g)	15.89	11.28	12.00	5.47	1.98	9.32	2.09	***
Shoot dry weight/plant (g)	4.30	2.84	2.97	1.56	0.54	2.44	0.65	***
Total dry matter weight/plant	6.33	4.58	4.91	2.35	0.93	3.82	0.94	***
Second experiment								
	Nitrogen levels (p.p.m.)							
	130	65	33	17				
Number of tillers/plant (12h)	6.44	4.89	2.22	2.00		4.00	0.74	***
Total dry matter wt/plant (12h) g	2.51	1.87	1.08	0.69		1.54	0.26	***
Number of tillers/plant (12h) g	4.89	4.33	2.67	1.33		3.35	0.73	***
Total dry matter wt/plant (12h) g	2.79	2.02	1.05	0.75		1.65	0.62	***
Grain weight/plant (12h) g	0.42	0.38	0.33	0.29		0.35	0.06	***

Confidence limits ** = 0.05 > a > 0.01; *** = 0.01 > d > 0.001
Source. Gani (1988).

Table 5.9 Average yield of *D. exilis* in different African countries in kg/ha

Country	*kg/ha*
Senegal	330
Mali	400
Niger	181
Guinea	500
Burkina Faso	366
Cote Ivoire	500
Benin	300
Togo	257

17 p.p.m.). He found that the levels 65 and 130 p.p.m. produced more grain weight per plant than others. This increase has been mainly caused by positive effect of nitrogen on grain weight. Spiertz and Van Keuler (1980) believed from their experiment that the grain weight per plant differed according to level of nitrogen and plant species. Gani (1988) concluded that the species *D. iburua* and *D. exilis* showed potential for increased crop development at higher nitrogen levels than *D. barbinodis* which tended to have more vegetative biomass. The apparent accumulation of resources to vegetative biomass (including roots) suggested that it is a strong competitor for water and mineral elements in the soil for growth rather than crop development.

5.4.4 **Yields**

The grain yield of *D. exilis* is variable, depending on the soil and the time of planting. Reported show yield from 170–225 kg/ha in very poor soils. But yields of over 1000 kg/ha have been achieved. In general, grain yield can be obtained within the range of 600–900 kg/ha. Table 5.9 shows the yield reported by several authors in various countries in West Africa.

5.4.5 **Pests and diseases**

Both species of fonio, *D. exilis* and *D. iburua*, have shown low susceptibility to pests and diseases. The fungi, *Phyllachora sphaerosperma* and *Sphaerodothis sphaerosperma*, have been seen to affect *D. exilis* (Cisse, 1975), as does *Paspalum scrorbiculatum*, and *Pennisetum polystachion* Schult. and *Pennisetum subangustum* Stapf.

Striga is often found in the field to affect the crop. This is particularly true of *Striga hermontica* Benth. and *S. senegalensis* Benth. *S. rowlandi* can cause serious damage to the crop in Guinea, Senegal and Gambia.

Jideani (1990) reported that no major fungus, insect or other pest cause much damage to *D. exilis* in Nigeria. Birds may necessitate scarring as with other cereals.

5.4.6 **Mixtures of fonios**

Henrard (1950) reported that *Digitaria* species are cultivated together in West Africa, particularly in Nigeria and Togo. It is possible that the species will compete for nutrients and light. But the relationship between the competitive ability of a particular plant with a mixture of different plants of the same species to the yield of that same plant in pure culture has been shown to be negative. This has been shown in barley cultivars (Suneson, 1949) and in rice (Jennings and de Jesus, 1968). This is an area, for fonio, which requires more research.

5.4.7 **Seed storage**

A large percentage of cereal grains (in Nigeria alone about 37%) is lost during storage. The farmer does not face a major problem in storing *Digitaria* grains. The grain is less susceptible to attack by pests or by fungus than other cereals.

Storage in airtight containers is found to be suitable for keeping grains for more than 24 months (Jideani, 1990). He suggested that the storage life could be longer if conventional siloes are used as for millet in China.

5.5 GENETICS

5.5.1 **Cytology**

As with most of the Paniceae, the basic chromosome number of the genus *Digitaria* is 9. The chromosomes are generally small and polyploidization has played a role in the evolution of the genus. The genus contains diploid ($2n = 2x = 18$), tetraploid ($2n = 4x = 36$), hexaploid ($2n = 6x = 54$), octoploid ($2n = 8x = 72$) and even dectaploid ($2n = 12x = 108$) species. There are some species that have given more than one chromosome count (Darlington and Wylie, 1955). *D. exilis* is hexaploid ($2n = 6x = 54$). *D. iburua* is also considered to be hexaploid.

5.5.2 Germplasm

A relatively large germplasm collection of species of *Digitaria* is held by CSIRO, Division of Tropical Crops and Pastures, Australia. This comprises about 400 accessions of 43 species in Henrard's sense. (This is about 30 species according to current views.) They have a list of 700 accessions but this includes lines that are either not viable or that have been lost before the project adopted a long-term maintenance role. Most of the accessions are perennial forage species. This germplasm is being utilized to aid the development of improved pasture (A. Kidd, pers. comm.). Most accessions originated in Africa. CSIRO is interested in any future collections of perennial forage types that can be made. ILCA, Addis Ababa, Ethiopia also hold about 40 accessions from about six different species. About half the accessions are of *D. exilis* from Togo. CIAT in Colombia also maintains a collection of *Digitaria* spp., particularly the forage types. Royal Botanical Gardens, Kew also has some collections of wild types of *Digitaria* spp. A small research collection (about 60 accessions) of *D. exilis* and *D. iburua* is held at the Department of Biology, Southampton University, UK.

ORSTOM, Center de Bondy, France, hold over 400 accessions of *D. exilis* collected by Mr. Jean Clement in Benin, Burkina Faso, Guinea, Mali and Togo. This is a very comprehensive collection from those areas sampled. However, the collection does not cover all the areas where fonios are cultivated, although it probably covers a large part of the genetic variability that exists. Unfortunately, Nigeria, which contains what could be at least a secondary centre of diversity, is not included. This collection remains taxonomically unverified (it may actually contain three species). There has been no characterization or evaluation of the germplasm to date. Recently Dr. Adam Toudou of University of Niamey, Niger and Mr. Halvadan Jakobsen of Terra Nova have collected some local materials and their evaluation and characterization will be most useful.

5.5.3 **Breeding**

Little effort has so far been put into crop improvement through breeding. The most recent breeding work was carried out by Gani (1988). He also reported ideotype breeding but no conclusion was reached on the results because of shortage of time. However, he discussed the concepts of using environment, particularly the nutrient level, crop density and planting arrangement of *D. exilis* and *D. iburua*. He also discussed the accumulation and utilization of dry matter and its relation to grain yield, vegetative parts development and photoperiod growth pattern and its relation to grain and straw yield.

5.6 GAPS IN KNOWLEDGE

The genetic diversity of the species needs to be studied as the information on species relationships and on eco-geographic distribution is limited. This is because of lack of attention to fonio due to drawbacks on yield-related characters, such as small seeds, seed shattering, seed losses due to non-viability, inefficient root structure, lodging and poor tillering, which result in low yields. Some other agronomic investigations, such as requirement of water and fertilizer, soil types, effect of day length and temperature, need to be carried out to improve existing yields. The apparent wide range of variability in the two crops *D. exilis* and *D. iburua* could be exploited for increased grain yield, higher nutritive quality, and the existing wide range of adaptations and disease resistances is valuable but not well documented.

The harvest index seems to be poor (Gani, 1988) but he suggested that a satisfactory ideotype would provide a reasonable harvest index (not less than 45%) as most C4 cereals accumulate their yield on the basis of similar physiology. (*Digitaria* has a C4 photosynthetic pathway.) Gani (1988) carried out hybridization between and within species of *D. exilis* and *D. iburua* and *D. barbinodes* and he reported authentic hybrids only within species. However, this is valuable for crop improvement.

Diets involving grain legumes and fonio are useful in developing countries. Jideani (1990) suggested that gluten extracted from *Digitaria* by modern techniques can be used as a by-product and can aid in the production of new nutritious food products. In Nigeria, the Plateau Foods Processing Company (PFPC) Ltd has identified fonio as a viable and beneficial food processing venture (Jideani, 1990). However, any such venture will depend on the supply of raw materials, but such an enterprise strengthens the case for further research on this crop.

REFERENCES

Al Onan (1927) *Notebooks 1342–1346*, translated by Demonbynes, Paris, Busson, F. (1965) *Plantes Alimentaires de L'Oest Africain*. Lecunte, Marseille.

Benito, O., De Lumen, R. Becker and Ryes, P.S. (1986) Legume and a cereal with high methionine/cystine contents. *J. Agric. Food Chem.*, **34**, 361–4.

Cisse, I.B. (1975) La culture de fonio et quelques aspects écophysiologiques de la plante. *Tropische Plantenteelt*, Wageningen.

Clayton, W.D. (1972) *Flora of West Tropical Africa: Gramineae.* Crown Agents, London.

Dalziel, J.M. (1937) *The Useful Plants of West Tropical Africa*, Crown Agents, London.

Darlington, C.D. and Wylie (1956) *Chromosome Atlas*, George Allen and Unwin Ltd, London.

Gani, A.M. (1988) Mineral nutrition and morphological studies as approaches to development of ideotypes for *Digitaria iburua* Stapf. and *D. exilis* Stapf. M.Sc. thesis, Wye College, University of London.

Haq, N. (1991) Development of underutilized crops and their role in sustainable production. Presented at Bio-diversity of Traditional and Underutilized Crops, Malta, June 1991.

Hayward and Hacker (1980) Genetic control of some esterase isozymes of *Digitaria* species, and their utility in the identification of hybrids.

Heckel, E.M. (1901) *Les graines grasses nouvelles ou peu connus des colonies françaises*. A. Challomel, Paris.

Henrard, T.H. (1950) *Monograph of the Genus Digitaria*, Leiden University Press, Leiden, Netherlands.

Hewitt, E.J. and Smith, T.A. (1975) *Plant Mineral Nutrition*, John Wiley and Sons, Inc., New York.

Johnson, V.A. (1984) New horizon in cereal production, in *Cereal Production*, (ed E.J. Gallacher), Butterworth and Co., England. pp. 1–13.

Jennings, P.R. and Jesus J. de (1968) Studies in competition in rice. 111. Mechanism of varieties. *Evolution*, **22**, 119–24.

Jideani, I.A. (1990) Acha-*Digitaria exilis* – the neglected cereal. *Agric. Inst.*, **42**(5), 132–43.

Malleshi, N.G. (1989) Processing of small millets for food and industrial uses, in *Small Millets in Global Agriculture*, (eds A. Seetharum, K.W. Riley, and G. Hannardyana), Oxford, IBH Pubs. Co. Put. Ltd, Delhi, pp. 325–39.

Muller, H.G. (1970) Traditional cereal processing in Nigeria and Ghana. *Ghana J. Agric. Sci.*, **3**, 187–95.

National Research Council (1993) *Lost Crops of Africa*. Natrul Academy Press, Washington D.C.

Oyenuga, V.A. (1968) *Nigeria's Foods and Feedstuffs. Their Chemistry and Nutritive Value* (3rd edn), University Press, Ibadan, Nigeria.

Portères, R. (1946) L'aire culturales de *Digitaria iburua* Stapf, céréale mineure de l'Ouest-Africain. *L'Agronomie Tropicale*, **1**(11–12), 389–92.

Portères, R. (1976) African cereals, *Eleusine*, fonio, black fonio, *Braciaria*, *Paspalum*, *Pennisetum*, and African Rice, in *Origins of African Plant Domestications*, (eds J.R. Harland, J.M.J. de Wet and A.B.L. Stemler), Mouton Publishers, the Hague, pp. 409–52.

Portères, R. (1955) Les céréales mineures du genre *Digitaria* en Afrique et Europe. *J. Agric. Trop. Bot.* Appl. 2.

Purseglove, J.W. (1985) *Tropical Crops. Monocotyledons*, Longman Group Ltd, London.

Murdock, G.P. (1959) *Africa: its people and their Cultural History*, McGraw Hill Book Co., New York.

Veldkamp, J.F. (1973) A revision of *Digitaria* Haller (Graminaea) in Malenia. *Blumea*, **21**, 1–80.

6

Intermediate wheatgrass (*Thinopyrum intermedium*) Development of a perennial grain crop

Peggy Wagoner

6.1 INTRODUCTION

The development of intermediate wheatgrass as a grain crop is unique due to the fact that it is perennial. All of our current cereal and pseudocereal crops are annuals. Given the proper conditions annual crops can produce high yields and good profits. However, on hilly or otherwise marginal land, annual crop production can lead to severe soil erosion and associated environmental degradation (Colacicco *et al.*, 1989; Larson *et al.*, 1983; Pimentel *et al.*, 1987). The development of perennial grain crops could provide valuable alternatives for erodible land that cannot sustain annual crop production. Perennial grains such as intermediate wheatgrass can provide year-round soil protection, as well as lower annual inputs of labour and materials (Jackson, 1985; Schatz and Sobolik, 1989).

Intermediate wheatgrass has many favourable agronomic culinary traits which make it a likely candidate for development into a perennial grain crop. Mechanical seed production of intermediate wheatgrass is already a standard practice in the Great Plains of the USA and also the intermountain west (Atkins *et al.*, 1967; Hennes, pers. comm.). Intermediate wheatgrass seeds mature synchronously, resist severe shattering and are easily threshed from the hulls to product naked, edible grain. Most importantly, seeds are highly palatable for use as a grain

Cereals and Pseudocereals. Edited by J.T. Williams. Published in 1995 by Chapman & Hall, 2–6 Boundary Row, London SE1 8HN. ISBN 0 412 46570 1.

for human consumption (Becker *et al.*, 1991). Commercial varieties are available such as cv. Luna and cv. Oahe.

6.2 CROP HISTORY

The use of intermediate wheatgrass as an agricultural crop is not new. It has been used extensively as an introduced cool-season forage in the USA and Canada during this century (Asay and Knowles, 1985; Hansen, 1972) and as a native forage in temperate regions of the Old World.

During Byzantine times, the use of this perennial grass species as a grain crop was alluded to in Turkey, Armenia and the south-western region of the Soviet Union (Derzhavin, 1960). More recently in the former Soviet Union, efforts were made to develop perennial wheat utilizing intermediate wheatgrass germplasm. In the 1920s Soviet researchers cross-bred annual wheat with intermediate wheatgrass, yielding partially fertile hybrids (Tsitsin, 1960). Interest in developing a perennial wheat through such wide hybridization waned due to problems with sterility and genetic incompatibility (Dewey, 1984). Instead, intermediate wheatgrass has since proven useful as a source of genes for disease resistance which have been transferred to annual wheat via classical breeding techniques (Cauderon, 1979; Cauderon and Saigne, 1961; Dewey, 1984).

Today in the USA efforts are focused on developing intermediate wheatgrass into a perennial grain crop. This approach of domesticating the perennial species will avoid problems associated with the wide hybridization with an annual grain. Since 1984, research at the Rodale Research Center (RRC) in Pennsylvania, USA, has been conducted to develop grain production techniques and to select and breed higher grain-yielding varieties of intermediate wheatgrass (Wagoner, 1990a; Wagoner, 1990b; Wagoner and Schauer, 1989).

6.3 TAXONOMY

Intermediate wheatgrass is a perennial, cool-season, rhizomatous grass (family Gramineae). As a member of the tribe Triticeae, intermediate wheatgrass is related to important annual grain such as wheat, rye and barley as well as 250 species of perennial grasses, many of which are important forage grasses (Dewey, 1984). Wheatgrasses are generally thought of as those grasses belonging to *Agropyron senso lato*.

The taxonomic classification of intermediate wheatgrass has been unsettled for some years. Until recently, North American botanists have classified intermediate wheatgrass as *Agropyron intermedium* (Host) Beavois. A close relative, pubescent wheatgrass, has been classified as

Agropyron intermedium var. *trichophorum*. Pubescent wheatgrass shows its pubescence on the seed-heads and leaves in contrast to intermediate wheatgrass which is glabrous (Hanson, 1972). Both are considered part of an intermediate wheatgrass species complex because they coexist in Old World habitats and are freely interfertile (Dewey, 1978; Mariam and Ross, 1972). The form introduced to the USA represents a forage type from the mountains of Turkey, Iran and Iraq usually considered

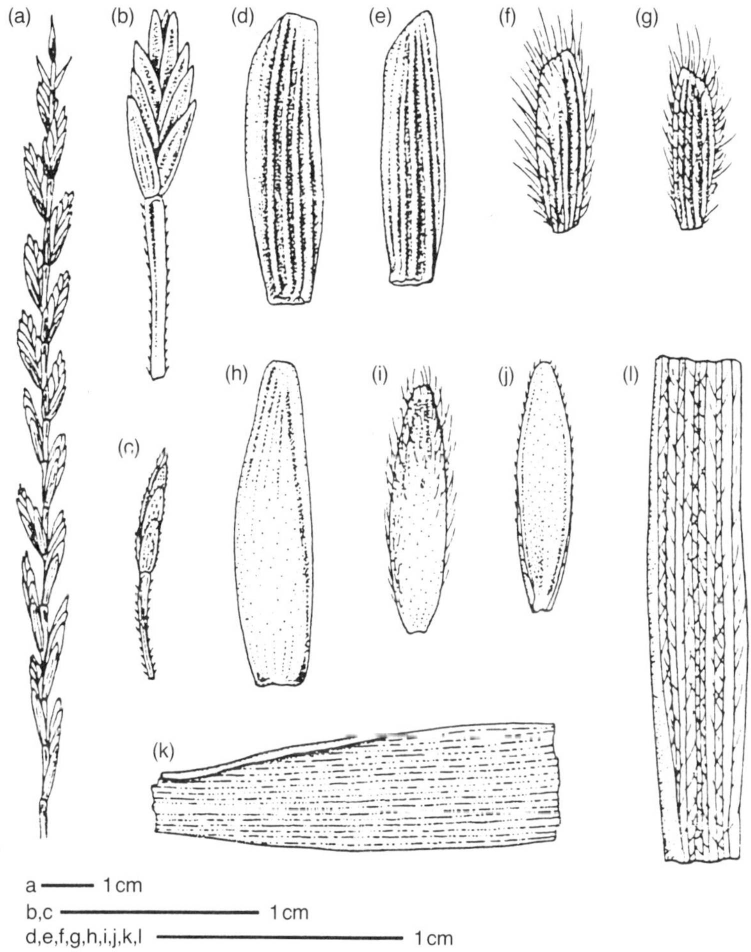

Figure 6.1 *Elytrigia intermedia* subsp. *intermedia*. (a) spikes; (b) spikelet of glabrous form; (c) spikelet of hirsute form; (d) upper glume of glabrous form; (e) lower glume of glabrous form; (f) upper glume of hirsute form; (g) lower glume of hirsute form; (h) lemma of glabrous form; (i) lemma of hirsute form; (j) palea; (k) blade of glabrous form; (l) blade of hirsute form. (Diagrams from J.K. Jarvie, Ph.D. thesis, Utah State University, 1990.)

subsp. *Podgerae* (Nabel.) Dewey; syn. *A. podgerae* Nabel. Dewey (1984) proposed a new taxonomic classification of the perennial Triticeae grasses based upon cytogenic information, phylogeny and biological relationships. Under this classification, both intermediate and pubescent wheatgrass are *Thinopyrum intermedium*. Other names under which intermediate wheatgrass has been identified include *Elytrigia intermedia* (Host) Nevski (Figures 6.1 and 6.2), *Elymus hispidus* (Opiz) Melderis and *Agropyron glaucum* (Desf. ex DC.) Roemer and Schultes (Dewey, 1984). The Triticeae include a number of genera of considerable interest to

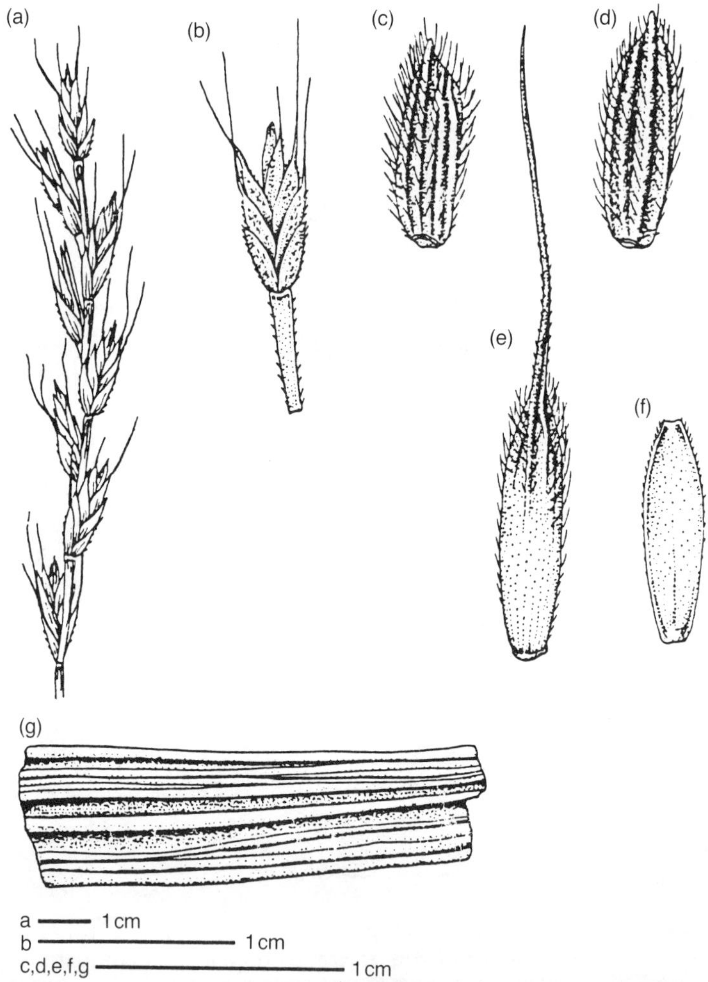

Figure 6.2 *Elytrigia intermedia* subsp. *pulcherrima*. (a) spikes; (b) spikelet; (c) upper glume; (d) lower glume; (e) lemma; (f) palea; (g) blade. (Diagrams from J.K. Jarvie, Ph.D. thesis, Utah State University, 1990.)

wheat breeders and *Thinopyrum* is one of these. Genomically it is designated genome J. The potential, in the future, of wide crossing to improve wheatgrass cannot be overlooked.

6.4 ORIGIN AND ECOLOGY

Intermediate wheatgrass is native to the Old World from southern Europe through the Middle East and southern parts of the former USSR to western Pakistan (Dewey, 1978). The natural distribution of intermediate wheatgrass overlaps the area where wheat originated.

In addition to its area of primary density, intermediate wheatgrass is widely adapted to inland sites in the temperate zone (Dewey, 1984). In North America, it grows throughout the intermountain west and the Great Plains from the Canadian prairie in Saskatchewan, south into New Mexico. It also grows in higher rainfall areas east of the Great Plains to Pennsylvania and South Carolina.

Differences in adaptation between pubescent and glabrous forms have been noted in the USA (Hafenrichter *et al.*, 1968; Hanson, 1972). The pubescent form appears to be better adapted to areas of low annual precipitation (200–400 mm), droughty soils and slightly saline soils (Hafenrichter *et al.*, 1968). The glabrous form is better adapted to areas with 400 mm or more annual precipitation (Hansen, 1972) and grows well even in relatively high rainfall areas of the eastern USA (1000 mm annual precipitation). More disease problems have been noted on pubescent plants in the high rainfall areas of the eastern USA (Schauer, 1990; Wagoner, 1991). Intermediate wheatgrass has been reported to grow well from sea level up to 3000 m elevation (Hanson, 1972).

6.5 CROP MORPHOLOGY

Although there is some variation in height, in general intermediate wheatgrass grows to about 1–1.5 m including seed-stalks and seed-heads (spikes).

Intermediate wheatgrass produces its first grain crop approximately one year after planting followed by one crop in each subsequent year. The plants require a period of vernalization before flowering is initiated. Anthesis is generally in late June or early July. Grain is mature in mid August to early September. Grain maturity is fairly synchronous within 4–7 days.

Intermediate wheatgrass grain is similar in appearance to that of wheat though the size is considerably smaller. Grain averages 5–6 mm long and 1–2 mm wide. Grain weight varies from 3.5 to 8.2 mg as compared to 40 mg for wheat. The colour of the grain is golden brown. Grain can be easily threshed from the lemma and palea (Wagoner, 1991).

6.6 NUTRITIONAL COMPOSITION AND FOOD USE POTENTIAL

Nutritional and food use evaluations of intermediate wheatgrass grain were conducted by the United States Department of Agriculture (USDA) Western Regional Research Center (WRRC) in Albany, California (Becker *et al.*, 1986; Becker *et al.*, 1991). Intermediate wheatgrass grain has higher levels of protein (20.8%), fat (3.21%) and ash (2.64%) than wheat. The protein in intermediate wheatgrass grain is nutritionally limiting in lysine as in wheat, but has higher levels than wheat of all other essential amino acids (Becker *et al.*, 1991).

In the samples of intermediate wheatgrass tested by the USDA, no functional gluten was found (Becker *et al.*, 1991). However, in an earlier study in the USSR, Shibaev (1937) found varying quantities and qualities of gluten in the various germplasm lines studied.

Milling tests demonstrated that endosperm recovery rates of intermediate wheatgrass grain were lower (40–50%) than wheat (70–75%) due to the smaller seed size. However, there was considerably higher bran fraction within intermediate wheatgrass, compared to wheat (Becker *et al.*, 1991).

Farinograph tests of whole grain flours made from several different varieties of intermediate wheatgrass demonstrate that differences exist among varieties (Becker *et al.*, 1991). These results indicate that there may be sufficient variability within the intermediate wheatgrass species complex to select and develop lines for different culinary purposes, just as there are different types of wheat for breads, pastas and pastries.

Studies to determine if anti-nutritive substances such as tannins and trypsin inhibitors are present indicate that intermediate wheatgrass is similar to wheat. No anti-nutritive substances have been found at harmful levels (Becker *et al.*, 1991).

Baking and cooking tests conducted at USDA-WRRC have shown that the flavour qualities of intermediate wheatgrass grain are very good. The grain was ground into flour and used in four different baked products to which it imparted a sweet nutty flavour. These products were judged by more than 50 trained taste panellists at WRRC. All baked products were rated good to excellent by a majority of judges. In addition, the seeds were cooked as a whole grain in mixture with brown rice to make an attractive, flavourful, cooked-grain product (Becker *et al.*, 1991).

6.7 AGRONOMY

Cultural techniques to produce intermediate wheatgrass seed have been developed by the forage seed industry. These techniques are being modified for the production of grain varieties. Studies have been conducted at Rodale Research Center since 1984 in an effort to develop environmentally sound techniques for establishing and maintaining stands of intermediate wheatgrass in order to maximize grain yields and maintain soil fertility, while minimizing weed, disease and insect problems (Wagoner and Schauer, 1989).

Experimental plots of intermediate wheatgrass ranging in size from 260–2000 m^2 have been established using a standard grain drill. Forage varieties are being used in these plantings as sufficient seed supplies are available for establishing plots of this size. Although these will not be the varieties used for perennial grain production, they provide baseline information on seed yield and production methods within this species. All plots are being managed using full-scale farm equipment to simulate currently available farming techniques in the USA. This research is being conducted in unreplicated plots in order to test a broad range of stand establishment and maintenance techniques (Wagoner and Schauer, 1989).

Results indicate that the best stands are established when wheatgrass is seeded into a weed-free seed-bed. Intermediate wheatgrass germinates quickly, within 3–4 days. The optimum planting date appears to be late August to early September as warm-season weeds are not competitive at that time of year. Seeding rates of 15–20 kg/ha with rows spaced 15–20 cm apart produce good stands in Pennsylvania (Wagoner and Schauer, 1989). A legume companion crop such as white clover (*Trifolium repens*) is also planted with the grass to provide biologically fixed nitrogen to the system in subsequent years. In western USA, where annual precipitation is lower, seed growers plant intermediate wheatgrass at 7 kg/ha in rows spaced 60–100 cm apart (Ross, 1962).

In addition to studying establishment techniques, several management techniques are being tested in an effort to maintain the health and vigour of wheatgrass stands for sustained seed production over a period of 5–7 years (Wagoner and Schauer, 1989). Management techniques which are being tested include fertility maintenance through the use of legume companion crops and manure application as well as thatch control by grazing, mowing or burning.

Grain is harvested from the Rodale Research Center experimental plots each August using a small-grain combine with a wheat header. The grain is cleaned further to remove hulls using a brush thresher.

Yields of clean grain obtained at the Rodale Research Center have been fairly similar to those observed by forage seed producers in western USA at 100–600 kg/ha. Other researchers in the northern Great Plains have obtained grain yields of 800 kg/ha (Ross, 1962) to 950 kg/ha (Gardner, 1991) from forage varieties of intermediate wheatgrass.

After 6 years of testing at Rodale Research Center, trends in grain yield became apparent. Yields from the first year of production are generally the highest (Wagoner *et al.*, 1990). Similar seed yield results have been observed among this and other perennial grass species (Hebblethwaite *et al.*, 1980; Majerus, 1988; Ross, 1962; Van Keuren and Canode, 1963). Where solid stands of intermediate wheatgrass are established by the first year of production in Pennsylvania, yields average around 500–600 kg/ha. As the stands age, yields tend to decline. However, grain yields can be manipulated through the use of appropriate management techniques. For example, when intermediate wheatgrass is grown in combination with white clover, yields are higher than when grown without this nitrogen-fixing legume. In addition, post-harvest grazing tends to increase grain yields during the succeeding year, especially among stands of 3 years or older (Wagoner *et al.*, 1990). Similar results were obtained in other perennial grasses (Majerus, 1988).

These yield data provide baseline information regarding the yield potential of intermediate wheatgrass germplasm which was developed primarily for forage production. In order to obtain economic yield levels over an extended period of time, higher yielding varieties must be developed through breeding.

6.8 ECONOMICS OF INTERMEDIATE WHEATGRASS GRAIN PRODUCTION

Information regarding the economics of production is useful in setting yield goals for the breeding programme. A preliminary economic analysis of intermediate wheatgrass grain production was conducted by Watt (1989). His analysis was based on data and information about intermediate wheatgrass seed production techniques from the seed industry, USDA Soil Conservation Service (SCS) and the Rodale Research Center. Watt concluded that if yields of intermediate wheatgrass grain could be

Table 6.1 Break-even price in US $ cents per harvested pound of intermediate wheatgrass grain

Years of production		*Average annual yield*					
	(lbs/acre):	100	200	300	400	500	600
	(kg/ha):	(112)	(224)	(336)	(448)	(561)	(673)
4		40.1	20.1	13.4	10.0	8.0	6.7
6		33.6	16.8	11.2	8.4	6.7	5.6
8		30.4	15.2	10.1	7.6	6.1[a]	5.1
10		28.5	14.3	9.5	7.1	5.7	4.8

[a]Numbers in bold represent break-even prices which are competitive with spring wheat.

Source: Watt (1989).

maintained at 500 lb/acre (561 kg/ha) over a period of at least 8 years, or 600/acre (673 kg/ha) for 4 years, the breakeven cost of the grain would be around US$0.06/lb (Table 6.1). This cost of production is competitive with spring wheat production in the northern Great Plains and could penetrate existing markets (Watt, 1989).

Watt's analysis demonstrates that yield levels in this perennial grain crop do not have to be at the same high level as in annual grains to be economically viable. Reduced input costs make net profits possible at lower yield levels. This information is encouraging in light of the fact that yield levels at and above 500 lb/acre (560 kg/ha) were achieved in the first year of production in intermediate wheatgrass germplasm which was developed for forage, not grain production. Germplasm which consistently produces 560 kg/ha or more over the course of several years should now be selected and bred to develop economically viable grain varieties of intermediate wheatgrass.

6.9 BREEDING AND VARIETAL DEVELOPMENT

In order for intermediate wheatgrass grain production to be commercially viable, productive, well-adapted germplasm must be selected and bred. The breeding biology of intermediate wheatgrass is such that techniques of mass selection, maternal line selection, recurrent phenotypic selection and polycrossing have been employed to improve forage and seed yields. Intermediate wheatgrass is a wind-pollinated, essentially obligate cross-pollinated hexaploid (Heinrichs, 1953) with a chromosome number of 2n = 42 (Dewey, 1984).

A number of studies suggest that there is ample variation within intermediate wheatgrass germplasm to enable selection for improved grain yield (Dewey, 1978; Heinrichs, 1953; Hunt and Miller, 1965).

Since 1987, germplasm evaluations have been conducted at the Rodale Research Center (Wagoner, 1989; Schauer, 1990; Wagoner and Schauer, 1990; Wagoner, 1991). A total of 350 accessions representing breeding lines as well as collections from the species area of diversity have been evaluated for growth and seed yield characteristics including seed weight, seed-head fertility, seed quality, seed-head/area, vigour, resistance to disease and lodging. Seed weights ranged from 3.5–8.2 mg. Seed-head fertility as measured by the seed-set rating method (Trupp and Slinkard, 1965) ranged from 3–66%.

It has been interesting to note the performance of individual accessions over the course of this study. Some of the accessions which had high seed weight, seed-head fertility ratings and yield potential during their first year of seed production continued to exhibit good yield potential in subsequent years, while others dropped dramatically. Selections are being made among only those accessions which have high seed yield ratings over the course of several years to ensure the selection of consistent high yielding lines. Other favourable grain production characteristics such as synchronous maturity and resistance to shattering, lodging, disease and insects are also being considered.

In order to incorporate the favourable characteristics being identified in accessions at the Rodale Research Center into well-adapted high-yielding varieties, a breeding programme was initiated in collaboration with the USDA-SCS Plant Materials Center in Big Flats, New York. Progeny resulting from intercrossing 20 high yielding accessions are being evaluated in a 400 individual plant nursery at Big Flats. Selections of the best-adapted, highest-yielding individuals from this nursery will provide the basis for a recurrent phenotypic selection programme and ultimately lead to the development of intermediate wheatgrass grain varieties.

6.10 MARKET DEVELOPMENT

As with many new grains, intermediate wheatgrass will initially enter the market on a small scale as a speciality ingredient. Grain supply will have to be balanced with market development and demand. Intermediate wheatgrass grain is poised to enter a limited market at the present time.

A supply of grain is available from intermediate wheatgrass seed producers in the south-west USA. During the cleaning process, to provide seed to the forage seed industry, some intermediate wheatgrass seeds lose their hulls, producing naked hull-less grain. The hull-less seed tends to lose its viability more quickly and is therefore not sold to farmers and ranchers who wish to plant the seed for establishment of pastures. Previously, this naked grain represented a loss to the seed industry as it was not sold for planting. The cleaned, naked seed by-product is now being sold as grain by Southwest Seed Company, Dolores, Colorado (Hennes, pers. comm.).

Small speciality food companies and bakeries are becoming interested in using intermediate wheatgrass grain in their cereal and baked products. Marketing strategies will emphasize the favourable protein qualities and high bran content of the grain as well as its uniqueness in being the only perennial grain available. Increasingly, consumers are concerned about the environment. A perennial grain such as intermediate wheatgrass has obvious environmental advantages over annuals. These advantages should be emphasized when targeting the environmentally conscious consumer segment of the market.

Another important marketing consideration is the naming of this new grain ingredient. The name 'intermediate wheatgrass' is too long and may be confusing to consumers who know of wheatgrass as sprouted wheat kernels. At the Rodale Research Center Bookstore, intermediate wheatgrass is being sold under the name 'Wild Triga'. 'Triga' is a shortened form of the genus name *Elytrigia*.

The marketing of this new grain is in its infancy. The next several years will be important for the development of markets and appropriate supply levels.

6.11 FUTURE POTENTIAL

Intermediate wheatgrass grain has a lot to offer consumers, the food industry and growers. The flavour, protein and bran content are attractive to consumers, bakeries and other grain products' manufacturers. Because this new grain crop is a perennial, not requiring annual replanting, it is attractive to farmers. The production of intermediate wheatgrass grain on hilly erodible land that is currently in annual crops would help to reduce soil erosion and could therefore benefit the environment and ultimately society as a whole.

More effort must be put into the development of markets and higher-yielding varieties before these benefits can be fully realized. However, at this point, intermediate wheatgrass is poised to be the first perennial grain commercially marketed in our 10 000 year history of grain culture.

Intermediate wheatgrass was selected for further development after almost 100 species of perennial grasses were evaluated at the Rodale Research Center. Selection of the most promising was based on:

1. production of seeds with favourable flavour;
2. production of easily threshed seeds;
3. manageable seed size;
4. resistance to shattering;
5. strong non-lodging seed-heads;
6. seed-heads held above foliage canopy;
7. drydown of seed-heads at maturity;
8. high potential for mechanical harvest.
9. vigorous perennial growth.

This selection process resulted in focus on *Thinopyrum intermedium* which has all the attributes of a new crop (Wagoner, 1990).

Germplasm is available at the USDA-ARS Living Perennial Triticeae Collection at Evan's Farm, Logan, UT; Rodale Research Center, Kutztown PA and the N.I. Vavilov Institute, St Petersburg, Russia. These collections include related genera such as *Lophopyrum* based on the E genome (some *Thinopyrum* species combine genomes J and E) and *Trichopyrum* also including E (Jarvie and Barkworth, 1990). These species may be of value in crop improvement in the future.

REFERENCES

Asay, K.H. and Knowles, R.P. (1985) The wheatgrasses, in *Forages; The Science of Grassland Agriculture,* (eds M.E. Heath, R.F. Barnes and D.S. Metcalfe), Iowa State University Press, Ames.

Atkins, M.D. and Smith, J.E. (1967) *Grass Seed Production and Harvest in the Great Plains,* Farmer's Bulletin No. 2226, United States Department of Agriculture, Washington D.C.

Becker, R., Hanners, G.D., Irving, D.W. and Sanders, R.M. (1986) Chemical composition and nutritional qualities of five potential perennial grains. *Food Sci. Technol.,* **19**, 312–15.

Becker, R., Wagoner, P., Hanners, G.D. and Saunders, R.M. (1991) Compositional, nutritional and functional evaluation of intermediate wheatgrass (*Thinopyrum intermedium*). *J. Food Proc. and Pres.,* **15**, 63–77.

Colacicco, D., Osborn, T. and Alt, K. (1989) Economic damage from soil erosion. *J. Soil and Water Cons.,* **44**, 35–9.

Cauderon, Y. (1979) Use of *Agropyron* Species for wheat improvement, in *Broadening the Genetic Base of Crops,* (eds A.C. Zeven and A.M. van Harten), Pudoc, Wageningen, Netherlands.

Cauderon, Y. and Saigne, B. (1961) New interspecific and intergeneric hybrids, involving *Agropyron. Wheat Inf. Serv.,* **12**, 13–14.

Derzhavin, A.I. (1960) The theory and practice of producing perennial rye varieties, in *Wide Hybridization in Plants,* (ed. N.V. Tsitsin), Israel Program for Scientific Translation, Jerusalem.

Dewey, D.R. (1978) Intermediate wheatgrasses of Iran. *Crop Science,* **18**, 43–8.

Dewey, D.R. (1984) The genomic system of classification as a guide to intergeneric hybridization with the perennial Triticeae, in *Gene Manipulation in Plant Improvement,* (ed. J.P. Gustafson), Plenum Press, New York, pp. 209–79.

Gardner, J.C. (1991) Herbaceous perennials for grain production, in *Alternative Crop and Alternative Crop Production Research,* North Dakota State University, Fargo.

Hafenrichter, A.L., Schwendiman, J.L., Harris, H.L., MacLauchlan, R.S. and Miller, H.W. (1968) *Grasses and Legumes for Soil Conservation in the Pacific Northwest and Great Basic States,* Agricultural Handbook 339, United States Department of Agriculture, Washington D.C.

Hanson, A.A. (1972) *Grass Varieties in the United States,* Agricultural Handbook No. 170, United States Department of Agriculture, U.S. Government Printing Office, Washington D.C.

Hebblethwaite, P.D., Wright, D. and Noble, A. (1980) Some physiological aspects of seed yield in *Lolium perenne* L. (perennial ryegrass), in *Seed Production,* (ed. P.D. Hebblethwaite), Butterworths, Boston.

Heinrichs, D.H. (1953) Methods of breeding *Agropyron intermedium. Can. J. Agric. Sci.,* **33**, 470–93.

Hunt, O.J. and Miller, D.G. (1965) Coleoptile length, seed size and emergence in intermediate wheatgrass (*Agropyron intermedium* (Host) Beauv.). *Agron. J.,* **57**, 192–5.

Jackson, W. (1985) *New Roots for Agriculture,* University of Nebraska Press, Lincoln, Nebraska.

Jarvie, J.K. and Barkworth, M.E. (1990) Isozyme similarity in *Thinopyrum* and its relatives (Triticae: Gramineae). *Genome,* **31**, 885–91.

Larson, W.E., Pierce, F.J. and Dowdy, R.H. (1983) The threat of soil erosion to long-term crop production. *Science,* **219**, 458–65.

Majerus, M.E. (1988) Stimulating existing grass stands for seed production, in *Proceedings of Northern Plains Grass Seed Symposium*, (eds J.R. Johnson and K.K. Beutler), Northern Plains Grass Seed Symposium Committee, Pierre, SD.

Miriam, E.K. and Ross, J.G. (1972) Intermediate pubescent wheatgrass complex in native collection from Eastern Turkey. *Crop Science*, **12**, 472–4.

Pimentel, D., Allen, J., Beers, A., Guinand, L., Linder, R., McLaughlin, P., Meer, B., Musonda, D., Perdue, D., Poison, S., Siebert, S., Stoner, K., Salazar, R. and Hawkins, A. (1987) World agriculture and soil erosion. *Bioscience*, **37**, 277–83.

Ross, J.G. (1962) *'Oahe' Intermediate Wheatgrass*, Bull. No. 506, Agric. Exp. St. South Dakota State College, Brookings.

Schatz, B.G. and Sobolik, F. (1989) Contrasting production practices of annuals versus perennials, in *Grass or Grain? Intermediate Wheatgrass in a Perennial Cropping System for the Northern Great Plains*, (eds P. Wagoner, J. Gardner, B.G. Schatz, F. Sobolik and D. Watt), North Dakota State University, Fargo/ Rodale Institute Research Center, Kutztown, PA.

Schauer, A. (1990) *Evaluation of Intermediate Wheatgrass – 1989 Summary*, Rodale Press, Emmaus, PA.

Shibaev, P.N. (1937) Grain quality of couch grass and wheat-couch grass hybrids. *Cereal Chem.*, **14**, 437–9.

Trupp, C.R. and Slinkard, A.E. (1965) Seedset rating as a measure of fertility in grasses. *Crop Science*, **5**, 599–600.

Tsitsin, N.V. (1960) *Wide Hybridization in Plants*, Israel Program for Science Translation, Jerusalem.

Van Keuren, R.W. and Canode, C.L. (1963) Effects of spring and fall plantings on seed production of several grass species. *Crop Science*, **3**, 122–5.

Wagoner, P. (1989) *The Study of Intermediate Wheatgrass as a Perennial Grain Crop 1988 Research Summary*, Rodale Press, Emmaus, PA.

Wagoner, P. (1990a) Perennial grain development: past efforts and potential for the future. *Crit. Rev. in Plant Sci.*, **9**, 381–408.

Wagoner, P. (1990b) Perennial grain, new use for intermediate wheatgrass. *J. Soil and Water Cons.*, **45**, 81–2.

Wagoner, P. (1991) *Evaluation of Intermediate Wheatgrass Germplasm – 1990 Summary*, Rodale Press, Emmaus, PA.

Wagoner, P. and Schauer, A. (1989) *Intermediate Wheatgrass Grain Production Trials at the Rodale Research Center – 1988 Summary*, Rodale Press, Emmaus, PA.

Wagoner, P. and Schauer, A. (1990) Intermediate wheatgrass as a perennial grain crop, in *Advances in New Crops*, (eds J. Janick and J.E. Simon), Timber Press, Portland, OR, pp. 143–5.

Wagoner, P., Schauer, A. and Janke, R.R. (1990) Perennial grain development: the potential of intermediate wheatgrass. Presented at Agronomy Society of America Conference, October 1990, San Antonio, TX.

Watt, D. (1989) Economic feasibility of a perennial grain: intermediate wheatgrass, in *Grass or Grain? Intermediate Wheatgrass in a Perennial Cropping System for the Northern Great Plains*, (eds P. Wagoner, J. Gardner, B.G. Schatz, F. Sobolik and D. Watt), North Dakota State University Press, Fargo/Rodale Institute Research Center, Kutztown, PA.

7

Wild rice (*Zizania palustris*)

M.R. Duvall

7.1 HISTORY

Wild rice (*Zizania* spp.) is an aquatic grass cultivated in temperate regions of North America and Asia as a minor cereal crop. The species commonly known as 'northern wild rice' has probably grown in North America since at least 135 AD, based on the age and identity of fossil pollen in sediment cores taken from a US lake (McAndrews, 1969). Wild rice was an important food of native Americans for over 1000 years and may have been used considerably earlier by aboriginal peoples. The cereal was also eaten by the early European explorers. Natural stands of northern wild rice are still common in Canada and Minnesota and until about 1970 were the source of most of the wild rice grain harvested on the continent. To protect natural stands and the native American industry Minnesota state law since 1939 has provided that wild rice grown in public waters can be harvested only by hand. Efforts to domesticate northern wild race as a paddy-grown crop were begun in the 1960s and paddy-grown wild rice, produced on a maximum of about 8500 ha in northern Minnesota, was the primary source of the cereal in North America until 1985 (Aiken *et al.*, 1988). Wild rice was introduced into northern California around 1977 with striking success. By 1991 the Sacramento Valley of California had become the foremost producer of the grain, with approximately 8000 ha in cultivation. It produced 2495 t compared with 2405 t from Minnesota. Canadian production was about half of the Minnesota production (Oelke, 1993).

In Asia northern wild rice is found as an introduced plant. There is also a native species called 'Manchurian water-rice'. In the former USSR both species are used as fodder on livestock, fish and game farms. In

Cereals and Pseudocereals. Edited by J.T. Williams. Published in 1995 by Chapman & Hall, 2–6 Boundary Row, London SE1 8HN. ISBN 0 412 46570 1.

China the indigenous wild rice has been cultivated since ancient times as a cereal and as a vegetable (see below).

7.2 TAXONOMY AND ECOLOGY

Zizania was first described taxonomically in 1743. The name is taken from *zizanion*, a Greek name for a weed of grain fields. Modern taxonomists recognize four species of wild rice. *Zizania latifolia* Turcz, ex Stapf (Manchurian water rice) is a perennial species that propagates vegetatively by means of underground stems. This species has a unique chromosome count for the genus (2n =) 34, all other species have (2n =) 30 chromosomes. Hybrids between *Z. latifolia* and the annual species of wild rice are thus sterile. Manchurian water rice is susceptible to infection by an edible smut fungus (*Ustilage esculenta*) that suppresses flowering but stimulates enlargement of the stem bases into large, succulent growths that are eaten as a vegetable by the Chinese (Terrell and Batra, 1982).

A second species, *Z. texana* Hitchcock ('Texas wild rice'), propagates vegetatively by tillering and is the only North American wild rice that is perennial. This type of perenniality is inherited by hybrids between Texas wild rice and 'interior wild rice' (the latter is described below). Texas wild rice is restricted in range to several kilometres of the headwaters of the San Marcos River in south-central Texas and the species has been officially listed as endangered. Texas wild rice grows in relatively deep waters with strong currents and exhibits several adaptations to that environment including rooting of the lower nodes, and decumbent or trailing stems.

The remaining two species are annuals indigenous to North America and are well-distinguished from each other by: 1) a unilateral barrier to interspecific hybridization (Duvall and Biesboer, 1988a); 2) the anatomy (Duvall and Biesboer, 1988b) and morphology (Aitken *et al.*, 1988) of the female flowers; and 3) isoenzyme profiles (Warwick and Aiken, 1988b). *Zizania aquatica* L. is a weedy species of two varieties. The typical variety, var. *aquatica* ('southern wildrice'), can be found from the shores of the Great Lakes down through the estuaries and marshes of the eastern seaboard states into Florida and west to southern Louisiana. There is little commercial interest in this taxon. A dwarf variety, var. *brevis* Fossett ('estuarine wild rice'), is found only along the tidal flats of the St Lawrence River estuary about 80 km on either side of Quebec City (Duvel, 1906). The stands of var. *brevis* furthest downstream are exposed to brackish water and exhibit unique features including a somewhat succulent or rubbery texture and an unusual tallness for this otherwise dwarfed variety.

A third variety of *Z. aquatica*, var. *subbrevis*, from south-eastern

Canada, cannot be reliably distinguished from var. *aquatica* and is no longer recognized.

Zizania palustris L. is the species of greatest commercial value as a cereal. There has been confusion in the past between Z. *aquatica* and Z. *palustris* but they are now clearly recognized and isozyme analysis also confirms that the species are separate (Warwick and Aiken, 1986). Two varieties are recognized. The typical variety, var. *palustris*, syn. Z. *aquatica* var. *angustifolia* Hitchcock (northern wild rice), is widely found on lake and stream margins across central and southern Canada and the north-eastern USA. The variety from which the cultivated wild rice was developed is most probably var. *interior* (Fassett) Dore (interior wild rice – so called because of its distribution in the interior of the northern USA). Compared to var. *palustris*, interior wild rice is a more robust plant with larger and more numerous grains. The account which follows below is restricted solely to Z. *palustris*.

7.3 MORPHOLOGY AND ANATOMY

Z. *palustris* varies in height, leaf size and panicle type. The florescence is a terminal panicle, the lower branches are spreading with male spikelets that hang down and upper branches usually are upright with female spikelets. Both male and female spikelets are borne on short persistent pedicels. Shattering of the pedicels occurs just below a collar-like tissue which probably represents vestigial glumes. The two varieties are characterized by the following morphologies.

var. *palustris*: 0.5–1.5 m high; leaf blades 4–12(–15) mm wide; panicle slender, few flowered (Figure 7.1); male spikelets normally fewer than 15 on a branch and female spikelets usually 2–6 per branch.
var. *interior*: 0.9–3.0 m high; leaf blades 10–40 mm wide; panicle spreading, branches divergent and many flowered; male spikelets (20–) 20–60 on a branch, female usually 10–30 on a branch. When the two varieties come together many intermediate forms develop and these can be exploited in breeding.

7.3.1 **Development**

During germination the wild rice shoot emerges before the root, unlike the typical pattern of germinating seeds. The development of the young plant depends on the depth of water in which it grows (Weir and Dale, 1960). Seedlings in deep water will have relatively long stem segments and produce two or three streaming, ribbon-like submerged leaves. Much of the interior of the submerged leaf is occupied by large longitudinal air spaces partitioned by transverse cellular diaphragms. In

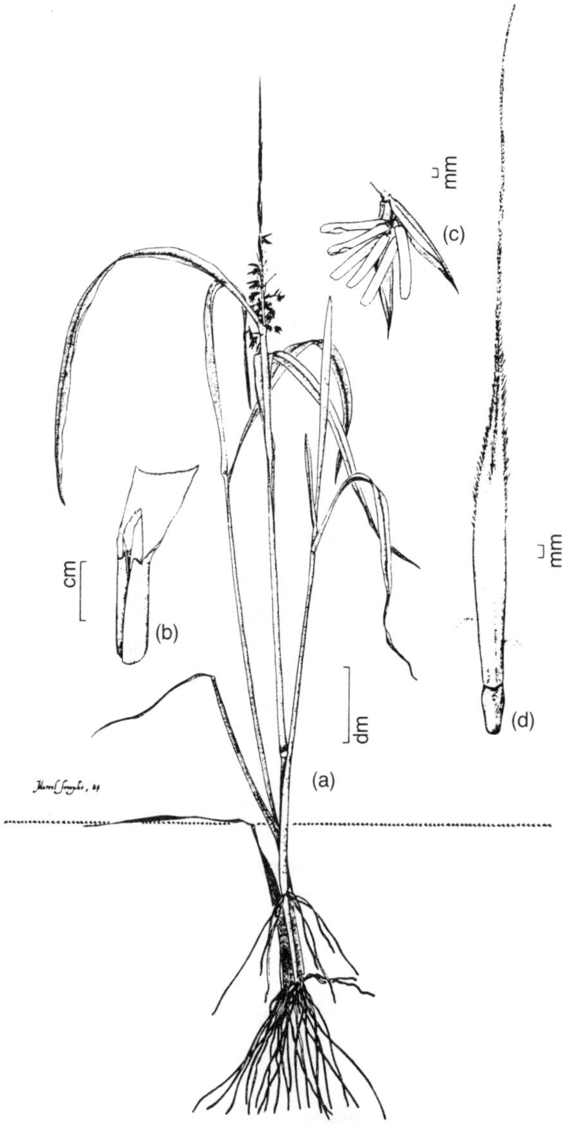

Figure 7.1 Wild rice (*Zizaria palustris*). (a) a flowering plant approximately 1.5 m high; (b) the junction between leaf sheath and blade showing ligule. Scale bar = 1 cm; (c) pendulous male spikelet at anthesis, showing six anthers. Scale bar - 1 mm; (d) erect female spikelet with long awn. Scale bar = 1 mm. (Courtesy of N.C. Press Ltd., Canada. From Aiken, Lee and Stewart (1988).)

shallow water, seedlings will produce fewer submerged leaves and have shorter stem segments. Within a few weeks of germination one or two linear floating leaves are produced at the surface of the water. The cells of the upper leaf surface are covered with a waxy cuticle and possess numerous minute hairs and projections that trap air bubbles. These features, also found on mature leaves but not on the submerged leaves, repel water and promote buoyancy.

The flowering head emerges from within the sheath of the uppermost (flag) leaf on the mature plant. Wild rice is monoecious (i.e. the male and female flowers are borne on the same plant). Unlike the arrangement of flowers on many other monoecious grasses, the uppermost flowers of wild rice are female. Thus, female flowers emerge before male flowers on the same head and cross-pollination is more likely although wild rice is self-fertile. Like most grasses wild rice is wind-pollinated. Pollen grains that become lodged in the feathery stigmas may germinate pollen tubes and effect fertilization in a matter of hours. Within 24 h of fertilization the stigmas wither. The kernels mature over the next 2 or 3 weeks from the tip of the head towards the base and fall from the head in the sequence that they mature, a condition known as shattering.

The anatomical changes accompanying the abscission of grains have been studied in some detail (Hanten *et al.*, 1980). Across the stalk of a female wild rice flower are two transverse cell layers together called the 'separation layer'. After fertilization these cells dehydrate and die. As the grain matures 'separation pockets' form randomly between the two cell layers while the walls of the cells surrounding the separation layer thicken and accumulate lignin, a strengthening polymer. At maturity the separation pockets merge and the exposed cell walls of the separation layer break down.

Once abscised the grain drops and plunges through the water, pulled by the greater weight at the base and steered by the bristle-like awn at the tip, and usually lodges in the soft substrate near the base of the parent plant.

7.4 PHYSIOLOGY OF GROWTH AND DEVELOPMENT

Germination of wild rice usually occurs only after 3 to 6 months of storage in cold water. The mechanism of dormancy in wild rice is not completely known. Damage to the pericarp (outer wall of the grain) inflicted by pricking, scraping, mechanical abrasion, or ultrasonic vibration will terminate dormancy especially if the damage is immediately above the embryo. One explanation of this observed response is that water and dissolved oxygen that enter a damaged (scarified) grain induce germination. However, grains stored in cold water are not found to increase in moisture or oxygen content while naturally overcoming dormancy. A second explanation, that scarification permits leaching of inhibitors from the grain, is supported by the detection of high levels of endogenous abscisic acid, a growth inhibitor, in dormant embryos. However, leaching of abscisic acid through the waxy pericarp of a dormant wild rice kernel has not been observed, implying that reduction of the inhibitor is normally by other, perhaps metabolic, means. Finally, the need for chilling may be explained by the observation that low temperatures are required by some plants for the biosynthesis of growth-promoters. In wild rice, it is likely that an optimal ratio of promoters to inhibitors must be achieved for germination (Albrecht *et al.*, 1979).

The grains of wild rice are intolerant of drying. When stored dry for 6 weeks germination rates fall nearly 70% although many grains die within a few days of drying (Aiken *et al.*, 1988; Probert and Longley, 1989). Moreover, tolerance of desiccation does not increase with increasing storage nor as dormancy is overcome. The storage life of grain that is properly stored in near-freezing water is limited to only 18–24 months because of rot.

The typical submerged environment of the germinating wild rice seedling has a low oxygen tension. Seedlings thus cannot derive energy for early growth from aerobic metabolic pathways. Alcohol dehydrogenase is an enzyme that catalyzes the fermentation of sugar. Considerable activity of this enzyme has bee reported in wild rice seedlings implicating fermentation as the source of energy for development (Campiranon and Koukkari, 1977). Root growth, which is stimulated by materials translocated from photosynthetically active leaves, is inhibited by low oxygen concentrations, an explanation for the unusual pattern of development in germinating wild rice.

7.5 BIOCHEMICAL COMPOSITION

The chemical composition of the wild rice grain has been analysed with reference to its nutritive value (Anderson, 1976; Swain *et al.*, 1978; Terrell and Wiser, 1975). The bulk of the processed wild rice grain is carbohydrate (70–80%), of which 60–65% is starch.

Protein content is estimated at 12–15%, greater than that of most cereal including barley, maize, sorghum, rye, white rice and wheat but less than that in oat. Of the amino acids found in wild rice protein, glutamic acid is the most abundant, accounting for 18–19% of the total protein. The essential amino acid lysine is often limited in cereal proteins. The lysine content of wild rice is 3.8–4.2% of total protein, again higher than most cereals (including high-lysine maize) except oat.

The lipid content of processed wild rice (less than 1%) is lower than that of brown rice, oat, wheat, and corn but higher than that of.

polished white rice. Most of the lipid is in the form of linoleic acid (37.7% of total fat) or linolenic acid (30.0% of total fat), the former being a vital component of the human diet.

With respect to vitamins the quantities of riboflavin (vitamin B_2) and niacin in wild rice are higher than in oat, wheat and maize but thiamin (vitamin B_1) content is lower. Wild rice has a mineral content comparable to that of other cereals although it has less calcium and more iron and zinc than brown rice, polished white rice, maize or wheat.

Analyses of trace metals in wild rice show a level of cadmium comparable to that in maize and wheat, lead content similar to that of maize but less than half that of wheat, and an amount of copper over twice that of maize but comparable to that of wheat.

7.6 AGRONOMY, YIELD AND PRODUCTION

Wild rice is most commonly produced on flooded fields around which low dykes have been constructed to form a paddy. Optimal soils are soft with a high organic content and are often peat soils. A most effective method of seeding new paddies is to broadcast seed from an aeroplane directly into the flooded field at a rate of 30–35 kg wet seed per hectare. A paddy established in a cool wet climate will re-seed itself so that annual seeding is not necessary. On the contrary, to maintain optimal yield, airboats are routinely used to thin older fields so that densities do not exceed about 10 plants/m^2.

The character and chemistry of individual soils necessitate different nutrient and fertilization levels for maximal production of wild rice. The chemistry of flooded, oxygen-poor soils differs from that of well-aerated soils since minerals in the former are found in chemically reduced states. So, for example, reduced forms of iron do not bind phosphates into compounds that cannot be used by plants. By one estimate, the most productive soils for wild rice have the following minimum nutrient levels per hectare: 30 kg calcium; 20 kg nitrogen; 30 kg potassium; 30 kg magnesium; 15 kg phosphorus; 5 kg zinc; and 750 kg iron. The metals iron, zinc, copper and manganese may be found in productive wild rice paddies at levels toxic to non-aquatic crops (Aiken *et al.*, 1988).

The water depth in wild rice paddies should be maintained between 0.3–0.6 m. Deep or very turbid water reduces light penetration and photosynthetic rates in new seedlings at the critical time when energy from fermentation is exhausted. Abrupt changes in water levels or sudden strong current early in the growing season uproot and kill seedlings, a principal reason for crop failure in natural stands of wild rice.

The grain can be harvested using rice combines or other mechanical harvesters. Yield is maximized when harvest is conducted at a time when about one-third of the kernels are dark-coloured and have a moisture content of about 40%. Average yields in commercial paddies range from 480–1250 kg of unprocessed grain per hectare (Hayes *et al.*, 1989).

7.7 DISEASES AND PESTS

Wild rice hosts a number of fungal parasites that cause diseases known as fungal brown spot (*Bipolaris oryzae* and *B. sorokiniana*), stem rot (*Sclerotium* sp. and *Helminthosporium sigmoidium*), crown and root rot, stem and leaf smut (*Entyloma lineatum*) and ergot (*Claviceps zizaniae*) (Kernkamp *et al.*, 1976). Sanitation and a fungicide can control brown spot; sanitation is the only method for control of stem rot. Stem smut is not a major problem and ergot bodies are removed from grain by screening or flotation. In Minnesota fungal brown spot, caused by *Bipolaris oryzae* and other fungi, can reach epidemic proportions in cultivated stands because of the extraordinary susceptibility of commercial cultivars (Kohls *et al.*, 1987). Attempts to control this blight involve carefully timed use of fungicides and removal of wild rice stubble after harvest. Experimental applications of calcium silicate fertilizer to peat–soil paddies produced plants resistant to fungal brown spot. The fertilizer is applied as fly ash, from coal combustion, or slag, from phosphate fertilizer production, at 15 t/ha. Fungal brown spot has not been detected in California but another pathogen, *Phytophthora erythroseptica*, has been recently identified there as the cause of a severe crown and root rot (Gunnell and Webster, 1988).

Wheat streak mosaic virus is the only known pathogenic virus to attack wild rice, and the only important bacterial disease may be bacterial leaf streak.

Among animal herbivores, blackbirds, waterfowl, muskrats and carp have all been implicated in localized destruction of wild rice. The most common and devastating insect pest is a moth, *Apamea apamiformis* (Aiken *et al.*, 1988). Moth larvae (called 'rice worms') hatch in the female flowers where they feed on the immature grains. The rice worm is effectively controlled by insecticides.

7.8 POST-HARVEST TECHNOLOGY

Once harvested the wild rice is unloaded into open-air windrows where it is allowed to ferment for from 4 to 7 days. During this time the hulls partly degrade and the kernels develop a uniformly dark colour and a characteristic flavour variously described as 'earthy', 'nutty', or 'tea-like'. The kernels are stirred and water is added daily to prevent rot and accumulation of aflatoxin (Frank *et al.*, 1975). The fermented grain is next parched in a rotating oven for about 2 h at a temperature rising to over 125 °C. During parching moisture is reduced to less than 10%, the kernel shrinks and becomes loose in the hull and the starch gelatinizes. The parched grain is threshed and foreign matter is separated from the kernels. The outer layer of the kernels may next be partly removed before the wild rice is packaged to reduce cooking time.

7.9 GENETICS AND BREEDING

In 1963 a few shatter-resistant spontaneous mutant wild rice plants were observed in a commercial paddy in Minnesota. These plants served as the germplasm base for the seven shatter-resistant cultivars released since 1972. Two complementary genes have been proposed to account for the shattering trait (Elliot and Perlinger, 1977). Shattering susceptibility occurs when there is at least one dominant allele for each gene. All other genetic combinations, i.e. those in which at least one gene has two recessive alleles, are shatter-resistant.

Within the shatter-resistant mutants, improvements have also been effected for a number of other traits of agronomic bearing. Cultivars with earlier maturation are exposed for a shorter period to the hazards of damaging winds, disease and insect pests. Other traits of significance are synchronized maturation of main stem and tiller heads, high yield and short-seededness (preferred by the processing industry since the short wild rice kernels cook in about the same amount of time as the white and brown rice with which they are often blended). In spite of the restricted germplasm base of domesticated wild rice there is a considerable amount of genetic variation in all five of the above-mentioned traits (Hayes *et al.*, 1989).

The combined factors of prolonged dormancy, desiccation intolerance, and brief seed-storage life are serious drawbacks to the perpetuation of desirable strains which currently must be replanted annually.

7.10 FUTURE POTENTIAL

The difficulties associated with large-scale commercial production of wild rice are attributable in large part to its rather short history of domestication (30 years), especially when compared to the centuries of human selection that have been practised on most common seeds. Shattering is the most conspicuous character of wild grasses that has long been eliminated from domesticated wheat, rice, barley and other cereals. Yet significant crop losses are still attributed to shattering in the contemporary shatter-resistant cultivars of wild rice. Furthermore, paddies that are allowed to re-seed will gradually revert to greater shatter-susceptibility. In spite of some of the inconvenient aspects of the cereal, worldwide production of wild rice has risen continuously since 1960 to

Table 7.1 Key events in the domestication and development of *Zizania* rice

1950	First attempts to cultivate wild rice in USA and Canada
1960	Strain with low shattering identified
late 1960s	Expansion of cultivation
1967	A Development Corporation established
1970	A Wild Rice Growers' Association formed
	University of Wisconsin initiates a research programme
1971	Co-operative formed for marketing and processing
1972	University of Minnesota initiates a research programme
	Cultivar K2 released
1974	Minnesota Paddy and Wild Rice Research and Promotion Council formed
	A seed co-operative established
	Cultivar M_3 released
1982	International Wild Rice Association formed
1983	A third co-operative established
1985	Cultivar with shorter height released
1987	Some growers establish a marketing and product company

a remarkable level of 13.7 million kg unprocessed grain in 1985 (Hayes *et al.*, 1989). Increases have been achieved by putting more land under cultivation and by increasing productivity.

The key to the success exhibited in such a short time span is probably due to the fact that soon after the first non-shattering cultivar became more widely known, a series of development and grower associations were established so that production and marketing interests were stimulated in parallel with the research programmes (Table 7.1).

REFERENCES

Aiken, S.G., Lee, P. and Stewart, J. (1988) *Wild Rice in Canada*. N.C. Press Ltd., Toronto.

Albrecht, K.A., Oelke, E. and Brenner, M. (1979) Abscisic acid levels in the grain of wild rice. *Crop Science*, **19**, 671–6.

Anderson, R.A. (1976) Wild rice: nutritional review. *Cereal Chem.*, **53**, 949–55.

Campiranon, S. and Koukkari, W. (1977) Germination of wild rice, *Zizania aquatica*, seeds and the activity of alcohol dehydrogenase in young seedlings. *Physiol. Plant.*, **41**, 293–7.

Duvall, M.D. and Biesboer, D. (1988a) Nonreciprocal hybridization failure in cross between annual wild rice species (*Zizania palustris* × *Z. aquatica*: Poaceae). *Syst. Bot.*, **13**, 229–34.

Duvall, M.D. and Biesboer, D. (1988b) Anatomical distinctions between the pistillate spikelets of the species of wild rice (*Zizania*, Poaceae). *Amer. J. Bot.*, **75**, 157–9.

Duvel, J.W.T. (1906) *The Storage and Germination of Wild Rice Seed*. Bulletin 90, USDA Bureau of Plant Industry.

Elliott, W.A. and Perlinger, G. (1977) Inheritance of shattering in wild rice. *Crop Science*, **17**, 851–3.

Frank, J.F., Torrey, G., Marth, E., Stuiber, D., Lindsay, R. and Lund, D. (1975) Control of aflatoxin production during fermentation of wild rice. *J. Milk Food Technol.*, **38**, 73–5.

Gunnell, P.S. and Webster, R. (1988) Crown and root rot of cultivated wild rice in California caused by *Phytophthora erythroseptica* sensu lato. *Plant Disease*, **72**, 909–10.

Hanten, H.B., Ahlgren, G. and Carlson, J. (1980) The morphology of brain abscission in *Zizania aquatica*. *Can J. Bot.*, **58**, 2269–73.

Hayes, P.M., Stucker, R. and Wandrey, G. (1989) The domestication of American wildrice (*Zizania*, Poaceae). *Econ. Bot.*, **43**, 203–14.

Kernkamp, M.F., Kroll, R. and Woodruff, W. (1976) Diseases of cultivated wild rice in Minnesota. *Plant Dis. Rep.*, **60**, 771–5.

Kohls, C.L., Percich, J. and Huot, C. (1987) Wild rice yield losses associated with growth-stage-specific fungal brown spot epidemics. *Plant Disease*, **71**, 419–22.

McAndrews, J.H. (1969) Paleobotany of a wild rice lake in Minnesota. *Can. J. Bot.*, **47**, 1671–9.

Oelke, E. (1993) Wild rice: Domestication of a native North American genus, in *New Crops*, (eds J. Janick and J.E. Simon), John Wiley and Sons, New York, pp. 235–43.

Probert, R.J. and Longley, P. (1989) Recalcitrant seed storage physiology in three aquatic grasses (*Zizania palustris*, *Spartina anglica* and *Porteresia coarctata*). *Ann. Bot.*, **63**, 53–63.

Swain, E.W., Wang, H. and Hesseltine, C. (1978) Notes on vitamins and minerals of wild rice. *Cereal Chem.*, **55**, 412–14.

Terrell, E.E. and Batra, L.R. (1982) *Zizania latifolia* and *Ustilaga esculenta*, a grass-fungus association. *Econ. Bot.*, **36**, 20–6.

Terrell, E.E. and Wiser, W. (1975) Protein and lysine contents in grains of three species of wild-rice (*Zizania* Gramineae). *Bot. Gaz.*, **136**, 312–16.

Warwick, S.I. and Aiken, S. (1986) Electrophoretic evidence for the recognition of two species in annual wild rice (*Zizania*, Poaceae). *Syst. Bot.*, **11**, 464–73.

Weir, C.E. and Dale, H. (1960) A developmental study of wild rice, *Zizania aquatica* L. *Can. J. Bot.*, **38**, 719–39.

Postscript

N. Haq

Since the accounts in this book are so diverse it was thought that it might be helpful to draw together a number of principles and important considerations and constraints to future development of minor cereals and pseudocereals which have been mentioned in the text.

The accounts illustrate crops which have become 'new' to particular countries or regions but which are already established crops at some other location (e.g. quinoa, amaranth and buckwheat); crops which in their areas of traditional cultivation are being researched for new market products (e.g. amaranth); and one which has the potential to yield well with low inputs on substandard soils and which is being inserted into the array of staple crops as an insurance for consistency of yield in such conditions (triticale).

It is therefore difficult to make sweeping generalizations about the role of the many minor cereals and pseudocereals in world agriculture. The accounts of the limited number covered in this book, in effect, provide an initial identification of crops with potential for further exploitation. Most of them have undergone preliminary testing in sufficient places to warrant further research and development based on productivity and utilization.

Limits to increased production and use of these crops can be imposed by climatic, edaphic and biological constraints affecting adaptation and productivity; by socio-economic constraints or the fitting of a traditional crop into a more intensive farming systems pattern.

It cannot be stressed enough that the continued research and development has to be part of an integrated multidisciplinary crop improvement programme where the minor crop clearly has a place within the farming systems. Furthermore, these under-utilized crops have to be seen as complements to, and not substitutes for, major staples.

Cereals and Pseudocereals. Edited by J.T. Williams. Published in 1995 by Chapman & Hall, 2–6 Boundary Row, London SE1 8HN. ISBN 0 412 46570 1.

Several considerations emerge:

1. The development of minor crops, when geared to commercialization and product development and market research is a different scenario from development of the crops in many developing countries. In the latter, financial resources are more limited and the growers, often resource-poor smallholders, are unwilling to take risks. Put another way, development of new crops in countries where less than 5% of the population is involved in agriculture, and where there are all the commercial trappings – mechanization, processing and markets – and trends to lower production costs and better quality produce, differs from the development of the same crops in countries where 60–70% of the population are emergent or subsistence farmers.
2. There will continue to be limited focus on minor crops (whatever scientists may think), not only because of financial considerations but because attempts to revitalize interest in a crop which has declined in production in its traditional areas must be based on clear socio-economic analysis and resource management research combined with clear needs for crop diversification.
3. The changes in agriculture in developing regions of the world seen in recent decades relate not only to inputs but also to intensification of production over time, with often more than one crop per year and better farming systems which may include multiple cropping. Knowledge and application of technology generated on minor crops can play an important role in inherently less fertile and marginal environments with low-input systems. They can also play a role in high input systems to aid sustainability. This is because sustainability must be balanced with productivity objectives and wise use of the resource base.
4. The development of alternative crops in the affluent world is centred on crop surpluses of major staple, the needs of industries, government subsidies and more exotic consumer tastes. Scientists working in this background must avoid patronizing attitudes of telling people elsewhere what to grow. Additionally they must be sensitive to the export needs of poorer countries so as not to cut them out without alternatives being provided. However trite is the term 'global village', it is more and more a reality with interdependence between peoples. Major fluctuations in export revenue can lead to foreign exchange crises and an adverse effect on the capacity to import needed commodities.
5. A stimulus to applied crops research has been the increasing demands for animal feedstuffs. Several of the accounts in this book have shown the value of grains in compound feed rations. In many cases this may be a stronger justification for enhanced local production of the under-utilized crop than for human food.

Current financial support to agricultural research and development relies heavily on a previously accepted high-payoff input model in which agriculture becomes a productive source of economic growth. Although this is changing towards a more complete model of induced innovation, it is unlikely to result in substantial funding for innovative research apart from that directly concerned with food and poverty alleviation. We would counsel those researchers who have produced so much new information on minor crops to integrate their research into a wider economic context: by doing this, major application of new technologies can be assured.

Index

Afghanistan 85
Agriculture
  alternative xii–xiii
  policies xiv
  smallholder xviii, 109, 137, 154
  subsistence 5, 37, 39, 40, 48, 58
  traditional xviii, 87, 152, 155, 237
Agriculture Canada Research Station 114
Agropyron 248–50
  A. glaucum 250
  A. intermedium 249
  A. podgerae 250
Algeria 189
Amaranth Institute (USA) 163, 178
Amaranthus 27, 199
  A. australis 176
  A. caudatus var. *atropurpurea* 139
  A. caudatus subsp. *caudatus* 129, 131–3, 138, 139, 142
  A. caudatus subsp. *mantegazzianus* 133, 138, 139, 147, 170
  A. cruentus 129, 130, 134–5, 138, 139, 142
  A. edulis 133
  A. flavus 139
  A. frumentaceous 139
  A. hybridus 129, 138, 139, 140, 142, 147, 157
  A. hypochondriacus 129, 136, 138, 142
  A. leucocarpus 139
  A. powellii 138, 139, 146, 169
  A. quitensis 138, 139, 169
  A. retroflexus 170
  see also Grain amaranth
Archaeology
  amaranths 133, 136
  quinoa 3
Argentina 35, 129, 131, 133, 189
Armenia 248
Arrowhead Mills Inc. 163
Asian Vegetable Research and Development Center (AVRDC) 175
Australia xv, xx, 85, 189
Austria 85, 189

Bangladesh xiii, xv, xviii
Barley xii, 4, 19, 21, 27, 110, 154, 188, 197, 206, 242, 248, 266, 269
Belgium 189, 198
Benin 225, 231, 243
Bhutan 85, 87, 118, 119
Biodiversity conservation xii
Bolivia 3, 5, 12, 27, 110, 154, 188, 197, 206, 242, 248, 266, 269
Bolivian Institute of Andean Crops 58
Brazil xv, 85, 119, 189, 195, 196
Buckwheat
  amino acids 100–1, 102
  breeding
    heterosis 115–16
    mass selection 114–15
    polyploidy 116–18
  carbohydrates 101
  cytology 92
  cultivars 103, 105, 109, 112, 114, 115, 116, 117, 118
  diseases 108–9
  ecology 93
  ecotypes 97
  electrophoresis 89
  fats 99, 101, 112
  fertilizers 105–7
  germination 97–8

Buckwheat (continued)
 growth forms 89, 93, 97, 104–5, 115
 incompatibility 113, 115
 landraces 87, 107
 markets 119–20
 pests 108
 photoperiodicity 97
 pollination 95, 98, 108, 113
 proteins 99, 100
 rutin 88, 101–2, 112, 113
 uses 87–8, 110–12
 vitamins 101–2, 118
 yield 98, 104, 105, 114
Bulgaria 85, 189
Burkino Faso 231, 243

Canada xx, 85, 103, 108, 112, 113, 114, 118, 119, 130, 189, 248, 251, 261, 263
Cape Verde 225
Cassava xiii
Centre for Economic and Social Studies for the Third World (Mexico) 168
Centres of diversity
 fonio 230, 233, 243
 grain amaranth 131–6, 137
 intermediate wheatgrass 251
 quinoa 3
Centro Internacional de Agricultura Tropical (CIAT) Colombia 243
Centro Internacional de Mejoramiento de Maiz y Trigo (CIMMYT) xii, xix, xx, 188, 190, 193, 194, 196, 203
Cereals
 perennial 247
 rotation 3, 7
 yields xvi
Chenopodium 11, 15, 48
 C. album 11, 15, 48
 C. ambrosioides 11
 C. berlanderi 11
 C. hircinum 11, 13, 14
 C. pallidicaule 10
 C. quinoa, see Quinoa
 C. quinoa subsp. *milleanum* 13
Chile 3, 5, 11, 13, 14, 17, 38, 189
China xv, 85, 86, 88, 110, 118, 119, 129, 131, 136, 168, 189, 262
Colombia 3, 5, 14, 36, 69, 141, 176
Colorado State University, USA 61
Commonwealth Science Council, London 88
Commonwealth Scientific Industrial Research Organisation (CSIRO), Australia 242, 243
Cote d'Ivoire 231
Crop diversification xiii, xviii, 20, 110, 273
Crop genepools xi
 see also Germplasm collections
Crop rotations 20
Czechoslovakia 189, 198, 199

Denmark xv, 7, 64
Dickey-John Corporation 162
Digitaria
 D. atrofusca 132
 D. barbinodis 231, 239, 244
 B. cruciata 229
 D. exilis, see Fonio
 D. fuscesens 229
 D. laeviglumis 231
 D. longiflora 229, 231
 D. macroblephara 228
 D. milanjiana 228, 232
 D. pentzii 228
 D. smutsii 228
 D. ternata 231
 D. tricustulata 231
Distribution of cultivation
 buckwheat 85
 fonio 225
 grain amaranth 129–30
 quinoa 4, 5–6
 triticale 188–90
Domestication, in Sudanian Africa 225
 of buckwheat 86
 of fonio 225
 of grain amaranth 133, 134, 139
 of intermediate wheatgrass 269
 of quinoa 4, 11–12

Ecuador 5, 11, 50, 58, 59, 68, 69, 129, 133, 141, 176, 195
Egypt 129
Eleusine, see Millets
Elymus hispidus 250
Elytrigia intermedia 249, 250
European Economic Community (EEC) 70
Fagopyrum
 F. cymosum 86, 89, 91, 92, 113

F. esculentum, see Buckwheat
var. *emarginatum* 89, 91, 92, 93
var. *himalianum* 87
F. kashmirianum 89, 97, 98, 102
F. tartaricum 85–6
see also Buckwheat
Finland 7
Fonio
breeding 243
botanical varieties 232–3
cytology 232, 242
diseases 241
drought resistance 225, 228
ecology 233–4, 238
fertilizers 239
fodder and feed uses 228, 236–7
food uses 227–8
germination 238
gluten 244
growing season 237–8
harvest index 239
isozymes 232
protein 235
pests 241
yield 241
France 85, 89

Gambia 231, 241
Genetic resources, *see* Germplasm
Germany 7, 85, 86, 118, 188, 189
Germplasm collections
buckwheat 118
fonio 242–3
grain amaranths 175–7
intermediate wheatgrass 257
quinoa 58, 62
triticale 211
Ghana 230, 231
Grain amaranths
amino acids 147–9
antinutritional factors 153, 169
breeding
crossing 172–3
gene markers 173–5
male sterility 172, 173
breeding systems 170
carbohydrates 149
cultivars 131, 158, 177
cultural uses 146–7
cytology 169
diseases 161–2
ecology 139–41, 145, 154
electrophoresis 177
fatty acids 149, 151
fertilizers 158–9
food uses 137, 165–8
germination 146
grain colour 131, 142, 143, 163
grain processing 164–8
growth 145, 157–8, 174
harvesting 163–4
landraces 144, 159, 174, 176, 177
mechanical production 154, 155, 163, 164, 177
minerals 151, 152, 154
morphological types 143–4, 154, 162
pests 160–1
photoperiodicity 146
pigments 152–3
pollination 171
protein 147–9, 153–4
seed anatomy 144–5
varieties 157
vitamins 151, 152, 154
yield 145, 156, 158, 174
Greece 7, 129
Groundnut 154
Guatemala xx, 129, 134, 136, 141, 166, 178
Guinea 225, 231, 232, 238, 241, 243

Hedeoma mandosiana 48
Himalayas 11, 18, 20, 85, 86, 130, 133, 136
Hungary 85, 189

India xiii, xv, xvi, xix, 7, 11, 85, 86, 87, 88, 108, 110, 112, 114, 119, 129, 131, 136, 158, 160, 161, 162, 189, 190, 195
Indian Council of Agricultural Research 131
Institute of Nutrition of Central America and Panama 168
Instituto Boliviano de Tecnologia Agropecuaria (IBTA) 61
Instituto Nacional de Investigaciones Agropecuaria (INIAP) Ecuador 59
Inter-American Institute for Cooperation on Agriculture (IICA) 176
Intercropping 5, 110, 160
Intermediate wheatgrass
breeding

Intermediate wheatgrass (continued)
  selection 255–6
  selection criteria 257
  intercrossing 256
 cytology 255
 economics of production 254
 flour 252
 forage use 252
 germplasm 255
 gluten 252
 grain composition 252
 growth cycle 251
 harvesting 253
 hybridization 248
 market development 256–7
 mechanical seed production 247
 pollination 255
 protein 252
 varieties 248
 yield 253–4
International Board for Plant Genetic Resources (IBPGR) xii, 88, 118, 175, 176
International Buckwheat Research Association 88, 121
International Crops Research Institute for the Semi-Arid Tropics (ICRISAT) xv, 228
International Center for Agricultural Research in the Dry Areas (ICARDA) 212
International Development Research Centre (IDRC), Canada xx, xxi, 176
International Institute for Tropical Agriculture (IITA) 228
International Livestock Centre for Africa (ILCA) 228, 243
International Network for Bamboo and Rattan (INBAR) xvi
International Wild Rice Association 270
Iran 85, 91, 249
Iraq 91, 249
Italy 7, 85, 189

Jalaharlal Nehru University, Delhi 174
Japan 7, 85, 87, 88, 108, 110, 112, 113, 118, 119

Kenya xv, xx, 7, 131, 136, 189, 195, 231
Korea, PDR 85, 118
Korea, RO 108, 110, 120
Kyoto University, Japan 91

Landraces 12, 87, 107, 144, 159, 174, 176, 177, 232
Latinreco SA 58, 59
Liberia 231
Lophopyrum 157

Maize xii, 5, 21, 22, 99, 100, 101, 109, 154, 164, 236, 266, 267
Mali 228, 230, 231, 233, 243
Market development xvi, xvii, 3, 68–9, 70, 178, 256–7, 270
Marginal lands/soils xiv, xviii, 3, 5, 19, 20, 93, 121, 177, 213, 226, 247
Mayaguez Institute of Tropical Agriculture, Puerto Rico 175
Mexico xvi, xx, xxi, 5, 11, 129, 134, 136, 137, 160, 161, 164, 166, 178, 189, 190
Millets xiii, 235
 barnyard xv
 common or proso xv
 finger xiii, xv, 226, 235, 236, 237
 foxtail xv, 87
 kodo xv
 pearl xii, xiii, 233, 236, 237
 teff xv
Minnesota Paddy and Wild Rice Research and Promotion Council 270
Mongolia 85, 103
Morocco 189
Multiple cropping 110, 237, 273
Myanmar 118

National Academy of Sciences (NAS), USA 88, 130, 131, 225
National Bureau of Plant Genetic Resources (NBPGR), India 118, 131, 175
National Institute of Agrobiological Resources (NIAR), Japan 118
Nepal 85, 87, 110, 113, 118, 119
Netherlands xvi, 7, 64, 189
New Zealand 130, 189
Niger 230, 243
Nigeria xvi, 131, 161, 225, 227, 228, 230, 231, 242, 243

North Central Regional Plant Introduction Station (USDA), USA 176

Oats 21, 197, 266, 267

Pakistan 7, 118, 161, 251
Paspalum scrobiculatum 241
Pennisetum species 241
Peru xx, xxi, 3, 5, 12, 14, 19, 20, 38, 44, 50, 65, 68, 69, 129, 131, 133, 137, 141, 166, 168, 176
Philippines xix
Plateau Foods Processing Co., Nigeria 244
Poland 85, 86, 87, 113, 189, 213
Polish Genebank, Radzikow 212
Portugal 189
Potato 4, 5, 19, 20, 27, 37, 44, 150

Quinoa
amino acids 22
breeding
male sterility 53, 56
marker genes 60
methods 59 61
objectives 58
breeding systems 51–4
carbohydrates 25–7
cytology 11
diseases 43–5
domesticated types 12
drought resistance 19–20
ecotypes 12–13, 17
electrophoresis 11, 13, 14
fertilizers 20, 39–40
flowering 18
frost resistance 19
life cycle 12–13, 15–17, 18, 57
lipids 23, 24, 35
mechanical harvesting 38
minerals 23–4, 34, 35
pests 20, 45–8
pigmentation 12, 54–5
protein 22–3, 31–5, 36
protein efficiency rate 32–3
races 12
saponins 12, 13, 21, 27, 30, 31, 55–6, 64–5
trypsin inhibitor 30
vitamins 23, 24
yield 36, 39, 41–3, 70

Rape seed 20
Rice xii, xviii, 21, 22, 27, 87, 100, 101, 150, 228, 235, 236, 238, 242, 267, 269
Rockefeller Foundation 175
Rodale Research Center (RRC), USA xx, xxi, 131, 138, 157, 171, 173, 175, 248, 253, 254, 255, 257
Romania 85, 189
Royal Botanic Gardens Kew, UK 243
Russia xv, 7, 85, 86, 87, 108, 112, 113, 118, 133, 188
Rye 21, 99, 248, 266
see also Triticale 187 *et seq.*

Senegal 231, 241
Sierra Blanca Associates 61
Sierra Leone 231
Soil Conservation Service (USDA) 254, 256
Sorghum xii, 233, 236, 237, 266
South Africa xv, 85, 118, 119, 189, 231
Southampton University, UK 243
Southwest Seed Co., USA 256
Soy bean 154
Spain 20, 189
Spinach 35
Sri Lanka xv
Striga
S. hermontica 241
S. rowlandi 241
S. senegalensis 241
Sudano–Sahelian climatic zone 237
Sugarbeet 20
Sustainability xvi, xviii, 121, 274
Sweden 7, 188, 198, 199
Switzerland 85, 189

Tanzania 189
Thailand xii, xx, 129, 131, 136, 161
Thinopyrum intermedium, *see* Intermediate wheatgrass
Togo 225, 230, 231, 232, 242, 243
Trichopyrum 257
Triticale
amino acids 199
breeding lines 190, 193, 196
breeding methods 210–11

Triticale (continued)
carbohydrates 200
diseases 203–4
dough 188, 212
drought resistance 194
feed and forage use 189, 206–7, 210
fertilizers 202
food use 189, 190
genetic base 212
germination 197
gluten 188, 205, 206
growth characteristics 195
isozymes 248
lipids 200
lodging resistance 193
malting 206
meiotic instability 209
pre-harvest sprouting 190, 196
pests 202, 204
photoperiodism 195
protein 190, 198
repetitive DNA 195
salinity tolerance 195
starch 198
types
primary 191, 208
secondary 192, 206
substituted 192, 203
varieties 193, 196, 201, 202, 203, 210
yield 203, 210, 213
Triticum, see Triticale
Tunisia 189
Turkey 248, 249

Uganda 226
UK xvi, 7, 11, 12, 18, 20, 38, 44, 45, 68, 69, 85, 188, 189, 196, 198, 199, 213
USA xv, xvii, xx, xxi, 7, 54, 68, 69, 85, 103, 112, 113, 116, 117, 118, 119, 129, 130, 131, 134, 136, 158, 161, 162, 164, 168, 178, 189, 247, 248, 251, 253, 256, 261, 263
US Agency for International Development (AID) xxi, 131
US Department of Agriculture (USDA) 252, 254, 257
US National Plant Germplasm System 173, 175, 176
USSR 115, 116, 119, 189, 190, 248, 251, 261
Universidad Nacional Agraria, Peru 58, 61
Universidad Nacional del Centro, Peru 61
Universidad Nacional Tecnica del Altiplano, Peru 58
University of California, Davis, USA 177
University of Chapingo, Mexico 176
University of Concepcion, Chile 58
University of Manitoba, Canada 188
University of Minnesota, USA 270
University of Niamey, Niger 243
University of San Carlos, Guatemala 176
University of Wisconsin 270

Vavilow Institute, Russia 257
Venezuela 131, 141

Western Regional Research Center, USDA 252
Wheat xiii, xvi, 4, 21, 22, 26, 27, 99, 100, 101, 145, 150, 154, 248, 251, 252, 255, 266, 269
see also Triticale 187 *et seq.*
Wild rice
cultivars 269, 270
cytology 262
diseases 268
isozymes 262, 263
lipids 266–7
minerals 267
morphology 263–5
paddy production 267
pests 268
protein 266
seed shattering 269
soils 267
vitamins 267
yield 268
Wild Rice Growers' Association, USA 270
Wild triga, *see* Intermediate wheatgrass

Yugoslavia 85, 113, 118

Zizania
Z. aquatica 262, 263
Z. latifolia 262
Z. palustris, see Wild rice
Z. texana 262